Lecture Notes in Geoinformation and Cartography

Series Editors: William Cartwright, Georg Gartner, Liqiu Meng, Michael P. Peterson

Robert J. Peckham • Gyozo Jordan
(Eds.)

Digital Terrain Modelling

Development and Applications in a Policy Support Environment

With 126 Figures

Springer

Editors:

Dr. Robert Joseph Peckham
4 Rossie Lodge
Dores Road
Inverness
IV2 4HB
U.K.

E-mail: rjpeckham@hotmail.com

Dr. Gyozo Jordan
Geological Institute of Hungary (MAFI)
Stefania ut 14
Budapest 1143
Hungary

E-mail: jordan@mafi.hu

ISBN 10 3-642-43309-x **Springer Berlin Heidelberg New York**
ISBN 13 978-3-642-43309-2 **Springer Berlin Heidelberg New York**
ISSN 1863-2246

Springer is a part of Springer Science+Business Media
springeronline.com

Softcover re-print of the Hardcover 1st edition 2007

Cover design: deblik, Berlin
Production: A. Oelschläger
Typesetting: Camera-ready by the Editors
Printed on acid-free paper 30/2132/AO 54321

Preface

This book presents a set of papers on Digital Terrain Modelling for Policy Support which aims to be informative and stimulating for both developers and users of digital terrain models. It should also be useful for professionals who are interested in the wider aspects of the applications of terrain models in support for policies and decision making.

There is an increasing demand for regional and continental scale data for use in environmental modelling and spatial analysis to support environmental policy development and implementation. Important environmental processes act at regional and continental scales, for example climatic change inducing floods and soil erosion, and management of these processes for the sustainable development of society requires policies to be applied at the corresponding scales. One of the most important factors influencing these environmental processes which act on the earth's surface is the surface topography, hence the need for extensive, harmonised digital terrain models.

Despite the steadily increasing literature on the developing field of digital terrain modelling, there is a lack of information about their regional and policy applications. Some reasons for this are that, on one hand, policies are only recently being developed for the regional and continental scales, and are still largely limited to developed parts of the world and, on the other hand, large-scale DEMs and the required computational power have only become widely available relatively recently. This book therefore aims to give an inspiring insight into the problems, methods and some of the applications of DEMs in policy support. It is practice-oriented by providing descriptions of algorithms, databases and information sources oriented to the environmental modeller and GIS expert working on regional planning and policy support applications.

The first six papers describe recent advances in the theories and foundations as well as the practical techniques for developing and tailoring the digital terrain models for their intended purposes. The topics addressed include the mathematical and numerical development of digital terrain analysis for geomorphometric and surface transport modelling, the use of mathematical morphology and image analysis to describe terrain features, optimisation of interpolation parameters using cross-validation, and the influence of input data design on the accuracy of the resulting elevation models. Also reported are projects which are making use of data from the Shuttle Radar Topography Mission (SRTM). Modern techniques for data capture based on space technology now allow us to obtain elevation information on a European and even global scale, but even with these methods for extensive data capture a lot of work still needs to be done in processing, refining and adapting the data for its specific purpose. These techniques are brought out in papers describing development of European databases for use in soil applications and hydrology.

The following five papers focus more closely on applications. Three are devoted to applications of digital terrain models related to flooding, including a decision support system for flood control, flood risk mapping, and the design of flood emergency reservoirs. The next paper describes the development of physiographic units for the European sector of the World Soil and Terrain (SOTER) Digital Database, which has a very broad range of applications in the management of agricultural and environmental resources. The final application is solar energy resource modelling, for which elevation, aspect and slope are key input data.

Two further technical notes describe statistical tools for analysis of Digital Elevation Models, and the definition and calculation of the Potential Drainage Density Index which is useful for characterising landscapes in hydrogeological applications.

We are especially grateful to the European Commission's Joint Research Centre (JRC) which has contributed extensively to this volume. Most of the authors of the presented papers are either based at the JRC's Environment Institute, or have spent some time there working on digital terrain models in a policy support context. The JRC is concerned with the application of the innovations in science and technology to support policies for the whole of Europe and it is therefore particularly appropriate that they are involved in the development of the methods and applications for Digital Terrain Modelling over this extensive region. In order to provide sound, fair and well balanced policy support for all European countries it is essential that the underlying scientific and technical data be accurate, reliable, and harmonised for their intended purpose. The work on harmonising geospatial information in general is of special concern for the European

Commission, as it is needed to sustain many European Policy areas, and the JRC is strongly supporting the INSPIRE initiative (Infrastructure for Spatial Information in Europe) which is working towards this goal. Digital Terrain Models are an excellent example of a type of geographic information where, in the past, different approaches, methods and standards applied in different countries have lead to discrepancies arising at country borders when information is put together. These discrepancies can no longer be tolerated when we need to model and simulate, for example, flooding in rivers which cross the borders. The work described in this book can therefore be seen as a contribution to the INSPIRE initiative, but also an essential step in making Digital Terrain Models applicable for a range of policy applications over a wide region. The scientific and technical foundations of this, including the theories, methods, and algorithms, are the daily work of the authors of these papers and we sincerely thank them all for their substantial contributions and patience during the editing phase. Additional thanks are due to Pierre Soille for his excellent assistance with indexing.

We hope that this book will provide a useful input to the research field of Digital Terrain Modelling, as well as a helpful support, and indeed an inspiration to people working in this area.

Robert Peckham
Gyozo Jordan

January 2007

Contents

TECHNICAL NOTES

Digital Terrain Analysis in a GIS Environment. Concepts and Development

Gyozo Jordan

1. Introduction

Digital models of topographic elevation data form an integral part of geographic information systems (GIS) and are most often used for (1) hydrological modelling including flood simulation, delineation and analysis of watersheds and drainage networks, (2) soil erosion and sediment transport modelling, (3) delineation and study of physiographic units, (4) soil and ecological studies, (5) geomorphological evaluation of landforms, (6) civil engineering and military applications such as site and route selection, landslide hazard assessment, visibility analysis (viewshed analysis), and (7) remotely sensed image enhancement for 3D analysis. Groundwater and climatic models also use digital topographic data as essential components. Digital elevation models provide an opportunity to characterise quantitatively land surface in terms of slope gradient and curvature and yield digital terrain information not blurred by land cover features which is often a problem in stereo-aerial photograph interpretation and remotely sensed image analysis.

Systematic digital terrain analysis is hampered by: (1) the lack of such studies in literature and (2) the non-uniform description and use of relevant digital methods in different fields of the Earth Sciences. Essentially identical methods are often used in these different fields with different names and for different purposes that makes their application difficult. For example, calculated gradients are used as edge enhancement filters in digital image processing while they have direct geometric implications in geomorphometry. Numerical operators have different names in the different fields of study. For example, Prewitt operators (Prewitt 1970) of image analysis are called 'Sharpnack and Akin (1969) gradient filter' or 'six-term quadratic model' (Evans 1980) in the terrain modelling literature, all being equal to a least-squares linear trend fit to the nine grid neighbours. Analysis of directional aspect data is crucial in image processing, hydrology, and climatic studies. Image analysis methods are often applied to grey-scale terrain models, thus losing the inherent geometric information of digital terrain models (Simpson and Anders 1992). Accordingly, available software designed for the specific needs of each field of study in many cases does not offer all the operations required for consistent digital terrain analysis. GIS software can easily perform most of the analyses but some procedures may be very difficult to implement. Digital terrain analysis requires the use of an integrated system of many analytical and software tools. The objective of this study is to present some basics for consistent use of digital terrain analysis methods in a GIS environment.

2. Digital Terrain Analysis in a GIS Environment

GIS is software that processes geo-referenced data and provides four basic functionalities: (1) data input, (2) data management (database management system; DBMS), (3) data analysis, and (4) data output (Aronoff, 1989). Thus, digital terrain modelling in a GIS environment expands from (1) data input by topographic elevation data acquisition, digitization and geo-referencing, to (2) management of elevation data in the form of vector data (random points, survey lines, contour lines and TINs) or raster (grid) data, to (3) analysis of digital terrain data, and to (4) generation of outputs such as contour maps and various colour-shaded terrain models. Terrain data input and output are primarily related to the field of digital cartography (El-Sheimy et al. 2005). There is no sharp boundary between data management and data analysis. For example, interpolation from point or contour elevation data to raster elevation model can be regarded as vector-to-raster transformation between data formats in the DBMS, or as an essential tool for the analysis of the studied terrain. It is the results of terrain analysis that are used as input into various environmental models and engineering applications. In a policy support environment these models and applications are used, in turn, to provide scientific bases for planning and decision making. Under these conditions the policy making organisation or its contractors rarely have the time and resources to generate the necessary terrain models and most often they obtain a terrain model readily available in a given data format derived from elevation data acquired by a given method. Based on these considerations, this study focuses on digital terrain data analysis and data input, output and management are considered only to the extent that influence the analysis of terrain data.

By definition, digital terrain models (DTM) are ordered arrays of numbers that represent the spatial distribution of terrain attributes (Moore et al. 1991). A digital elevation model (DEM) is defined as an ordered array of numbers that represent the spatial distribution of elevations above some arbitrary datum in a landscape. DEMs are the most basic type of DTMs. Digital terrain analysis (DTA) is implemented on digital elevation models in order to derive digital terrain models of various terrain attributes. Topographic attributes, such as slope and aspect can be derived from contour, TIN and grid DEMs, however, the most efficient DEM structure for the estimation and analysis of topographic attributes is generally the grid-based method. Regular grid data structure is needed also for spatial data manipulation by GIS (Aronoff 1989). In order to make maximum advantage of well-established methods of GIS technology, digital image processing and DTA, the present discussion on terrain analysis is confined to the most widely-used grid-based DEMs.

Topographic attributes can be classified into primary and secondary (or compound) attributes according to their complexity (Moore et al. 1993). *Primary attributes* are directly calculated from a DEM and include (1) point attributes of elevation, slope, aspect and curvatures, and (2) area properties such as integrals (e.g. area or surface integrals such as catchment area, or line integrals such as flow path length), and statistical properties of elevation over an area (e.g. mean and standard deviation of elevation or slope). Fractal dimension of terrain can also be regarded as a primary attribute. *Compound attributes* involve combination of primary attributes and can be used to characterise the spatial variability of specific processes occurring in the landscape, such as surface water saturation (wetness index) and sheet erosion (Moore et al. 1991). This classification originates from geomorphology. Topographic attributes can also be classified according to their spatial character. *Point attributes* result from spatial (also called local) operations that modify each pixel value based on neighbouring pixel values. Examples are gradients calculated for each pixel in a moving kernel. *Spatial attributes* result from point (also called global) operations that modify pixel values independently from neighbouring pixels. Line and area integrals, and overall or local statistical calculations yield spatial attributes. This classification originates from digital image processing. In this study focus is on geometric characteristics of the terrain that influence surface matter transport processes such as water and sediment transport processes. Thus, discussion is limited to primary point attributes in this paper.

3. Mathematical Development

The basis of digital terrain analysis is provided by mathematical vector analysis. Three types of functions are investigated by vector analysis: (1) vector-scalar functions, (2) scalar-vector functions, and (3) vector-vector functions. Common in these functions is that they use vector variables. In vector-scalar functions the independent variables are scalars such as the x, y, z co-ordinates and the resulting dependent variable is a vector. The graph of the vector end point is a spatial curve or a surface. Thus the analysis of this type of function is often called differential geometry because it describes the geometry of curves and surfaces and calculates curvatures, for example. In scalar-vector functions the independent variable is a vector and the resulting dependent variable is a scalar measured at the end point of the vector, such as the temperature in a room or soil organic matter content on the topographic surface. Thus the analysis of this type of function is often called scalar-field analysis because the central interest is the change of the scalar attribute in various directions. With this function type the gradient vector can be derived that defines the maximum change at a given point. If the dependent scalar variable is the z co-ordinate itself, the scalar-vector function defines a geometric surface. In vector-vector functions both the independent and dependent variables are vectors. Thus the analysis of this type of function is called vector-field analysis because it describes the spatial character of vector attributes at a point, such as slope vectors at a point on the topographic surface. The vector-vector function can be regarded as a tensor and calculations involve tensor algebra. In digital terrain modelling literature all the three function types are superficially called differential geometry. Differential geometry deals with parameters that are rotation invariant. For the mathematical analysis of each of the three function types both differentials and integrals are calculated. In this paper vector-vector functions and vector fields are not discussed. In the present discussion only differentials are discussed to the depth necessary for DTA.

3.1 Vector-scalar Functions: Spatial Curves, Curvature and Surfaces

3.1.1 Univariate Vector-scalar Function: Spatial Curves and Curvature

A *spatial curve* is the line obtained along a section of surface. If $\mathbf{R}$ is the set of real numbers, $T \subset \mathbf{R}$, $t \in T$, and $\mathbf{R}^3$ is the three-dimensional space, then the function $\mathbf{r}(t)$: $T \mapsto \mathbf{R}^3$ is a single-variable vector-scalar function, where $\mathbf{r}$ is a vector. The graph of $\mathbf{r}(t)$ is a spatial curve (Figure 1a). If the co-ordinates of vector $\mathbf{r}(t)$ are x, y, and z, then its co-ordinates are also functions of t: $\mathbf{r}(t) = x(t)\mathbf{i} + y(t)\mathbf{j} + z(t)\mathbf{k}$, which is the vector form of a spatial curve. Thus, defining a vector-scalar function is equivalent to defining three single-variable functions: $x = x(t)$, $y = y(t)$, $z = z(t)$, which is the scalar form of a spatial curve.

The *limit* at t_0 of $\mathbf{r}(t)$ vector-scalar function is vector $\mathbf{r}_0$, if for any $\varepsilon > 0$ there is a $\delta > 0$ value, that $|\mathbf{r}(t)-\mathbf{r}_0| < \varepsilon$, whenever $0 < |t-t_0| < \delta$:

$$\lim_{t \to t_0} \mathbf{r}(t) = \mathbf{r}_0 \,.$$

It can be shown that a vector-scalar function has limit at point t_0 if all its three co-ordinates have limits there. The co-ordinates of the limit vector are identical to the limits of the co-ordinates.

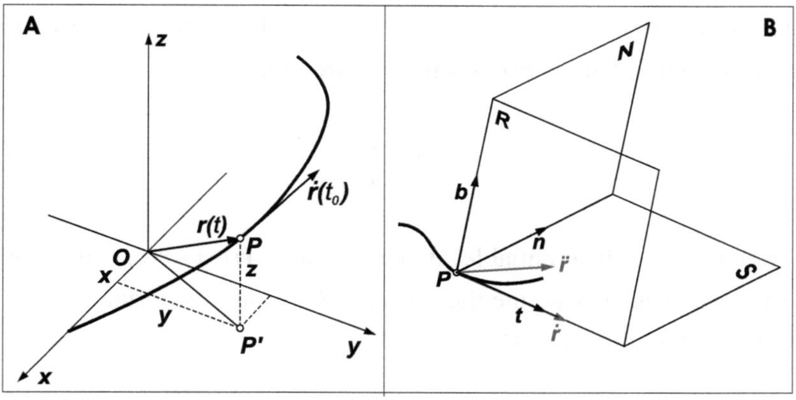

Fig. 1. A. Spatial curve as the graph of $\mathbf{r}(t)$ univariate vector-scalar function. First-derivative vector tangent to the curve is also shown. B. Some fundamental vectors and planes for spatial curves ('Frenet frame'). Note that the first- and second-order derivative vectors define the tangent plane to curve S and its normal, the binormal vector **b**

The $\mathbf{r}(t)$ vector-scalar function is *continuous* at $t_0 \in T$, if it has a limit here, a function value, and these are equal:

$$\lim_{t \to t_0} \mathbf{r}(t) = \mathbf{r}(t_0) \,.$$

The $\mathbf{r}(t)$ function is differentiable at $t_0 \in T$, if the limit

$$\lim_{t \to t_0} \frac{\mathbf{r}(t) - \mathbf{r}(t_0)}{t - t_0} = \dot{\mathbf{r}}(t_0)$$

exists (Figure 1a). This limit is the *differential* of $\mathbf{r}(t)$ at point t_0. It can be shown that the function $\mathbf{r}(t)$ is continuous at point t_0 if it is differentiable there (the reverse is not true). The differential is equal to the differential of its co-ordinates:

$$\dot{\mathbf{r}}(t) = \frac{dx}{dt}\mathbf{i} + \frac{dy}{dt}\mathbf{j} + \frac{dz}{dt}\mathbf{k} = \dot{x}\,\mathbf{i} + \dot{y}\,\mathbf{j} + \dot{z}\,\mathbf{k}\,.$$

The tangent of a spatial curve at point P is the limit line of transects across points P and Q of the curve, when Q $\mapsto$ P. It can be shown that if $\mathbf{r}(t)$ vector-scalar function is differentiable at point P corresponding to parameter value t_0 and the differential is not zero vector, then the tangent at point P exists and it is equal to $\dot{\mathbf{r}}(t_0)$. The equation of the tangent line at point P is $\mathbf{R} = \mathbf{r}(t_0) + \dot{\mathbf{r}}(t_0)\tau$, $-\infty < \tau < +\infty$. The $\mathbf{t}$ *unit tangent vector* is defined as the unit length $\dot{\boldsymbol{r}}(t)$ vector (Figure 1b):

$$\boldsymbol{t} = \frac{\dot{\mathbf{r}}}{|\dot{\mathbf{r}}|}\,.$$

If $\mathbf{r}(t)$ is twice-differentiable at point P and $\dot{\mathbf{r}}(t_0) \times \ddot{\mathbf{r}}(t_0) \neq \mathbf{0}$, then the $\dot{\mathbf{r}}(t_0)$ and $\ddot{\mathbf{r}}(t_0)$ vectors define the *tangent plane*, and its normal vector, the *binormal vector* (Figure 1b):

$$\boldsymbol{b} = \frac{\dot{\mathbf{r}} \times \ddot{\mathbf{r}}}{|\dot{\mathbf{r}} \times \ddot{\mathbf{r}}|}\,.$$

The equation of the tangent plane at point P is given by

$$(\mathbf{R} - \mathbf{r}(t_0))(\dot{\mathbf{r}}(t_0) \times \ddot{\mathbf{r}}(t_0)) = 0\,.$$

The tangent plane S (Figure 1b) at point P is a limit plane of the planes that go through three points of the curve as the points converge to point P. The curve best smoothes into the tangent plane of all planes fitting to the tangent line. The *unit normal to the curve* is defined as $\mathbf{n} = \mathbf{b} \times \mathbf{t}$ (Figure 1b).

Curvature measures the deviation of the curve from a straight line. The larger the change of the direction of the tangent at a point, the more curved is the curve there.

Based on Figure 2, the curvature is defined as

$$k = \lim_{\Delta s \to 0} \frac{\Delta \alpha}{\Delta s}\text{, when Q} \mapsto \text{P.}$$

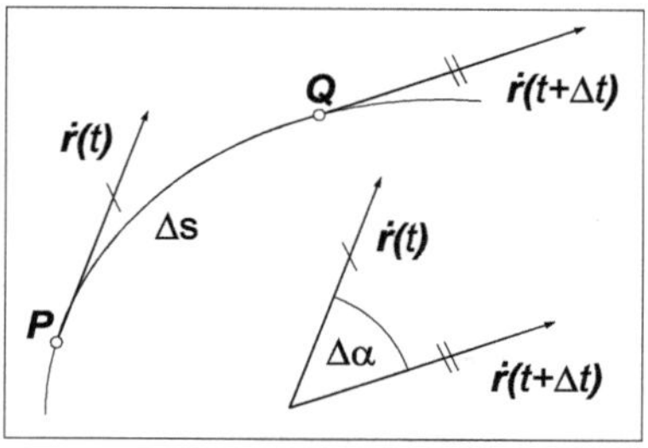

Fig. 2. Definition of curvature as the angular change between tangents along a spatial curve

It can be shown that $k = \frac{|\dot{\mathbf{r}} \times \ddot{\mathbf{r}}|}{|\dot{\mathbf{r}}|^3}$. The curvature of a spatial curve is non-negative. The curvature of a straight line is $k = 0$. Curvature can be derived by the *tangent circle* at the given point P when three points of the curve defining a circle approach point P. The radius of the tangent circle, the *curvature radius* is given by R = 1/k. Curvature radius is measured in metres [m] and curvature is measured as m^{-1}. It can be seen from the above that curvature of a curve is measured in the tangent plane S (Figure 1b).

3.1.2 Bivariate Vector-scalar Function: Surfaces, Tangent Plane and Normal Vector

A *surface* is the boundary of bodies resulting from dissection of space. If set $T \subset \mathbf{R}^2$ consisting of pairs of real numbers, $(u,v) \in T$, and $\mathbf{R}^3$ is the three-dimensional space then the function $\mathbf{r}(u,v)$: $T \mapsto \mathbf{R}^3$ is a two-variable vector-scalar function, where **r** is a vector. The graph of function $\mathbf{r}(u,v)$ is a surface (Figure 3a). If the co-ordinates of vector $\mathbf{r}(u,v)$ are x, y, and z, then its co-ordinates are also functions of u and v: $\mathbf{r}(u,v) = x(u,v)\mathbf{i} + y(u,v)\mathbf{j} + z(u,v)\mathbf{k}$, which is the vector form of a surface. Thus, defining a vector-scalar function is equal to defining three two-variable functions: $x = x(u,v)$, $y = y(u,v)$, $z = z(u,v)$, which is the scalar form of a surface. When parameters u and v are eliminated from these equations by setting x = u, y = v, the form of the surface can be written as $F(x,y,z) = 0$ (isosurface of scalar-vector function) or as $z = f(x,y)$ (Euler-Monge form). The vector form of the $z = f(x,y)$ surface can be written as $\mathbf{r} = \mathbf{r}(x,y) = \mathrm{x}\mathbf{i} + \mathrm{y}\mathbf{j} + f(\mathrm{x,y})\mathbf{k}$, where the parameters are chosen as x = u and y = v.

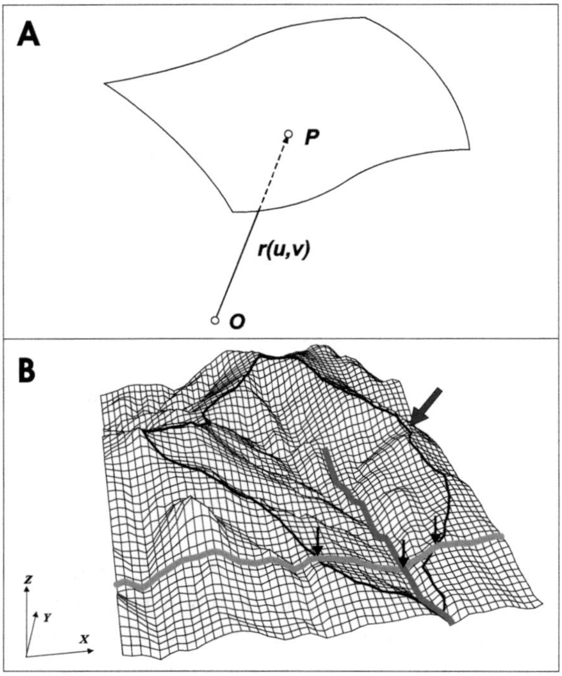

Fig. 3. A. Surface as the graph of $\mathbf{r}(u,v)$ bivariate vector-scalar-function. B. Surface curves. Parameter lines are parallel to x and y axes defining a grid in the xy plane. Dark grey curve: valley line (drainage line); black curve: ridge line (watershed divide); light grey curve: an $\mathbf{r} = \mathbf{r}(x,y_0)$ parameter line; light arrows show local extremes along the $\mathbf{r} = \mathbf{r}(x,y_0)$ parameter line corresponding to ridge- and valley lines; heavy arrow shows local minimum on the surface curve of ridge line (saddle point)

The *limit* at (u_0,v_0) of $\mathbf{r}(u,v)$ vector-scalar function is $\mathbf{r}_0$, if for any $\varepsilon > 0$ there is a $\delta > 0$ value, so that $|\mathbf{r}(u,v)-\mathbf{r}_0| < \varepsilon$, whenever $0 < \sqrt{(u-u_0)^2+(v-v_0)^2} < \delta$:

$$\lim_{(u,v)\to(u_0,v_0)} \mathbf{r}(u,v) = \mathbf{r}_0 .$$

It can be shown that a univariate vector-scalar function has limits at a point (u_0 v_0) if all its three co-ordinates have limits there. The co-ordinates of the limit vector are identical to the limits of the co-ordinates.

The $\mathbf{r}(u,v)$ vector-scalar function is *continuous* at $(u_0,v_0) \in$ T, if it has a limit here, a function value, and these are equal:

$$\lim_{(u,v)\to(u_0,v_0)} \mathbf{r}(u,v) = \mathbf{r}(u_0,v_0).$$

The *partial derivative* of function $\mathbf{r}(u,v)$ (Figure 4) according to u is

$$\lim_{\Delta u\to 0} \frac{\mathbf{r}(u+\Delta u,v)-\mathbf{r}(u,v)}{\Delta u} = \frac{\partial \mathbf{r}}{\partial u} = \mathbf{r}'_u = \mathbf{r}_u,$$

and according to v is

$$\lim_{\Delta v\to 0} \frac{\mathbf{r}(u,v+\Delta v)-\mathbf{r}(u,v)}{\Delta v} = \frac{\partial \mathbf{r}}{\partial v} = \mathbf{r}'_v = \mathbf{r}_v.$$

It can be shown that the function $\mathbf{r}(u,v)$ is continuous at point (u_0 v_0) if it is differentiable there (the reverse is not true). The differential is equal to the differential of its co-ordinates:

$$\mathbf{r}'_u = \mathrm{x}'_u\,\mathbf{i} + \mathrm{y}'_u\,\mathbf{j} + \mathrm{z}'_u\,\mathbf{k},$$

$$\mathbf{r}'_v = \mathrm{x}'_v\,\mathbf{i} + \mathrm{y}'_v\,\mathbf{j} + \mathrm{z}'_v\,\mathbf{k}.$$

An $\mathbf{r} = \mathbf{r}(u,v)$ surface is a *smooth surface* if $\mathbf{r}'_u$ and $\mathbf{r}'_v$ are continuous and $\mathbf{r}'_u \times \mathbf{r}'_v \neq \mathbf{0}$. It can be shown that the tangent lines of surface curves going through a given P point of the surface are found in a plane. The plane fitting to point P and parallel to $\mathbf{r}'_u$ and $\mathbf{r}'_v$ vectors is called the *tangent plane to the surface* at point P (Figure 4). A smooth surface has tangent plane, and its normal vector is: $\mathbf{n} = \mathbf{r}'_u \times \mathbf{r}'_v$ (Figure 4). If the surface is given as $z = f(x,y)$, then it can be written that $\mathbf{r}$ = x**i** + y**j** + f(x,y)**k**. Thus $\mathbf{r}'_x = \mathbf{i} + f_x\mathbf{k}$ and $\mathbf{r}'_y = \mathbf{j} + f_y\mathbf{k}$ ($f_x = \partial z/\partial x$, $f_y = \partial z/\partial y$), so

$$\mathbf{n} = \mathbf{r}'_u \times \mathbf{r}'_v = \mathbf{r}'_x \times \mathbf{r}'_y = \begin{vmatrix} \mathbf{i} & \mathbf{j} & \mathbf{k} \\ 1 & 0 & f_x \\ 0 & 1 & f_y \end{vmatrix} = -f_x\mathbf{i} - f_y\mathbf{j} + \mathbf{k}\,.$$

The absolute value of normal vector **n** is given as $|\boldsymbol{n}| = \sqrt{f_x^2 + f_y^2 + 1}$.

If the normal vector **n** = (A,B,C) and point $P(x_0,y_0,z_0)$ are given, then the equation of the tangent plane (Figure 4) is

$$A(x\text{-}x_0) + B(y\text{-}y_0) + C(z\text{-}z_0) = 0.$$

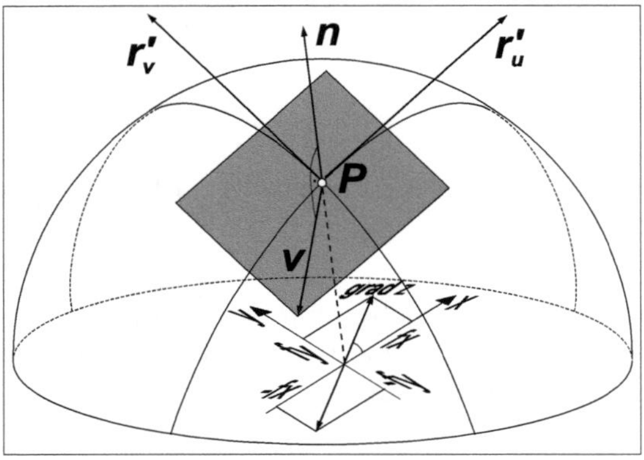

Fig. 4. Tangent plane to the surface (grey shading), its normal vector **n**, the partial derivative vectors $\boldsymbol{r}'_u$ and $\boldsymbol{r}'_v$, and the slope vector **v**. The gradient vector and the partials are also shown for the case $u = x$, $v = y$, and $z = f(x,y)$

3.1.3 Surface Curves, Surface Curvature, Surface Convexity, Inflexion and Special Curvatures

If u and v depend on a common parameter t, u=u(t) and v=v(t), then a *surface curve* is defined: **r** = **r**(u(t),v(t)) = x(u(t),v(t))**i** + y(u(t),v(t))**j** + + z(u(t),v(t))**k**. If $x = u$, $y = v$ and $z = z(x,y)$, then **r** = **r**($x(t),y(t)$) = x(t)**i** + + y(t)**j** + z(x(t),y(t))**k**. A surface curve is also defined if u and v are linked, for example v = v(u): **r** = **r**($u,v(u)$) = x((u),v(u))**i** + y(u,v(u))**j** + z(u,v(u))**k**. If $x = u$, $y = v$, $z = z(x,y)$ and $y = y(x)$, then **r** = **r**($x,y(x)$) = x**i** + y(x)**j** + + z(x,y(x))**k**. This means that x is the single parameter of the surface curve. If one of the parameters is fixed and only the other one remains variable we receive *parameter lines*. If $x = u$, $y = v$, $z = z(x,y)$ and y = y_0 (or x = x_0)

then these lines are parallel to the y (or x) axis and are infinite in number. Projection of parameter lines to the xy plane at discrete intervals are 'grid-lines' and in the above case the **r** = **r**(x,y_0) surface curve is the projection of the 'gridline' to the surface (Figure 3b).

If the surface is given by **r** = **r**(u,v), **m** is the unit normal vector of the surface at a given point (**m** points to the same side of the surface everywhere), **n** is the unit normal vector to the surface curve passing through a given point (Figure 5), and the Gaussian principal quantities are given as $\boldsymbol{E} = \boldsymbol{r}'_u\boldsymbol{r}'_u$, $\boldsymbol{F} = \boldsymbol{r}'_u\boldsymbol{r}'_v$, $G = \boldsymbol{r}'_v\boldsymbol{r}'_v$, $\boldsymbol{L} = \boldsymbol{r}''_{uu}\boldsymbol{m}$, $\boldsymbol{M} = \boldsymbol{r}''_{uv}\boldsymbol{m}$, $\boldsymbol{N} = \boldsymbol{r}''_{vv}\boldsymbol{m}$, then the curvature of the surface curve is given as

$$\boldsymbol{k} = \frac{1}{\boldsymbol{mn}} \frac{\boldsymbol{L}\dot{u}^2 + 2\boldsymbol{M}\dot{u}\dot{v} + \boldsymbol{N}\dot{v}^2}{\boldsymbol{E}\dot{u}^2 + 2\boldsymbol{F}\dot{u}\dot{v} + \boldsymbol{G}\dot{v}^2}.$$

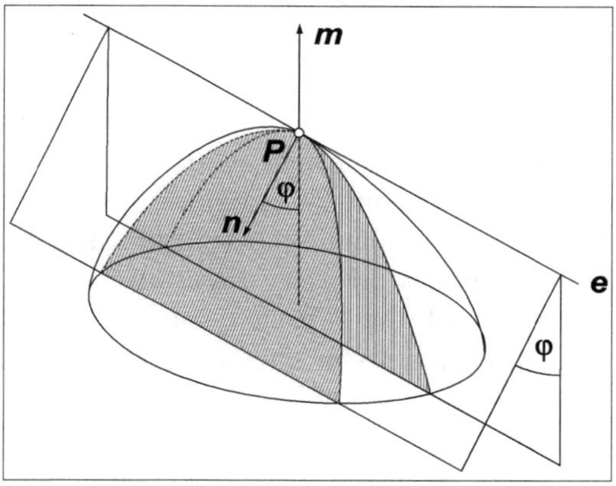

Fig. 5. Curvature on surface defined by curvature of curves in planar sections (compare to Figure 1b). See text for details

In this formula only **n**, $\dot{u}$ and $\dot{v}$ depend on the surface curve, all other parameters are related to the surface. It can be shown that $\dot{u}$ and $\dot{v}$ define the direction of the tangent of the surface curve $\dot{\boldsymbol{r}}(t) = \dot{\boldsymbol{r}}(u(t), v(t))$ therefore the curvature k of the surface curve is defined by its tangent and the **n** unit normal vector to the curve, i.e. by the tangent plane to the curve (see section 3.1.1 on spatial curves: the tangent plane S is defined by the unit tangent **t** and normal **n** vectors) (see Figure 1). This means that surface curves with the same tangent plane at a point have identical curvature, which is the same as the curvature of the curve at the intersection of the surface and the tangent plane S to the curve. Therefore, it is sufficient to study the curvature of curves in planar sections.

Let **m** be the unit normal vector to the surface at point P, and *e* one of the tangent lines at point P (Figure 5). The plane of **m** and *e* is called *normal plane*. The curve intersected by the normal plane from the surface is the *normal section* with curvature k_n. It can be shown that the curvature k of the surface curve intersected from the surface by a plane at point P having angle φ to the normal plane ($0 \le \varphi \le \pi/2$) is given by (Meusnier theorem)

$$k = \frac{k_n}{\cos\varphi} = \frac{1}{|\mathbf{mn}|} k_n, \text{ and the curvature radius is } R = \frac{1}{k_n}.$$

It can be shown that there are two perpendicular normal sections (surface curves) at point P, with k_1 and k_2 curvatures at P, that the curvature of any normal section across point P is (Euler theorem)

$$k_n = k_1 \cos^2\varphi + k_2 \sin^2\varphi .$$

The quantities k_1 and k_2 are the *principal curvatures* that form at the section of the corresponding principal normal sections. Thus, knowing k_1 and k_2 principal curvatures, the curvature of any normal section can be calculated, hence the curvature of any plane section of the surface. Curvature on a surface can be calculated in any arbitrary direction.

Assigning the curvature of the normal sections positive sign when $\mathbf{n} = \mathbf{m}$, and negative sign when $\mathbf{n} = -\mathbf{m}$, then the above definition of curvature facilitates the classification of surface points based on the sign of k_1 and k_2 principal curvatures as follows:

1. Elliptic point:
   - General case: $k_1k_2 > 0$ $LN\text{-}M^2 > 0$ $f_{xx} f_{yy} - f^2_{xy} > 0$
     (e.g. points of ellipsoid, sphere, paraboloid)

   - Spheric: $k_1 = k_2$ $LN\text{-}M^2 > 0$ $f_{xx} f_{yy} - f^2_{xy} > 0$
     (all points of sphere)

2. Hyperbolic point:
   - General case: $k_1k_2 < 0$ $LN\text{-}M^2 < 0$ $f_{xx} f_{yy} - f^2_{xy} < 0$
     (e.g. all points of saddle)

3. Parabolic point:
   - General case: $k_1k_2 = 0$ $LN\text{-}M^2 = 0$ $f_{xx} f_{yy} - f^2_{xy} = 0$
     (e.g. all points of cylinder)

- Planar: $k_1 = k_2 = 0$ $L = N = M = 0$ $f_{xx} = f_{yy} = f_{xy} = 0$ (e.g. all points of plane).

The '*surface curvature*' at a point is defined by curvature of surface curves, specifically by the principal curvatures. $K = k_1k_2$ is the total curvature and $H = k_1 + k_2$ is the mean curvature. Sometimes the $k = (k_1 + k_2)/2$ curvature is also used.

There are other types of curvature. A complete discussion of surface curvatures is provided by Shary (1995).

The convexity of the surface at a point is also described by curvature of surface curves. Obviously, '*surface convexity*' can be evaluated only if the surface point is elliptic ($LN\text{-}M^2 > 0$, $f_{xx} f_{yy} - f^2_{xy} > 0$) in which case *all* normal sections have the same sign (i.e. both k_1 and k_2 principal curvatures have the same sign). In this case, the surface is concave upward if $L > 0$ ($f_{xx} > 0$) and $N > 0$ ($f_{yy} > 0$) (i.e. signs of principle curvatures are positive), and it is convex if $L < 0$ ($f_{xx} < 0$) and $N < 0$ ($f_{yy} < 0$) (i.e. signs of the principle curvatures are negative).

By giving sign to curvatures as above, the *inflexion point* of a surface curve can be defined as the point separating convex and concave parts of the curve. Thus, point of inflexion is given where the curvature of the surface curve changes sign, i.e. it becomes zero. Inflexion of a surface at a point can only be discussed by specifying the surface curve, similarly to defining curvature and convexity at a point.

Finally, three important curvatures are defined. Let the curvature k of a plane section through a point P on a surface be given as follows (Rektorys 1969):

$$k = (f_{xx} \cos^2\beta_1 + 2 f_{xy} \cos\beta_1\cos\beta_2 + f_{yy} \cos^2\beta_2)/\sqrt{q}\cos\varphi,$$

and let $p = |\boldsymbol{grad}\ z|^2 = f_x^2 + f_y^2$ and $q = |\boldsymbol{n}|^2 = f_x^2 + f_y^2 + 1$, φ is the angle between the normal to the surface at a given point and the section plane; β_1, β_2 are angles between the tangent of the given normal section at a given point (Figure 5) and axes x, y, respectively. It can be shown that this formula is identical to the curvature defined with the Gaussian quantities.

Profile curvature, the curvature of normal plane section in gradient vector direction is defined as

$$k_p = \left(f_{xx} f_x^2 + 2 f_{xy} f_x f_y + f_{yy} f_y^2\right) / p\sqrt{q^3},$$

where $\cos\varphi = 1$, $\cos\beta_1 = f_x/\sqrt{pq}$, $\cos\beta_2 = f_y/\sqrt{pq}$.

Plan curvature, measured in the horizontal plane section in the direction perpendicular to the gradient vector (the curvature of contour lines) is given by

$$k_h = f_{xx} f_y^2 - 2 f_{xy} f_x f_y + f_{yy} f_x^2 / \sqrt{p^3} \ ,$$

where $\cos\varphi = \sqrt{p/q}$, $\cos\beta_1 = f_y/\sqrt{p}$, $\cos\beta_2 = -f_x/\sqrt{p}$.

Tangential curvature, measured in the normal plane section in a direction perpendicular to the gradient vector is given by

$$k_t = f_{xx} f_y^2 - 2 f_{xy} f_x f_y + f_{yy} f_x^2 / p\sqrt{q} \ ,$$

where $\cos\varphi = 1$, $\cos\beta_1 = f_y/\sqrt{p}$, $\cos\beta_2 = -f_x/\sqrt{p}$.

As it will be seen in the following sections these curvatures are fundamental parameters linking surface geometry and mass flow on topographic surfaces.

3.2 Scalar-vector Functions: Gradient Vector, Slope and Aspect

If $\mathbf{R}^3$ is the three-dimensional space and $T \subset \mathbf{R}^3$, then the function $u(\mathbf{r})$: $T \mapsto \mathbf{R}$ is a (three-variable) scalar-vector function, where $\mathbf{r}$ is a vector. If the co-ordinates of vector $\mathbf{r}$ are x, y, and z so $\mathbf{r} = x\mathbf{i} + y\mathbf{j} + z\mathbf{k}$, then the scalar-vector function can be written as $u = u(\mathbf{r}) = f(x,y,z)$. The scalar-vector function defines a *scalar field* and it renders a scalar value (u) to each point of space by vector $\mathbf{r}$.

An *equipotent surface* is defined by points where the value of the measured scalar is constant. If $u = f(x,y,z)$ then the equation of one equipotent surface is given by $u = C$, i.e. by the $C = f(x,y,z)$ equation. If the studied space is two-dimensional we have equipotent curves, i.e. *contour lines*. If the two-variable scalar-vector function has the form $z = f(x,y)$ then the scalar z is measured in the plane of x and y co-ordinate axes. Writing the above two-variable scalar-vector function as $C = z - f(x,y)$ shows that a surface can be regarded as the $C = 0$ equipotent surface of the three-dimensional $u = z - f(x,y)$ scalar field. This leads to the definition of a surface in the $F(x,y,z) = 0$ form.

A is the *limit* value of the $u(\mathbf{r})$ scalar-vector function at point $\mathbf{r}_0$, if for any $\varepsilon > 0$ there is a $\delta > 0$ value, that $|u(\mathbf{r}) - A| < \varepsilon$, whenever $0 < |\mathbf{r} - \mathbf{r}_0| < \delta$:

$$\lim_{\mathbf{r}\to\mathbf{r}_0} \mathrm{u}(\mathbf{r}) = A .$$

The u(**r**) scalar-vector function is *continuous* at point $\mathbf{r}_0$, if

$$\lim_{\mathbf{r}\to\mathbf{r}_0} \mathrm{u}(\mathbf{r}) = \mathrm{u}(\mathbf{r}_0) .$$

The u(**r**) scalar-vector function is *differentiable* at point $\mathbf{r}_0$, if vector **g** exists when

$$\lim_{\Delta\mathbf{r}\to\mathbf{0}} \frac{|u(\mathbf{r}_0+\Delta\mathbf{r})-u(\mathbf{r}_0)-\mathbf{g}\Delta\mathbf{r}|}{|\Delta\mathbf{r}|} = \lim_{\Delta\mathbf{r}\to\mathbf{0}} \frac{|\Delta u-\mathbf{g}\Delta\mathbf{r}|}{|\Delta\mathbf{r}|} = 0 .$$

The definition of the derivative of scalar-vector function has to consider not only the magnitude but the direction of derivation a well. Vector **g** is the derivative of the u(**r**) scalar-vector function at point $\mathbf{r}_0$ and it is called the *gradient vector*:

$$\mathbf{g} = \left(\frac{du}{d\mathbf{r}}\right)_{\mathbf{r}_0} = grad\ u(\mathbf{r}_0) .$$

If u(**r**) = $f(x,y,z)$, it can be shown that

$$grad\ u = \frac{\partial f}{\partial x}\mathbf{i} + \frac{\partial f}{\partial y}\mathbf{j} + \frac{\partial f}{\partial z}\mathbf{k} = f_x\mathbf{i} + f_y\mathbf{j} + f_z\mathbf{k} .$$

The magnitude of the gradient vector, called *gradient*, is given by $|grad\ u| = \sqrt{f_x^2 + f_y^2 + f_z^2}$. It can be shown that the gradient vector is perpendicular to the equipotent surfaces: **n** = *grad u*, where **n** is the normal vector of the tangent plane of the equipotent surface. If the surface is given by $z = f(x,y)$, then the surface is the u = 0 equipotent surface of the scalar field $\mathrm{u} = z - f(x,y)$. Thus

$$\mathbf{n} = grad\ u = -\frac{\partial f}{\partial x}\mathbf{i} - \frac{\partial f}{\partial y}\mathbf{j} + \frac{\partial f}{\partial z}\mathbf{k} .$$

If **e** is a unit vector at point $\mathbf{r}_0$, then the straight line at point $\mathbf{r}_0$ in the **e** direction can be given as $\mathbf{r}(t) = \mathbf{r}_0 + t\mathbf{e}$ ($t \in$ R). In this case the scalar value u along the line depends only on t: u(**r**) = u(**r**(t)). The *directional derivative* of the scalar u in direction **e** is given as (using the chain rule)

$$\frac{du}{dt} = \frac{du}{d\mathbf{r}}\frac{d\mathbf{r}}{dt} = (grad\ u) \cdot \mathbf{e} .$$

Since $\mathbf{e} \cdot grad\ u(\mathbf{r}_0) = |\mathbf{e}||grad\ u(\mathbf{r}_0)| \cos\alpha$, where α is the angle between vectors **e** and grad u($\mathbf{r}_0$), it can be seen that the directional derivative is the maximum in the gradient direction. Therefore, the gradient vector points from the given point towards the largest increase of u scalar value.

If a surface is given by $z = f(x,y)$, then the gradient vector is given by $grad\ z(f_x + f_y)$ and its magnitude by $|grad\ z| = \sqrt{f_x^2 + f_y^2}$ (Figure 4). This shows that the gradient vector for surface function $z = f(x,y)$ lies in the plane of x and y co-ordinates. Since the magnitude (length) of the gradient vector is the maximum change at a point, it is the steepest slope at a point on the topographic surface defined by $z = f(x,y)$ (Figure 4):

$$\text{gradient} = |grad\ z| = \sqrt{f_x^2 + f_y^2}\ \text{, and}$$

$$\text{slope} = \arctan|grad\ z| = \arctan\sqrt{f_x^2 + f_y^2}\ .$$

The direction of the gradient vector (angle between the gradient vector and the x axis) measured in the xy plane is given by simple trigonometric considerations (Figure 4):

$$\text{aspect} = \arctan\left(\frac{f_y}{f_x}\right),$$

where aspect = 0 is in west direction with uphill gradient direction, or

$$\text{aspect azimuth} = 180^0 - \arctan\left(\frac{f_y}{f_x}\right) + 90^0\left(\frac{f_x}{|f_x|}\right),$$

where azimuth is in degrees and measured clockwise from north.

3.3 The Link between Surface Geometry and Surface Flow: Gradient and Curvature

3.3.1 First Derivatives (Gradient): Surface Flow Velocity and Direction

The flow on the surface is defined by the *slope vector* which is parallel to the two dimensional gradient vector, it points into the opposite direction (i.e. down-slope in the flow direction), and it is tangent to the surface. This means geometrically that the slope vector lies in the tangent plane of the surface and thus it is perpendicular to the surface normal vector (Figure 4):

$$\mathbf{v} = -f_x\mathbf{i} - f_y\mathbf{j} - (f_x^2 + f_y^2)\mathbf{k}\ .$$

If mass transport is driven by gravitational force then it is easy to calculate the force vector driving mass movement parallel to the surface based on the slope vector (assuming that gravitational force is perpendicular to the xy plane):

$$\vec{\mathbf{F}}_{slope} = \mathrm{m} \cdot \mathrm{g} \cdot \frac{\mathbf{v}}{|\mathbf{v}|} \cdot \sin\gamma$$

where m is mass, g is gravitational acceleration, γ is slope angle, and $\vec{\mathbf{F}}_{slope}$ is the gravitational force vector component parallel to the slope vector **v**, and $\frac{\mathbf{v}}{|\mathbf{v}|}$ is the unit slope vector.

Using the above consideration velocity and acceleration vectors for mass movement on the surface can be also calculated from the tangential component of gravitational force parallel to the slope vector. However, the complete description of the flow field on the surface requires vector-vector functions that are beyond the scope of the present discussion.

3.3.2 Second Derivatives (Curvature): Surface Convexity, Inflexion Points, Flow Convergence and Divergence

Krcho (1973) in his discussion regarded the two quantities of the gradient vector, magnitude (gradient) and direction (aspect angle), as two new scalar fields. With this view these scalar fields yield themselves again to directional derivative and gradient calculations. On this basis, it can be shown that the derivative of the gradient vector magnitude (or slope) in the direction of the gradient vector at a given point is identical to profile curvature. This derivative is the second derivative of elevation, thus profile curvature reflects the change in gradient along the slope line and controls the change of velocity of mass flowing down along the slope curve. It can be shown that at convex profile curvatures flow velocity increases ('erosion' occurs along upper slope sections on hill slopes) and at concave profile curvatures flow velocity decreases ('depositon' occurs along lower slope sections in valleys) (Krcho 1973; Shary 1995) (Figure 6). Since slope break is defined as a significant down-gradient change of slope, profile curvature is used to define slope-breaks where profile curvature value is above a specified threshold.

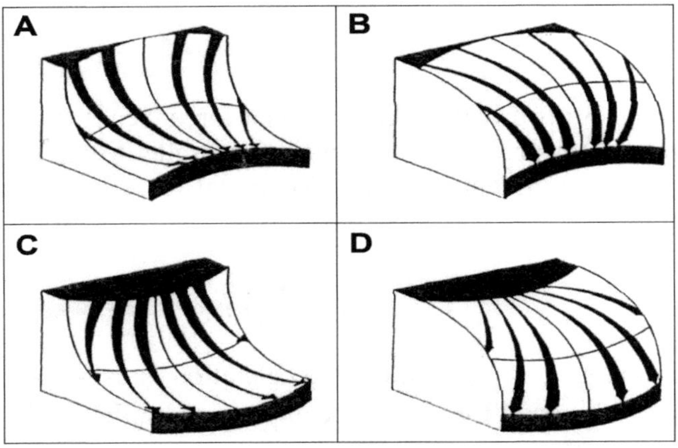

Fig. 6. Convexity of profile (arrows) and tangential (black areas) curvatures (k_p and k_t) defining velocity and convergence of surface flow, respectively. A. $k_p > 0$, $k_t > 0$. B. $k_p < 0$, $k_t > 0$. C. $k_p > 0$, $k_t < 0$. D. $k_p < 0$, $k_t < 0$. Positive and negative values define concave and convex curvature, respectively (after Mitesova and Hofierka 1993)

The inflexion point of the continuous slope line, along which profile curvature is measured, is calculated as $k_p = 0$, where it separates convex (negative) and concave (positive) curve sections. In the point of inflexion of the slope curves on the topographic surface, the value of slope acquires maximum or minimum values depending on whether we get from the convex part into the concave one or vice versa. Thus, the inflexion points of the slope curve are the local extreme values of the slope. Since slope curves on the topographic surface run through each other continuously, the inflexion points of slope curves form lines separating convex and concave profile curvature areas (Krcho 1973).

Tangential and plan curvature are both curvatures measured in a direction perpendicular to the gradient vector and hence reflect the change in aspect angle and influence the divergence/convergence (angular or directional change) of flow on a surface. Plan curvature measures the angular change of the two-dimensional gradient vector in the xy plane while tangential curvature measures the angular change of the three-dimensional slope vector in the tangent plane. Since actual flow on the surface is guided by the slope vector, tangential curvature is preferred to plan curvature in surface flow divergence/convergence calculations. However, plan and tangential curvatures are related via $\sin\gamma = k_t/k_h$ (where γ is the slope angle) and their computed distribution of convex and concave areas are the same (Figure 6). Inflexion points for these curvatures (surface curves) can be identified in the same way as for the slope line using profile curvature.

3.3.3 Singular Points: No Flow

For the purpose of topographic analysis two different types of points on a surface are distinguished: *regular points* with only one direction of flow, where the magnitude of gradient vector $|\text{grad } z| \neq 0$ ($f_x \neq 0$ or $f_y \neq 0$) and *singular points*, where $|grad\ z| = 0$ ($f_x = 0$ and $f_y = 0$) with undefined direction of flow. Singular points can be further classified. If $f_{xx} f_{yy} - f^2_{xy} > 0$ then the surface point is either a local maximum (peak) if $f_{xx} < 0$ and $f_{yy} < 0$, or a local minimum (pit) if $f_{xx} > 0$ and $f_{yy} > 0$. These are elliptic points, too ($LN\text{-}M^2 > 0$). If $f_{xx} f_{yy} - f^2_{xy} < 0$ then the surface point is a saddle point (pass). Geometrically, this is a hyperbolic point, too ($LN\text{-}M^2 < 0$). Undefined flow also occurs in a horizontal plane (flat) where $|grad\ z| = 0$ and $f_{xx} = f_{yy} = f_{xy} = 0$. This is a parabolic point (planar point, specifically), too ($L = N = M = 0$). Here we have a link between surface geometry and flow on the surface: local extremes, saddle points and flats with no flow can be found only at elliptic, hyperbolic and parabolic points, respectively.

In the case of bivariate functions of surfaces, similarly to single variable $y = f(x)$ planar curves, the first derivatives (gradients) are used to identify local extreme points and the second derivatives (curvatures) are used to identify inflexion points.

3.3.4 Ridge Line (Watershed Divide) and Valley Line (Drainage Line): Flow Boundaries

Finally, for the geometric characterisation of a topographic surface, valley lines and ridge lines have to be identified. Since valley lines are drainage lines and ridge lines are watershed divides their analysis requires, however, the analysis of surface flow by vector fields (vector-vector functions) and differential equations, which is beyond the scope of the present discussion. It can be said that valley lines (drainage lines) are lines across which there is no flow but flow along the line is permitted, and surface flow is convergent at drainage lines. Geometrically, at the points of a drainage line normal sections perpendicular to the gradient vector (flow direction) have local minima (Figure 3b). Similarly, ridge lines (watershed divides) are lines across which there is no flow, flow along the line is not permitted, and surface flow is divergent at watershed divides. Geometrically, at the points of a watershed divide normal sections perpendicular to the line have local maxima (Figure 3b). These definitions are sufficient to develop numerical methods for valley line and ridge line calculations from DEMs.

3.4 Vector Analysis and Digital Terrain Modelling: Geometric Characterisation of Topographic Surfaces

Based on point attributes, Evans (1972; 1980) developed geomorphometry using five basic parameters estimated from the elevation model: elevation, slope, aspect, profile and tangential curvatures that are calculated for regular points. Based on the preceding discussion, we further expand his system with surface specific points: (1) inflexion points, (2) singular points, and (3) valley lines and ridge lines. Inflexion points (lines) are added as the sixth parameter calculated for regular points because they define convex and concave areas, essential in surface flow modelling (Krcho 1973). Singular points calculated are (1) local minima (pits), (2) local maxima (peaks), (3) saddle points (passes) and (4) horizontal planar points (flats). Finally, we add flow boundaries: (1) valley line (drainage line) and (2) ridge line (watershed divide). In this way we receive a list of 12 parameters that provide a sufficient system for the complete geometric characterisation of topographic surface with surface flow:

1. Elevation,		
2. Local minimum (pit),		
3. Local maximum (peak),	*singular points*	
4. Saddle point (pass),		
5. Horizontal planar point (flat),		
6. Slope,	*gradient vector*	
7. Aspect,		
8. Profile curvature,		*regular points*
9. Tangential curvature,	*curvature*	
10. Inflexion points (convexity),		

11. Valley line (drainage line) } *flow boundaries*
12. Ridge line (watershed divide).

Singular points, inflexion points, valley lines and ridge lines are called 'surface specific points' because they are calculated only for certain isolated locations (points and lines). In contrast, elevation, gradient and curvature are calculated for points of areas.

In practice, first points of the topographic surface are classified as regular and singular points. Next, singular points are classified into peaks, pits, passes and flats, and are then excluded from further analysis. Third, the gradient vector (slope and aspect) and curvatures, together with inflexion points (convexity) are then calculated for areas of regular points. Finally, valley lines and ridge lines are calculated.

It is noted that length of surface curves (e.g. slope line or valley line length) or surface area (e.g. catchment drainage area) are given by line and surface integrals, respectively. Integrals are parts of exhaustive geometric characterisation of surfaces with flow, essential in surface hydrological and sediment transport modelling. However, due to space limitations here, integrals are the subject for another study.

From the above analysis it can be concluded that the estimation of the necessary 12 geometric parameters can be reduced to the estimation of five partial derivatives only:

$$f_x = \partial z/\partial x, \qquad f_y = \partial z/\partial y,$$
$$f_{xx} = \partial^2 z/\partial x^2, \qquad f_{yy} = \partial^2 z/\partial y^2, \qquad f_{xy} = \partial^2 z/\partial x \partial y.$$

One of the main results of all the above discussions is this surprising simplification of the problem. In the final section of the paper numerical methods are presented to estimate these five quantities from DEMs and to derive the 12 parameters of geomorphometry.

4. Numerical Methods

4.1 Digital Representation of Topographic Surface: Continuity and Smoothness

Grid DEMs are finite numerical representations of the infinite topographic surface. The finite grid data are used to estimate the five necessary partial derivatives for points of the infinite surface. One of the most important consequences of the preceding discussion is that the calculated geomorphometric parameters and thus the necessary partial derivatives depend on the character of the studied surface, i.e. the digital representation of the topographic surface. Two characteristics, surface continuity and smoothness are the most important limiting factors.

Surface continuity refers to the existence of limiting values at a point (see sections 3.1.1 and 3.1.2). The obvious physical continuity of the topographic surface is often mixed up with the mathematical definition of continuity in literature. DEMs represented by random points, survey lines, contour lines and grids have an infinite number of missing surface elevation values in between the data points, therefore they are discontinuous, too. In the special case of vertical walls in the topography, where the same (x,y) location can have two different z elevation values, the topographic surface cannot be represented by a mathematical function thus calculation of derivatives is not permitted.

DEMs represented by a limited number of small surface patches connected by a step are discontinuous along the connections, thus differentials cannot be calculated at these locations. Examples for discontinuous surfaces are horizontal planes connected by steps such as raster rectangles (squares) and Voronoi (Thiessien) polygons. Similar patches of horizontal planes result from the nearest neighbour and moving average interpolations, for example. In general, any surface interpolated with local (patchwise) interpolation producing patches of surfaces that does not ensure continuity at connections of patches poses limitations to differential calculations (Zevenbergen and Thorne 1987). Examples for continuous surfaces are TINs, polynomial trend surfaces calculated for the whole data set, sliding polynomials (Snyder et al. 1984), and splines such as bicubic and regularised splines (Mitesova and Hofierka,1993).

Surface smoothness refers to the existence of derivatives: a surface or a curve is smooth at a point if its first differential exists at the point. A smooth surface is differentiable everywhere. We may define the *degree of surface smoothness* according to the existence of higher-order derivatives. The higher the order of the derivative, the higher the order of smoothness at the given point. There are continuous surfaces that are not smooth. Examples are TINs and those patchwise interpolations that have ‘breaks’ or ‘peaks’ at connections of patches (triangle facets in TINs) where differentials cannot be calculated. Splines and multi-surface interpolation (Hardy 1971) make sure that patches are differentiable at the connections, thus they provide smooth surfaces. Apart from patch connections where differentials do not exist, degree of differentials, i.e. degree of smoothness at a point is also important. Surfaces resulting from linear interpolation such as TINs, bilinear interpolation, first-order trend surface (plane), or any other linear surface have only first-order derivatives, thus gradient vector can be calculated but curvatures requiring second-order derivatives cannot be estimated. Moreover, surfaces represented by horizontal planes, such as raster rectangles, Voronoi polygons and nearest neighbour interpolated surfaces have first derivatives equal to zero within patches, i.e. the gradient is zero resulting in undefined surface flow. Any function, such as second or higher order polynomials, including splines and bicubic interpolation, with higher order derivatives can be used for curvature calculations. According to Dubrule (1984), a surface fitted with kriging interpolation is twice differentiable under the conditions of trend order is k=1 and $K(h)=|h|^3$, where K(h) is the generalised covariance and h is the range.

The above discussed geometric smoothness that enables the calculation of derivatives of surfaces is often mixed up with the statistical (topographic) smoothness (or 'roughness') of the surface in literature. Statistical smoothness (or 'roughness') of the surface is the variation of elevation within a given area, often characterised by the standard deviation or range of elevation within the area. Geometric surface smoothness is a point attribute originating from the field of differential geometry, while statistical surface smoothness is an area attribute originating from geomorphometry. Thus, when describing an interpolation method producing a 'smooth surface', it should be made clear if the method produces a differentiable surface or an even surface. Also, smoothing of a DEM with moving average or other smoothing technique improves the evenness of the surface but not the geometric smoothness (existence of differentials).

In conclusion it can be stated that, from a mathematical point of view, a 'good' DEM for DTA should be continuous and smooth (continuity follows from smoothness, i.e. from the existence of differentials). In practice however, we consider a DEM 'good' if it represents the topographic surface accurately. In DTA we are often interested in break-lines, for example in the geomorphological investigation of tectonic faults (Jordan 2004). In such cases surface models representing sudden slope-breaks such as TINs are the most suitable (McCullagh 1988; Jordan et al. 2005). Some researchers argue that each landform may be considered to consist of smaller forms leading to the concept that land surface is not mathematically smooth and there are no actual values of any point attributes (slope, aspect, curvature) except elevation (Shary 1995). As it will be seen in the following section, numerical methods of partial derivative estimation from a grid DEM approximate the topographic surface at the point of interest with a locally smooth function, enabling the calculation of differentials.

4.2 Calculation of Partial Derivatives for Gradient and Curvature Estimation

Gradient calculation methods can be classified according to the number of neighbouring points used in the calculation in the 3x3 moving grid kernel as two-point, three-point, four-point and eight-point methods.

Two point methods do not use partial derivatives and estimate gradient vector slope and aspect directly from the elevation difference between the central point and one or more of the neighbours. The *maximum down-gradient* (or primary flow direction) method identifies the gradient direction towards the neighbour of the lowest elevation, and slope is calculated using the elevation difference of these two points. The *maximum gradient* (or steepest neighbour) method uses the neighbour with the highest elevation difference. Apart from the three-point forward (or backward) finite-difference scheme, these two methods are the only ones which use the elevation of the central point. Most of the hydrological watershed and drainage calculation methods, and erosion and sediment transport algorithms use these methods. Curvature requiring second-order derivatives cannot be estimated with the two-point gradient methods.

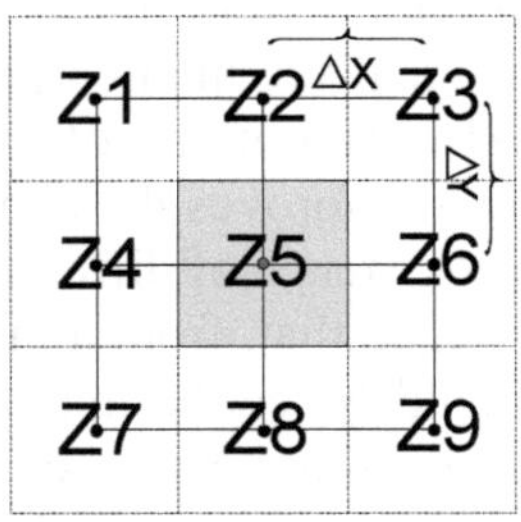

Fig. 7. The 3x3 moving grid kernel used for gradient estimation in the centre point Z5. Dotted lines and gey rectangle indicate corresponding 'pixels' or 'cells'

The *three-point method* proposed by O'Neill and Mark (1987) is similar to that of Onorati et al. (1992) and calculates slope and aspect from the plane determined by three neighbours (central, north and east) in the 3x3 kernel. This method provides good results for grid DEMs derived from TINs (Guth 1995). Since a plane is used, second-derivatives and thus curvature cannot be estimated with the three-point methods. The forward finite-difference scheme uses three-points for gradient vector estimation and calculates the partial derivatives as follows (Tuma 1979) (Figure 7):

$$f_x = (Z6\text{-}Z5)/\Delta X \text{ and } f_y = (Z2\text{-}Z5)/\Delta Y,$$

where ΔX and ΔY are grid spacing in the x and y axes direction, respectively.

The *four-point method* is the most widely used method for the estimation of the partial derivatives (Fleming and Hoffer 1979) (Figure 7):

$$f_x = (Z6\text{-}Z4)/2\Delta X \text{ and } f_y = (Z2\text{-}Z8)/2\Delta Y.$$

This method uses a simple central finite-difference scheme along the one-dimensional grid lines parallel to the x and y axes across the centre point in the 3x3 grid kernel (Tuma 1979). It can be shown that this scheme is identical to fitting a line to the three consecutive horizontal (Z4, Z5, Z6) or vertical (Z2, Z5, Z8) grid points (Figure 7) by the least-squares. This is also identical, in turn, to calculating the central averages for the middle half grid points (along the x axes: $Z_{-1/2(\Delta x)} = (Z4+Z5)/2$ and $Z_{1/2(\Delta x)} = (Z5+Z6)/2$) and using these averages for differencing at the central point Z5: $f_x = Z_{1/2(\Delta x)} - Z_{-1/2(\Delta x)})/\Delta X$. Slope and aspect calculations therefore use four neighbours. This method can be derived by fitting a full nine-term quadratic polynomial to the 9 kernel points by the least squares. Since this quadratic polynomial has nine terms, it passes exactly though the nine kernel elevations:

$$Z = Ax^2y^2 + Bx^2y + Cxy^2 + Dx^2 + Ey^2 + Fxy + Gx + Hy + I.$$

Since the f_x and f_y partials depend only on constants G and H, it can be easily seen that if a plane is fitted to the 9 kernel points by the least squares, i.e. $Z = Gx + Hy + I$, by setting $A = B = C = D = E = F = 0$ in the above nine-term quadratic polynomial, the same estimation is obtained for f_x and f_y (Sharpnack and Akin 1969). However, the plane fit does not allow for the estimation of the second-order derivatives. If the plane is fitted to the four cornel points of the kernel (Abramowitz and Stegun 1972) than

$$f_x = (Z3+Z9\text{-}Z1\text{-}Z7)/4\Delta X \text{ and } f_y = (Z3+Z1\text{-}Z7\text{-}Z9)/4\Delta Y.$$

The *eight-point method* of Zevenbergen and Thorne (1987) fits the above nine-term quadratic polynomial to the 9 kernel points by least squares providing the following expressions for the partials:

$f_x = (Z6\text{-}Z4)/2\Delta X$,
$f_y = (Z2\text{-}Z8)/2\Delta Y$,
$f_{xx}, = (Z4\text{-}2Z5+Z6)/ \Delta X^2$,
$f_{yy} = (Z2\text{-}2Z5+Z8)/\Delta Y^2$,

$f_{xy} = (Z3+Z7-Z1-Z9)/4\Delta X\Delta Y$.

The f_x and f_y first-order derivatives are identical to the four-point finite difference method.

In order to achieve some smoothing for the gradient, Evans (1980) suggested a six-term quadratic surface given as

$$Z = Dx^2 + Ey^2 + Fxy + Gx + Hy + I,$$

by setting A = B = C = 0 in nine-term quadratic polynomial. This yields the following expressions for the partials:

$f_x = (Z3+Z6+Z9-Z1-Z4-Z7)/6\Delta Y$,
$f_y = (Z1+Z2+Z3-Z7-Z8-Z9)/6\Delta Y$,
$f_{xx,} = (Z1+Z3+Z4+Z6+Z7+Z9-2(Z2+Z5+Z8)/3\Delta X^2$,
$f_{yy} = (Z1+Z2+Z3+Z7+Z8+Z9-2(Z4+Z5+Z6)/3\Delta Y^2$,
$f_{xy} = (Z3+Z7-Z1-Z9)/4\Delta X\Delta Y$.

The f_x and f_y partial derivative filters are known as the *Prewitt operators* (Prewitt 1971) in the image processing community. Note that the f_{xy} mixed second-order derivative is identical to the previous method.

Shary (1995) imposed the following condition: the six-term quadratic surface has to pass exactly through the Z5 middle point of the 3x3 kernel, giving the following for the partials:

$f_x = (Z3+Z6+Z9-Z1-Z4-Z7)/6\Delta Y$,
$f_y = (Z1+Z2+Z3-Z7-Z8-Z9)/6\Delta Y$,
$f_{xx,} = (Z1+Z3+Z7+Z9+3(Z4+Z6)-2(Z2+3Z5+Z8)/5\Delta X^2$,
$f_{yy} = (Z1+Z3+Z7+Z9+3(Z2+Z8)-2(Z4+3Z5+Z6)/5\Delta Y^2$,
$f_{xy} = (Z3+Z7-Z1-Z9)/4\Delta X\Delta Y$.

Note that f_x, f_y and f_{xy} are identical to the Prewitt operators and to Evans (1980) method above.

Horn's (1981) gradient estimation method also uses eight neighbours for gradient calculation and applies a third-order finite difference method which yields

$f_x = (Z3+2Z6+Z9-Z1-2Z4-Z7)/8\Delta X$ and
$f_y = (Z1+2Z2+Z3-Z7-2Z8-Z9)/8\Delta Y$,

where the unequal weights are proportional to the reciprocal of the square of the distance from the kernel centre. This weighting scheme is known as the *Sobel operator* in image analysis (Gonzalez and Woods 1993). Note that these weights can be obtained by first passing a four-point gradient method central finite-difference kernel followed by a smoothing kernel in the perpendicular direction (Gonzalez and Woods, 1993):

$$f_x = \frac{1}{8\Delta X}\begin{bmatrix} -1 & 0 & 1 \\ -2 & 0 & 2 \\ -1 & 0 & 1 \end{bmatrix} = \frac{1}{8\Delta X}\begin{bmatrix} 1 \\ 2 \\ 1 \end{bmatrix} * \begin{bmatrix} -1 & 0 & 1 \end{bmatrix} \text{ and } f_y = \frac{1}{8\Delta Y}\begin{bmatrix} 1 & 2 & 1 \\ 0 & 0 & 0 \\ -1 & -2 & -1 \end{bmatrix} = \frac{1}{8\Delta Y}\begin{bmatrix} 1 \\ 0 \\ -1 \end{bmatrix} * \begin{bmatrix} 1 & 2 & 1 \end{bmatrix}.$$

The Sobel operators therefore give higher weights to the points closer to the kernel centre than the Prewitt operators by applying smoothing in addition to differencing.

Second-order derivatives can be derived by simply applying the first-order derivatives twice. For example, if we apply the central finite-difference scheme of the four-point method in the x direction to the horizontal grid elevation series: $|Z_{-2}|Z_{-1}|\mathbf{Z_0}|Z_1|Z_2|$ (bold shows the central point of interest: $\mathbf{Z_{-1}}\equiv Z4$, $\mathbf{Z_0}\equiv Z5$, $\mathbf{Z_1}\equiv Z6$), then we get the following first derivative for the central point in the x direction (shown in bold):

$$|\times|(Z_0 - Z_{-2})/2\Delta X\ |\ \mathbf{(Z_1 - Z_{-1})/2\Delta X}\ |\ (Z_2 - Z_0)/2\Delta X|\ \times|,$$

which has been shown already above. Application of the $f_x = (Z6-Z4)/2\Delta X$ filter again to this series yields

$$|\times|\times|\mathbf{(Z_2 - 2Z_0 + Z_{-2})/4\Delta X^2}|\times|\ \times|,$$

i.e. second-order derivative in the x direction is $f_{xx} = (Z_2-2Z_0+Z_{-2})/4\Delta X^2$ for the central grid point. This estimation uses points lying two grid-spacings away from the central point outside of the 3x3 kernel. The result is obviously different from $f_{xx} = (Z4-2Z5+Z6)/\Delta X^2$ derived by the least-squares method applied to the nine-term quadratic polynomial. Since the latter gradient method uses points closer to the central grid point Z5 (labelled as Z_0 in the example) this method should be preferred.

Based on the above considerations, it is therefore relatively easy to design gradient filters by using the least-squares method (Papo and Gelbman 1984). Not only can the fitted function be changed, but the kernel size as well, achieving the representation of a larger area for the calculations. A great variety of gradient filters can be found in mathematical textbooks such as Abramowitz and Stegun (1972) or Tuma (1979). Another source of gradient filters is digital image processing literature where gradient methods are discussed under 'edge detection' and 'image segmentation' (e.g. Gonzalez and Woods 1993).

4.3 Which Gradient Calculation Method Should be Used?

The above discussion leads to the following conclusions. First, the same filter has various names depending on the field of origin and an understanding of the derivation of these filters helps to see clearly their similarities and differences. Second, before implementing gradient and derivative calculations by computer, it is worth knowing which type of filter or method is used by the applied software. Third, it is good to understand the original DEM surface (e.g. if it was generated from a TIN) and find the geometrically most suitable gradient filter.

The performance of a gradient algorithm is evaluated against the following criteria: (1) accurate representation of topographic reality, (2) resistance to random error in DEM, (3) resistance to systematic error in DEM, (4) response to horizontal and vertical DEM resolution, (5) introduction of random and systematic error by the filter, and (6) the objective of the investigation. Criteria 2-3 are related to the sensitivity of the gradient filter, i.e. its response to change in the input elevation signal.

Performance of a gradient algorithm depends on (1) source of elevation data, (2) the applied surface modelling (interpolation) method, (3) elevation data representation (such as TIN versus grid format, grid resolution and numerical resolution of height data), (4) whether they are used for low or high relief terrain, and (5) the objective of the investigation. Gradient methods are evaluated for these factors against the six performance criteria below.

As far as DEM *data source* is concerned, topographic elevation data can be acquired by means of direct field measurement with ground surveying or GPS equipment, or by means of remote sensing using orthophotos, LIDAR or RADAR technology. Direct field measurements produce irregular points (or strings of points along survey lines) that require interpolation to a grid, while LIDAR and RADAR height data are obtained already in grid format. Since LIDAR or RADAR elevation data are based on reflectance from earth surface cover including vegetation canopy and man-made structures, they are better called *digital surface models* instead of elevation models. Orthophoto interpretation produces either random points or contour lines. Random points (control points) from orthophotos are most often interpolated with TIN that is resampled to a grid in turn. Contour lines are the form of topographic surface representation in traditional topographic maps that are converted to grid DEM by interpolation after contourline digitisation. The various data acquisition methods, their accuracy for true topographic elevation representation, error sources and treatment are extensively discussed in El-Sheimy et al. (2005) and in Li et al. (2005).

Evaluating data sources against the performance criteria of gradient algorithms in general, it can be said that gradients derived from DEMs cannot be more accurate than the original elevation measurements. In other words, an inaccurately acquired DEM will produce inaccurate gradient estimations. It is noted however that topographic slope and aspect can be obtained by direct field measurement circumventing the problem of derivative calculations from an inaccurate DEM. LIDAR and RADAR data represent earth surface, therefore these data sources are inaccurate representations of the topographic surface especially in densely vegetated or built-up areas. Thus, some pre-processing is required to obtain topographic elevation before gradient calculations. Random error is more typical to DEMs derived from random points, from LIDAR and RADAR acquisition methods than to contour-based DEMs, thus gradient filters with smoothing effect such as the Prewitt operators should be preferred.

DEMs derived from contour data often display bias (systematic error) towards contour elevations (Carrara et al. 1997) and the curvature parameters, particularly profile curvature, can appear as waves over the landscape (Moore et al., 1993). These appear as sudden slope changes or even as bumps along contour lines and as flats in between. For distant and parallel running contour lines this problem is the most severe. Here, too, gradient filters with smoothing effect are preferred. Since the problem is related to contour line spacing of the original data, kernels larger than the average contour line spacing might treat the problem. Pre-smoothing of the DEM with kernel size matching contour line spacing can be a pre-processing step, or the DEM can be re-interpolated from contour lines extracted from the DEM with an interpolation method that avoids waviness (e.g. linear interpolation). Contours at regular intervals act as variable horizontal filters: in areas of steeper slopes small features are captured by the densely spaced contour lines while features of the same size are missed in flat terrains (Unwin 1981). In this way, contour data source impacts accuracy of topographic gradient calculation. Grid DEMs obtained from TIN-interpolated random points of orthophoto heights can also display systematic errors in the form of planar triangle facets at sparsely located control points, typically in flat terrains. Such areas should be either excluded from curvature (second-order derivative) calculations, or locally re-interpolated with smoother surfaces. LIDAR and RADAR heights data can display systematic error as sudden slope-breaks at the edges of forested or built-up areas thus biasing gradient estimation. The elimination of the edges is not a simple task and might require the use of ancillary data such as land-cover maps in a GIS.

The objective of the investigation determines the most appropriate data source. Ecological studies of vegetation canopy, for example, might prefer RADAR surface models. Contour-based methods have advantages in flow-modelling, for instance (Moore 1988; Moore and Grayson 1989). Also, waviness of contour-based DEMs influence most slope and profile curvature but aspect and plan curvature can be calculated with little distortion. When slope-breaks, ridge and valley lines are the subject of study, such as in morphotectonic studies, the data source should be able to capture the locations of sudden elevation changes. In such a case, random elevation points are required to be densely spaced at these locations, while curving lines of contours tend to smooth out slope-breaks. With the increasing availability of remotely sensed elevation data however, contour lines digitised from topographic maps tend to lose application. Often, the data availability is the limiting criteria for selection of elevation data source.

As far as DEM *surface modelling (interpolation) methods* are concerned, continuity and smoothness of surfaces in relation to gradient calculations have been discussed in section 4.1. Application of an interpolation method should be determined by the nature of input data, availability of computer resources, the characteristics of the modelled surface and the objective of modelling (Lam 1983). Therefore the effect of interpolation on gradient calculation should be viewed from the interaction between data source and interpolation method. For random points kriging is usually considered the most efficient method (El-Sheimy et al. 2005), however it requires a preliminary structural analysis by variogram estimation. In studies of terrain with complex relief where hilly areas meet flat valleys and basins, which is the most common case in fact, the autocorrelation property is spatially varying among low and high relief areas. Thus, application of a single variogram (autocorrelation function) is misleading and results in inaccurate modelling of the topographic surface, therefore biasing gradient calculations. If the characteristics of the modelled surface are known then the interpolation function best representing the surface should be selected. Interpolation from contour lines should avoid waviness around contour lines, therefore 'wavy' splines seem to be improper in most cases. Kriging also seems to be less appropriate for contour data source (El-Sheimy et al. 2005). Since contour lines capture the basic characteristics of the surface, selection of interpolation function does not need to be guided by surface geometric properties. Linear interpolation methods, such as TIN offer good results for contour source data. Triangulation inserts flats for closed contours at hill tops (peaks), however, and it can truncate valley bottoms and ridge tops where contours strongly curve. This feature disables the estimation of gradient at the truncated flat areas. Hill tops can be cured by adding spot heights or by interpolating locally for the flat tops. Flat valleys and ridges can be adjusted by using constraints in the algorithm ensuring that the three points of triangular facets do not fall on the same contour line. LIDAR and RADAR heights data do not require interpolation by default.

Spurious peaks and pits in DEMs often result from interpolation errors, however in natural landscapes pits are far less probable. While these features cause error in gradient calculations locally, the largest problem is artificial pits disabling surface flow routing. Artificial pits can be removed by construction of pit-free DEMs (Hutchinson 1989), smoothing and pit filling methods (Jenson and Domingue 1988; Tribe 1992; Martz and Garbrecht 1995). Atrificial flats, most often resulting from linear interpolation and pit filling algorithms, disable gradient estimation and flow routing, and can be treated by surface tilting such as the method of Martz and Garbrecht (1992). Areas should be marked where spurious pits and flats have been post-processed in order to note the locations of uncertain gradient and valley line calculations.

The objective of the investigation determines the most appropriate interpolation. In digital terrain analysis for morphostructural studies first- and second-order derivatives are frequently used (Florinsky 2000) and it is a requirement therefore that the fitted function should be continuous in the first two derivatives. This contradicts the idea that break-lines must also be included for tectonic analysis (McCullagh 1988). For example, TINs are efficient in the representation of areas with uniform slope or aspect, essential in identification of tectonically induced erosional morphology of spurs and pediments in mountain fronts (Riley and Moore 1993).

Considering *elevation data representation*, surface continuity and smoothness of various digital surface representation formats (Carter 1988), such as flat raster and Voronoi rectangles, planar TINs, and grids have been discussed in relation to gradient calculations in section 4.1. Digital elevation data often originate from tiles of digitised paper maps or of other data acquisition methods. Edge enhancement (gradient) methods and spectral methods can reveal systematic errors along tile boundaries that should be treated before gradient calculations. Differentiating over a rectangular grid causes quantification of slope and, especially, aspect data. All gradient estimation methods applied to grid data have undue clustering of aspect in the eight principal compass directions (Guth 1995; Jordan 2007), and the severity of this bias depends on the interpolation algorithm (Wilson and Gallant 2000). Use of gradient filters with smoothing effect can reduce systematic error in aspect, or aspect should be calculated from TIN DEM instead of grid DEM. The numerical resolution of heights defining vertical resolution has a strong impact on gradient calculations. Elevation data is often converted to the nearest integer in DEMs. Carter (1992) has shown that rounding results in increasing systematic error in slope and, particularly, in aspect calculations as slopes become gentle. Because of their smoothing effect eight-point finite difference methods yield the best approximation of gradient in this case (Carter 1992).

Experience demonstrates that a minimum slope value $\gamma_{min} > 0$ exists under which computation of aspect using $\alpha = \arctan(f_y / f_x)$, is inappropriate and the terrain should be classified as flat or as singular point (peak, pit or saddle point) with undefined aspect. The minimum slope value depends on the accuracy of the elevation data and the method of gradient calculation. The aspect computed in areas with slope less than this value reflects, more likely, 'noise' in the DEM rather than real structure of terrain (Mitasova and Hofierka 1993).

Grid spacing also has variable influence on slope and aspect calculations depending on the numerical algorithm (Chang and Tsai 1991; Guth 1995; Hodgson 1995). The largest differences in gradients calculated with various methods occur at pits, peaks, ridges and valleys, therefore algorithms tend to provide similar results when changing grid scale by smoothing (Guth 1995). Changing the scale by DEM aggregation (also called decimation or thinning) has most severe impact on the four-point finite difference and eight-point finite difference methods while the maximum gradient method is the most resistant (Guth 1995). This limits the comparison of slopes on different scales. Wilson and Gallant (2000) found that aspect estimations are relatively insensitive to DEM resolution. Evans (1980) noted the sensitivity of curvature to grid spacing. Plan curvature was found to be less sensitive to DEM resolution than profile curvature (Wilson and Gallant 2000). In general, terrain attributes measured on different scales should not be compared (Evans 1980).

Large pits can also result from insufficient grid spacing that cannot resolve pit outlet elevation and create artificial dams. Dams can be treated by burning blue lines into the DEM or by dam breaching algorithms (Garbrecht and Martz 1995).

With respect to modelling *low or high relief terrain*, the largest differences in calculated gradients occur at pits, peaks, valleys and ridges (Guth 1995). Jones (1998) compared seven different gradient algorithms based on their performance on a synthetic surface of known analytical derivatives. He concluded that the four-point finite central difference algorithm which computes gradient and aspect from only the nearest four elevation grid points in N-S and E-W directions was the most accurate. Skidmore (1989) used values of gradient and aspect manually calculated from paper maps as the true values against which to compare slope algorithms. His study concludes that the eight-point finite difference methods are more accurate in calculating slope and aspect because all grid points adjacent to the centre grid contribute to the calculation. Guth (1995) suggested in his comparative study that the eight-point unweighted Prewitt operators should be used for aspect calculation while the steepest neighbour method provides slope which is satisfying in most practical applications. Florinsky (1998) analytically derived errors (RMSE: root mean square error) of various gradient methods and concluded that (1) RMSEs in slope, aspect and curvatures are directly proportional to elevation RMSE, (2) these RMSE values increase with decreasing grid spacing, (3) gradient RMSEs increase with decreasing slope, (4) curvature RMSEs are more sensitive to grid spacing. For their smoothing effect he proposed the Prewitt operators for gradient calculations. Burrough and McDonnell (1998) suggested DEM error propagation with Monte Carlo simulation in order to estimate the resulting error of calculated topographic parameters. Since curvatures use second-order derivatives they are particularly sensitive to random error. According to Evans (1972) statistical properties of point attributes are more stable if the DEM is smoothed before analysis. Guth (1995) also suggests smoothing in order to reduce noise and discrepancies between various gradient methods.

The *objective of investigation* should eventually set the criteria for the selection of gradient calculation method. For instance, a higher smoothing effect can be desirable in drainage network modelling, while the fast computation of aspect (flow direction) in 45° bins is often satisfactory for hydrological or erosion and sediment transport modelling (Jordan et al. 2005).

4.4 Avoiding Second-order Derivatives: Break Lines and Inflexion Lines

Second-order derivatives extensively used in curvature and inflexion calculations significantly enhance error as discussed above, thus it is advisable to avoid the calculation of higher-order derivatives whenever possible. Slope-breaks defined by high profile curvature value, and inflexion lines separating convex and concave areas are very important parameters in geomorphology but it is almost impossible to identify these continuous lines with the point operations of gradient calculation methods. However, alternative methods for their analysis can be designed. The basic idea is based on the fact that continuous areas are much easier to find in digital images (DEMs in this case) than lines. If lines of interest can be defined as boundaries between areas, their continuity is easily achieved.

Jordan (2003) suggested the use of slope statistics and slope maps to find slope-break lines in a morphotectonic investigation in order to avoid the calculation of error-prone second-order derivatives. He calculated the cumulative percentage area-slope curve for the study area and identified break points in the curve (Figure 8). On this basis, areas were classified plain, hilly and mountainous where slope was ≤1°, 1°-3° and 3°<, respectively, corresponding to the break points in the curve. The obtained classified slope map displayed continuous linear sharp edges at the boundaries of slope classes corresponding to significant slope-breaks along morphotectonic lineaments (Figure 8). Unlike slope calculation, profile curvature using second-order derivatives failed to identify these lines (Jordan et al. 2003).

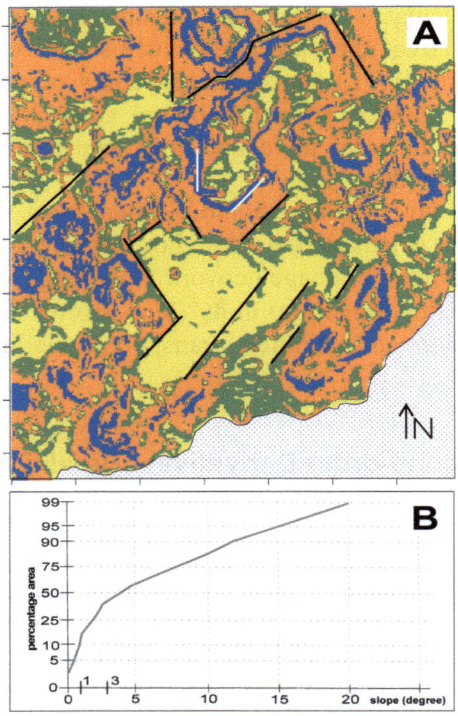

Fig. 8. A. Classified slope map. Slope angles according to colour shading: yellow ≤1°; green: 1°-3°; orange 3°-10°; blue: 10°<. Lines emphasise linear slope-breaks. B. Cumulative percentage area curve of slope. Slope is classified based on breaks in the curve (after Jordan et al. 2003)

Inflexion lines can be identified on a similar basis. Since points of inflexion lines separate convex and concave areas, first curvature values are divided into two classes (positive and negative) resulting in a binary image of convexity. Continuous inflexion lines are defined by the boundaries in the binary image that can be extracted either in the vector form by raster-to-vector conversion of the convexity polygons, or in the raster form by identifying the inner-boundary pixels of the uniform convexity areas with binary edge detection filters (Gonzalez and Woods 1993). Although this algorithm does need the calculation of second-order derivatives, it does not require the evaluation of the curvature $k = 0$ criteria for inflexion points and it also provides continuous lines.

Area-based line detection can be further enhanced by smoothing the DEM prior to derivative calculations, or by post processing of the obtained areas by majority filtering to eliminate small islands of pixels in order to obtain homogeneous and continuous areas with smooth edges.

4.5 Calculation of Singular Points

In this section, we use the term 'pixel' instead of 'grid node' because most of the discussed methods use image processing methods. However, 'pixel' is used in the same sense as 'grid node' in the following discussion.

While there is one unambiguous definition of horizontal flat surface points in vector analysis, several definitions of *flat* pixels are distinguished in numerical analysis of DEMs, based on the surface flow conditions at the pixel. A 'true flat' is a pixel surrounded by pixels of the same elevation. If the pixel has neighbours with equal and higher elevation only it is called an 'inflow-flat' pixel because it can receive inflow from the higher neighbours but flow cannot leave the pixel. This type is found at up-slope edges of flat areas. If the pixel has neighbours with equal and lower elevation only it is called an 'outflow-flat' pixel because it cannot receive inflow but flow can leave the pixel towards the lower neighbours. This type is found at down-slope edges of flat areas. These definitions indeed provide the algorithms to find flat points in grid DEM by applying these simple rules to the pixels in a 3x3 moving kernel. It is easy to see that the previously discussed numerical differential methods are quite inefficient in finding flats because they do not provide information on flow conditions at the central kernel point.

Localising surface specific points, i.e. local maxima (peaks), minima (pits), saddle points (passes) and inflexion points is fundamental in digital geomorphologic analysis (Peucker and Douglas 1975), however, the previously discussed numerical differential methods are quite inefficient for this purpose because they calculate the differences between elevations. For instance, locating a local maximum requires that the partials f_x and f_y are both zero. With this criterion, using the four-point finite difference method, for instance, would require that the local maximum present at the Z5 central kernel point is surrounded by N-S and E-W points of equal elevations, respectively (Z6–Z4 = 0 and Z2–Z8 = 0), which is a very unlikely situation in a real DEM. Thus, with methods based on vector analysis it is easy to miss identifying peaks. At the same time, grid DEMs often comprise flat areas represented by many pixels with the same elevation value due to insufficient grid resolution, numerical representation (integer elevations) or interpolation errors (such as DEM derived by TIN interpolation from contour lines). This leads to flat 'plateaus' instead of single pixel peaks. Therefore the analysis of local extremes requires methods which consider the spatial context of pixel points.

A simple algorithm for finding *local maxima* (peaks) first finds all pixels without higher elevation neighbours (i.e. single-pixel maxima, and outflow and true flats). Then, flat pixels having higher elevation neighbours (in-flow flat pixels) are located at the edges of the flats identified in the preceding step and are labelled as 'flat edge'. Next, 'flat edge' pixels are propagated in an iterative process to neighbour pixels that are flat pixels and have the same elevation as the given 'flat edge' pixels. This procedure marks all pixels that are found in flat areas with in-flow flat boundary pixels. Obviously, flat peak plateaus do not contain in-flow pixels only outflow and true flat pixels. If the marked 'flat edge' pixels areas are deleted then the remaining flat pixels are the sought peaks. Finally, flats at the DEM edges are deleted because their character cannot be decided. *Local minima* (pits) are found in the same way but having out-flow flat pixels propagated over flat areas and subsequently deleted.

Saddle points (passes) are the hardest to identify in DEM and there are no algorithms found for locating saddle points in literature. An exception is the method suggested by Peucker and Douglas (1975) that is based on parameter lines parallel to the x and y grid axes, benefiting from differencial geometric character of saddle points (see section 3.3.3). However, methods using transects reduce the two-dimensional problem to one-dimension thus they tend to fail in complex situations (Toriwaki and Fukumura 1978; Chorowitz et al. 1992). One way to find saddle points is based on the fact that saddle points lie on ridgelines defined by watershed divides, and saddle points are local minima on the surface curves of ridgelines (see Figure 3b). Therefore, ridgeline pixels are first identified by finding watershed boundaries as outer boundary pixels of watershed polygons. Then, by applying the above algorithm of local minimum identification locally to the watershed boundary pixels, saddle points can be found even if they consist of several flat pixels. The limitation of this simple method is that it depends on the success of the watershed extraction method that often fails to identify all sub-catchments and therefore all ridge lines (see below).

4.6 Digital Drainage Analysis: Ridge and Valley Lines

Extraction of single-pixel wide continuous valley and ridge line networks from DEMs is not an easy task and the problem of valley and ridge positioning has received great attention (e.g. Riazanoff et. al. 1988; Martz and Garbrecht 1992; Tribe 1992; McCormack et al. 1993; Mower 1994; Meisels et al. 1995). Valley and ridgeline extraction from a DEM is most often based on local processing of geometric properties using the 'higher than' concept (e.g. Band 1986; Peuker and Douglas 1975; Skidmore 1990; Chorowicz et al. 1992) or it is based on hydrological drainage direction information (O'Callaghan and Mark 1984; Martz and Garbrecht 1992). Some methods utilise the geometric property suggested by vector analysis, i.e. normal sections are concave with local minima at valley lines. If the DEM is sampled along x and y grid lines the obtained surface curves (parameter lines) should have local minima at valley lines (Chorowicz et al. 1992). Methods using local processing however need to be followed by connecting and thinning treatment (e.g. Band 1986). A polygon-breaking method proposed by Chang et al. (1998) tries to model human abilities of working on different scales at the same time in network analysis. Other methods apply hydrological drainage network algorithms that define the channels draining the landscape as those pixels that have an upstream drainage area greater than a threshold drainage area (critical source area) (Mark 1984; O'Callaghan and Mark 1984). Most of the models use the down-slope flow routing concept (two-point maximum down-gradient estimation method) (Douglas 1986). These methods provide a fully-connected, convergent and unidirectional down-slope drainage line network. In this approach, ridge lines are defined as boundaries between pixels of adjacent watersheds draining to drainage segments. Jenson and Dominigue's (1988) algorithm is used most often for removing spurious pits and the methods of Martz and Garbrecht (1992) or Soille et al. (2003) provide algorithms to treat flat areas that are obstacles for down-slope flow routing.

For the geometric analysis of the topographic surface we need the maximum number of valley lines extracted and to have them with their maximum length. Under these conditions however spurious multiple parallel channels are produced by hydrological drainage network algorithms due to insufficient DEM resolution where relief is insufficient to force flow lines to converge together (Tribe 1992). Spurious parallel channels are drainage channels that appear as double (multiple) pixel-wide lines. They occur in basin areas or in flat valley bottoms. Since ridges are defined as watershed boundaries between channels, spurious ridge lines appear in the middle of valleys. The formation of spurious channel segments is a fundamental problem in digital drainage analysis and many methods have been proposed to overcome it. In an attempt to solve the problem of extracting a single-pixel wide high-density continuous channel network, Jordan (2004; 2007) proposed a novel method that starts with applying the possible minimum critical source area threshold value in order to ensure high channel density. The resulting network still contains multiple pixel wide segments. Spurious segments can be reduced if relief is increased around streamlines in flat areas. Thus, the extracted channel network is reduced first to single-pixel width with the mathematical morphological method of skeletonisation (Soille 2003). The skeleton is the geometric middle line of the multiple channels. Next, the whole skeleton network is burnt into the DEM by subtracting a constant value from a pit and flat corrected DEM. The DEM is subsequently smoothed along the obtained one pixel-wide valleys with a moving average filter. This widens the one pixel-wide 'canyons' to the width of multiple channels thus creating smooth edges along the drainage line gradually descending towards the drainage pixels. The result is creating valley slopes around the middle skeleton line in the place of the multiple channel pixels dipping towards the skeleton line. The larger the specified moving average kernel size the wider the valley is along the drainage line. Subsequent application of the drainage network algorithm to the modified DEM results in single-line channels and watershed boundaries being correctly positioned on morphological ridgelines.

Garbrecht and Martz (1994) note that extracted drainage properties are dependent on the pixel size and drainage features at the scale of DEM cannot be accurately reproduced. Burrough and McDonnell (1998) notes that drainage networks and up-stream contributing areas are sensitive to errors in DEMs and they suggest the use of the Monte Carlo method to see the effect of random DEM errors on the extracted drainage network.

5. Conclusions

This paper discusses digital terrain analysis in a GIS environment in a systematic way which is not present in other literature. In accordance with the objective of geometric surface characterisation for surface flow modelling, the scope is limited to primary point attributes and to vector-scalar and scalar-vector functions. Vector-vector functions necessary for complete flow-field characterisation are not considered here. In this paper, after some basic definitions, mathematical derivation of parameters that are the most important for geometric surface characterisation was presented. The first part concluded with a novel system of geomorphometric parameters that is an extension of Evan's (1980) five-parameter method. The second part was dedicated to the development and discussion of numerical methods for the extraction of these parameters based on literature review.

It is noted that the presented univariate analysis of primary point attributes can be extended to bivariate analysis (e.g. secondary attributes and correlation analysis) and to texture analysis of surfaces (e.g. trend, autocorrelation and spectral analysis) using image processing techniques more extensively, as proposed by Jordan and Csillag (2001), Jordan and Schott (2005) and Jordan et al. (2005).

The consistent development of the subject also aimed at clarifying some of the misunderstandings around terrain analysis, such as surface continuity and smoothness, gradient and slope vectors, and curvature and 'surface curvature'. Understanding of underlying mathematics and assumptions of numerical methods, together with consistency of terminology and application of numerical methods, contribute to the efficient application of DTA to modelling in a decision support environment.

References

Aronoff S (1989) Geographic information systems: a management perspective. WDL Publications, Ottawa

Abramowitz M and Stegun IA (1972) Handbook of mathematical functions with formulas, graphs, and mathematical tables. Dover Publications Inc., New York

Band LE (1986) Topographic partition of watersheds with digital elevation models, Water Resources Research 22:15-24

Burrough PA and McDonnell RA (1998) Principles of geographical information Systems. Oxford University Press, Oxford

Carrara A, Bitelli G, Carla R (1997) Comparison of techniques for generating digital terrain models from contour lines. International Journal of Geographic Information Systems 11:451-473

Carter JR (1988) Digital representations of topographic surfaces. Photogrametric Engineering and Remote Sensing 54:577-1580

Carter JR (1992) The effect of data precision on the calculation of slope and aspect using gridded DEMs. Cartographica 29:22-34

Chang K, Tsai K (1991) The effect of DEM resolution on slope and aspect mapping. Cartography and Geographic Information Systems 18:69-77

Chang Y, Song G, Hsu S (1998) Automatic extraction of ridge and valley axes using the profile recognition and polygon-breaking algorithm. Computers and Geosciences 24:83-93

Chorowicz J, Ichoku C, Riazanoff S, Kim YJ, Cervelle B (1992) A combined algorithm for automated drainage network extraction. Water Resources Research 28:1293-1302

Dubrule O (1984) Comparing splines and kriging. Computers and Geosciences 10:327-338

Douglas DH (1986) Experiments to locate ridges and channels to create a new type of digital elevation models. Cartographica 23:2961

El-Sheimy N, Valeo C, Habib A (2005) Digital terrain modelling. Acquisition, manipulation, and applications. Artech House Inc., Boston

Evans IS (1972) General geomorphometry, derivatives of altitude, and descriptive statistics. In: Chorley RJ (ed) Spatial analysis in geomorphology. Methuen, London, pp 17-90

Evans IS (1980) An integrated system for terrain analysis for slope mapping. Zeitschrift für Geomorphologie 36:274-295

Flemming MD, Hoffer RM (1979) Machine processing of Landsat MSS data and DMA topographic data for forest cover type mapping. LARS technical report 062879, Laboratory for Applications of Remote Sensing, Purdue University, West Lafayette, Indiana, U.S.A.

Florinsky IV (1998) Accuracy of local topographic variables derived from digital elevation models. International Journal of Geographic Information Science 12:47-61

Florinsky IV (2000) Relationship between topographically expressed zones of flow accumulation and sites of faults intersection: Analysis by means of digital terrain modelling. Environmental Modelling and Software 15:87-100

Garbrecht J, Martz L (1994) Grid size dependency of parameters extracted from digital elevatin models. Computers and Geosciences 20:85-87

Garbrecht J, Martz LW (1995) Agricultural Research Service Publication NAWQL 95-3. In TOPAZ: An automated digital landscape analysis tool for topographic evaluation, drainage identification, watershed segmentation and subcatchment parameterisation: TOPAZ User Manual. U.S. Department of Agriculture, pp 95-3

Gonzalez RC, Woods RE (1993) Digital image processing. Addison-Wesley Publishing Company, New York

Guth PL (1995) Slope and aspect calculations on gridded digital elevation models: Examples from a geomorphometric toolbox for personal computers. Zeitschrift fur Geomorphologie N. F., Suppl.-Bd. 101:31-52

Hardy R (1971) Multiquadratic equations of topography and other irregular surfaces. Journal of Geophysical Research 76:1905-1915

Hodgson ME (1995) What cell size does the computed slope/aspect angle represent? Photogrammetric Engineering and Remote Sensing 61:513-517

Horn BK (1981) Hill shading and the reflectance map. Proceedings of the I.E.E.E. 69:14-47

Hutchinson MF (1989) A new procedure for gridding elevation and steam line data with automatic removal of spurious pits. Journal of Hydrology 106:211-232

Jenson SK, Domingue JO (1988) Extracting topographic structure from digital elevation data for geographic information system analysis. Photogrammetric Engineering and Remote Sensing 54:1593-1600

Jones KH (1988) A comparison of algorithms used to compute hill slope as a property of the DEM. Computers and Geosciences 24:315-323

Jordan G (2003) Morphometric analysis and tectonic interpretation of digital terrain data: a case study. Earth Surface Processes and Landforms 28:807-822

Jordan G (2007) Adaptive smoothing of valleys in DEMs using TIN interpolation from ridgeline elevations: an application to morphotectonic aspect analysis. Computers and Geosciences (in press)

Jordan G, Csillag G (2001) Digital terrain modelling for morphotectonic analysis: a GIS framework. In: Ohmori H (ed) DEMS and geomorphology. Special Publication of the Geographic Information Systems Association 1:60-61. Nihon University, Tokyo

Jordan G, Csillag G, Szucs A, Qvarfort U (2003) Application of digital terrain modelling and GIS methods for the morphotectonic investigation of the Kali Basin, Hungary. Zeitschrift fur Geomorphologie 47:145-169

Jordan G (2004) Digital terrain modelling with GIS for tectonic geomorphology. Numerical Methods and Applications. PhD Thesis. Comprehensive Summaries of Uppsala Dissertations, Faculty of Science and Technology, 1031. Acta Universitatis Upsaliensis

Jordan G, Schott B (2005) Application of wavelet analysis to the study of spatial pattern of morphotectonic lineaments in digital terrain models. A case study. Remote Sensing of Environment 93:31-38

Jordan G, Meijninger BML, van Hinsberen DJJ, Meulenkamp JE, van Dijk PM (2005) A GIS framework for digital tectonic geomorphology: case studies. International Journal of Applied Earth Observation and Geoinformation 7:163-182

Jordan G, van Rompaey A, Szilassi P, Csillag G, Mannaerts C, Woldai T (2005) Historical land use changes and their impact on sediment fluxes in the Balaton basin (Hungary). Agriculture, Ecosystems and Environment 108:119-133

Krcho J (1973) Morphometric analysis of relief on the basis of geometric aspect of field theory. Acta UC, Geographic Physica 1, Bratislava

Lam NS (1983) Spatial interpolation methods: a review. The American Cartographer 10:129-149

Li Z, Zhu Q, Gold C (2005) Digital terrain modelling. Principles and methodology. CRC Press, New York

Mark DM (1984) Automatic detection of drainage networks from digital elevation models. Cartographica 21:168-178

Martz LW, Garbrecht J (1992) Numerical definition of drainage networks and subcatchment areas from digital elevation models. Computers and Geosciences 18:747-761

Martz LW, Garbrecht J (1995) Automated recognition of valley lines and drainage networks from grid digital elevation models: a review and a new method. Journal of Hydrology 167:393-396

McCormack JE, Gahegan MN, Roberts SA (1993) Feature-based derivation of drainage networks. International Journal of Geographical Information Systems 7:263-279

McCullagh MJ (1988) Terrain and surface modelling systems: theory and practice. Photogrammetric Records 12:747-779

Meisels A, Raizman S, Karnieli A (1995) Skeletonizing a DEM into a drainage network. Computers and Geosciences 21:187-196

Mitasova H, Hofierka J (1993) Interpolation by regularized spline with tension: II. Application to terrain modelling and surface geometry analysis. Mathematical Geology 25:657-655

Moore ID (1988) A contour-based terrain analysis program for the environmental sciences (TAPES). Transactions of the American Geophysical Union 69:345

Moore ID, Grayson RB (1989) Hydrologic and digital terrain modelling using vector elevation data. American Geophysical Union 70:1091

Moore ID, Grayson RB, Ladson AR (1991) Digital terrain modelling: a review of hydrological, geomorphological and biological application. Hydrological Processes 5:3-30

Moor ID, Lewis A, Gallant JC (1993) Terrain attributes: estimation methods and scale effects. In: Jakeman AJ, Beek MJ, McAleer MJ (eds) Modelling change in environmental systems. John Willey and Sons, London

Mower JE (1994) Data-parallel procedures for drainage basin analysis. Computers and Geosciences 20:1365-1378

O'Callaghan JF, Mark DM (1984) The extraction of drainage networks from digital elevation data. Computer Vision, Graphics and Image Processing 28:323-344

O'Neill MP, Mark DM (1987) On the frequency distribution of land slope. Earth Surface Processes and Landforms 12:127-136

Onorati G, Poscolieri M, Ventura R, Chiarini V, Crucillá U (1992) The digital elevation model of Italy for geomorphology and structural geology. Catena 19:147-178

Papo HB, Gelbman E (1984) Digital terrain models for slope and curvatures. Photogrammetric Engineering and Remote Sensing 50:695-701

Peucker TK, Douglas DH (1975) Detection of surface-specific points by local parallel processing of discrete terrain elevation data. Computer Graphics and image processing 4:375-387

Prewitt JMS (1970) Object enhancement and extraction. In: Lipkin BS, Rosenfeld A (eds), Picture processing and psychopicotoics. Academic Press, New York, pp 75-149

Rektorys K (1969) Survey of applied mathematics. MIT Press, Combrige, MA, Iliffe Books Ltd., London

Riazanoff S, Cervelle B, Chorowicz J (1988) Ridge and valley line extraction from digital images. International Journal of Remote Sensing 9:1175-1183

Riley C, Moore McM (1993) Digital elevation modelling in a study of the neotectonic geomorphology of the Sierra Nevada, southern Spain. Zeitschrift fur Geomorphologie N. F., Suppl.-Bd. 94:25-39

Sharpnack D A, Akin G (1969) An algorithm for computing slope and aspect from elevations, Photogrammetric Engineering 35:247-248

Simpson DW, Anders MH (1992) Tectonics and topography of the Western United States - an application of digital mapping. GSA Today, 2:118-121

Skidmore AK (1989) A comparison of techniques for calculating gradient and aspect from a gridded digital elevation model. International Journal of Geographic Information Systems 3:323-334

Skidmore AK (1989) Terrain position as mapped from a gridded digital elevation model International Journal of Geographic Information Systems 4:33-49

Shary PA (1995) Landsurface in gravity points classification by a complete system of curvatures. Mathematical Geology 27:373-390

Snyder WM, Bruce RR, Harper LA, Thomas AW (1984) Two-dimensional sliding polynomials. Univ. Georgia Agric. Exp. Sta. Res. Bull. 320:58

Soille P (2003) Morphological image analysis: Principles and applications. Springer-Verlag, Berlin Heidelberg New York, 2nd edition

Soille P, Vogt J, Colombo R (2003) Carving and adpative drainage enforcement of grid digital elevation models. Water Resources Research 39:1366

Toriwaki J, Fukumura T (1978) Extraction of structural information from grey pictures. Computer Graphics and Processing 7:30-51

Tribe A (1992) Automated recognition of valley lines and drainage networks from grid digital elevation models: a review and a new method. J. Hydrology 139:263-293

Tuma J (1979) Engineering mathematics handbook. McGraw-Hill Book Company, New York

Unwin D (1981) Introductory spatial analysis. Metthuen, London

Wilson JP, Gallant JC (2000) Terrain analysis. Principles and applications. John Wiley & Sons, London

Zevenbergen LW, Thorne CR (1987) Quantitative analysis of land surface topography. Earth Surface Processes and Landforms 12:47-56

From Mathematical Morphology to Morphological Terrain Features

Pierre Soille

Spatial Data Infrastructures Unit, Institute for Environment and Sustainability
DG Joint Research Centre, European Commission

1. Introduction

Mathematical morphology can be defined as a theory for the analysis of spatial structures. It is called morphology because it concentrates on the form, shape, and size of the structures. It is mathematical in the sense that the analysis is based on set theory, integral geometry, and lattice algebra. When applied to digital spatial data, mathematical morphology offers a wide variety of methods and algorithms to solve scientific and practical problems.

The origin of mathematical morphology stems from the study of the geometry of porous media in the mid-sixties in France. Porous media are binary in the sense that a point of a porous medium either belongs to a pore or to the matrix surrounding the pores. This led Georges Matheron and Jean Serra to introduce a set formalism for analysing binary images (Matheron 1967, 1975; Serra 1982, 1988). Indeed, one may consider the matrix as the set of object points and the pores as the complement of this set. As a consequence, image objects can be studied by analysing their interaction with other sets of known shape and size called *structuring elements*. The interaction is defined in terms of basic set operations such as the union, intersection, and complementation. A 3-dimensional (3-D) set representation of 2-D grey scale images is obtained by considering the set of points lying above the image plane and below the image intensity surface. Hence, in mathematical morphology, every grey scale image is viewed as if it were a digital elevation model (DEM). This explains why numerous mathematical morphology terms stem from geomorphology and, conversely, why mathematical morphology is naturally suited for the analysis and processing of digital elevation models (Soille 1992).

This chapter presents an overview of the terrain features that can be derived from digital elevation models using mathematical morphology. Some of the most advanced techniques such as carving and optimal spurious pit removal have been motivated by the development of the pan-European river and catchment database detailed in this volume (Vogt et al. 2007).

This chapter is organised as follows. Background notions of mathematical morphology are presented in section 2. The generation of DEMs using a geodesic morphological interpolation technique is described in section 3. The automatic delineation of river networks from DEMs is detailed in section 4, the emphasis being put on the removal of spurious pits and the definition of flow directions on flat regions. The detection of watersheds using a flooding simulation and the extraction of crest lines using skeletonisation algorithms is briefly presented in section 5.

2. First Steps in Mathematical Morphology

Beyond the seminal books by Matheron (1967, 1975) and Serra (1982, 1988), a comprehensive introduction to the principles and applications to mathematical morphology can be found in (Soille 2003). In this section, we follow the definitions and notations of this latter reference.

2.1. Erosion and Dilation

The first question that may arise when we probe a set with a structuring element is *'Does the structuring element fit the set?'* The eroded set is the locus of points where the answer to this question is affirmative. In mathematical terms, the erosion ε of a grey scale image f at a given pixel $\mathbf{x}$ is equal to the minimum value of the image in the window defined by the structuring element B when its origin is at $\mathbf{x}$:

$$[\varepsilon_B(f)](\mathbf{x}) = \min_{\mathbf{b}\in B} f(\mathbf{x}+\mathbf{b}). \tag{1}$$

The dilation δ is the dual transformation (maximum filter):

$$[\delta_B(f)](\mathbf{x}) = \max_{\mathbf{b}\in B} f(\mathbf{x}+\mathbf{b}). \tag{2}$$

The erosion and dilation of a DEM by a disk structuring element are illustrated in Figs. 1b and c respectively. The small discrete disk structuring element B used in these figures is clearly visible in the eroded and dilated images. Indeed, a pixel near the upper left (resp. lower right) corner of the input image shown in Figure 1a has been arbitrarily set to a low (resp. high) value so that it generated a dark (resp. bright) disk of size and shape identical to that of the structuring element in the eroded (resp. dilated) image.

When the structuring element B is restricted to a pixel and its neighbours (e.g., 4- or 8-nearest neighbours for a 2-D image defined over a square grid), the corresponding erosion is called an elementary erosion and is denoted by $\varepsilon^{(1)}$. Similarly, the elementary dilation if denoted by $\delta^{(1)}$.

Erosions and dilations are the letters of the morphological alphabets in the sense that all other transformations are based on these two elementary transformations. For example, the notion of relief energy can be expressed in terms of erosions and dilations. Indeed, the relief energy of a given point of the terrain is usually defined as the maximal variation of the elevation values falling around a disk centred at this point. It is therefore equal to the arithmetic difference between the dilation and the erosion by this disk, see example in Figure 1d. This operation is called a morphological gradient because, in the case of a differentiable function, it is equal to the norm of the gradient vector when the radius of the disk tends to zero.

2.2. Opening and Closing

The geometric formulation of the morphological opening of a set in terms of structuring element fits is as follows: *'Does the structuring element fit the set?'* Each time the answer to this question is affirmative, the *whole* structuring element must be kept (for the erosion, it is the origin of the structuring element that is kept). It can be shown that the same result can be achieved by performing an erosion followed by a dilation. More precisely, the opening γ of an image f by a structuring element B is denoted by $\gamma_B(f)$ and is defined as the erosion of f by B followed by the dilation with the reflected structuring element $\check{B}$:

$$\gamma_B(f) = \delta_{\check{B}}[\varepsilon_B(f)], \tag{3}$$

where $\check{B} = \{-\mathbf{b} \mid \mathbf{b} \in B\}$. The morphological closing ϕ is the dual transformation:

$$\phi_B(f) = \varepsilon_{\check{B}}[\delta_B(f)]. \tag{4}$$

The opening/closing operations are illustrated on a sample DEM in Figs. 1E and f respectively. Note that the opening (resp. closing) levels down (resp. up) all ridges (resp. valleys) that are thinner than the structuring element until it fully fits the ridge (resp. valley).

2.3. Geodesic Transformations

The output of the elementary geodesic erosion of an image f (called the marker image) with respect to another image g (called the mask image) is denoted by $\varepsilon_g^{(1)}(f)$ and defined as the point-wise maximum between the mask image and the elementary erosion of f (assuming that the mask image g is less than or equal to the marker image f: $g(\mathbf{x}) \leq f(\mathbf{x})$ for all pixel $\mathbf{x}$):

$$\left[\varepsilon_g^{(1)}(f)\right](\mathbf{x}) = \max\{g(\mathbf{x}), \left[\varepsilon^{(1)}(f)\right](\mathbf{x})\}. \tag{5}$$

The geodesic erosion of size n of a marker image f with respect to a mask image g is obtained by performing n successive elementary geodesic erosions of f with respect to g:

$$\varepsilon_g^{(n)}(f) = \varepsilon_g^{(1)}[\varepsilon_g^{(n-1)}(f)], \tag{6}$$

with $\varepsilon_g^{(0)}(f) = f$. The reconstruction by erosion of a mask image g from a marker image f is defined as the geodesic erosion of f with respect to g iterated until stability is reached. It is denoted by $R_g^{\varepsilon}(f)$:

$$R_g^{\varepsilon}(f) = \varepsilon_g^{(i)}(f), \tag{7}$$

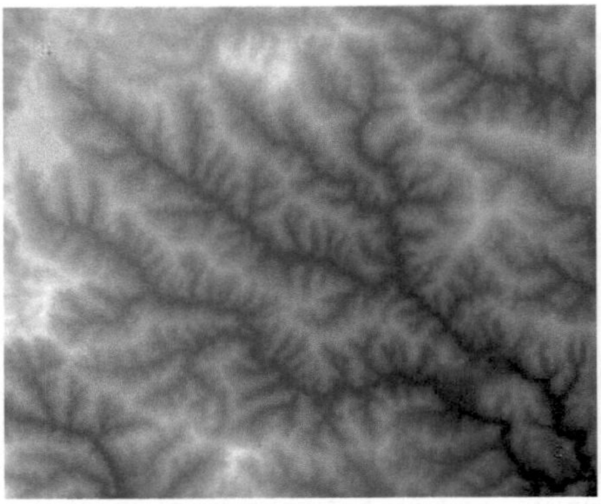

(A) Input image f

(B) Eroded image: $\varepsilon_B(f)$ with structuring element B materialised by the dark disk

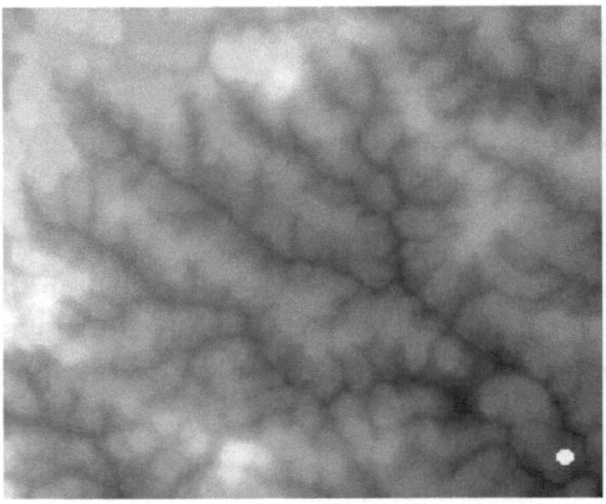

(C) Dilated image: $\delta_B(f)$ with structuring element B materialised by the bright disk

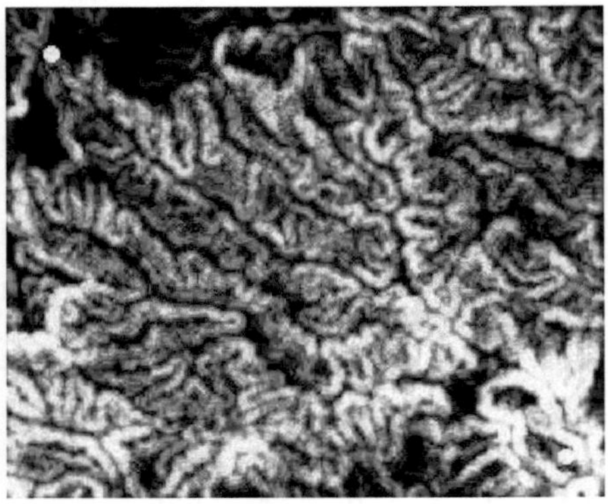

(D) Morphological gradient: $\delta_B(f) - \varepsilon_B(f)$

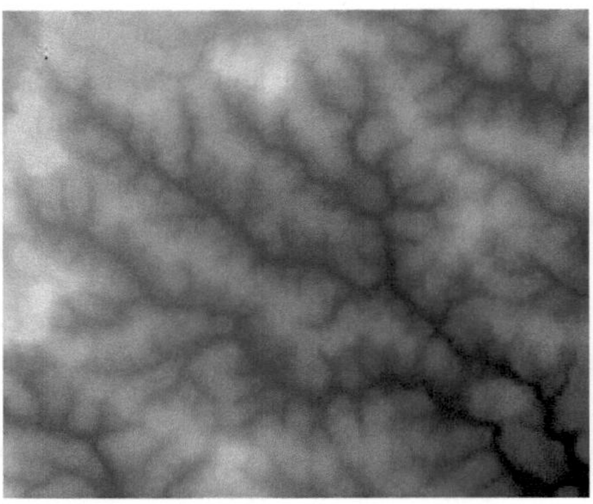

(E) Opened image: $\gamma_B(f)$

(F) Closed image: $\phi_B(f)$

Fig. 1. Basic morphological transformations applied to a 286×243 sample SRTM DEM of the Gardon valleys in France. The DEM has been projected in the ETRS-89 LAEA projection (Annoni et al. 2003) and gridded (Annoni 2005) at 100m where i is such that $\varepsilon_g^{(i)}(f) = \varepsilon_g^{(i+1)}(f)$. The dual transformation is called the reconstruction by dilation is denoted by $R_g^{\delta}(f)$. It is obtained by replacing the maximum with the minimum operator and the erosion by the dilation in Eqs. 5–7. For the reconstruction by dilation, it is assumed that the mask image g is greater than or equal to the marker image f: $g(\mathbf{x}) \geq f(\mathbf{x})$ for all pixel $\mathbf{x}$.

The reconstruction operators are illustrated on Figure 2. In this figure, the marker

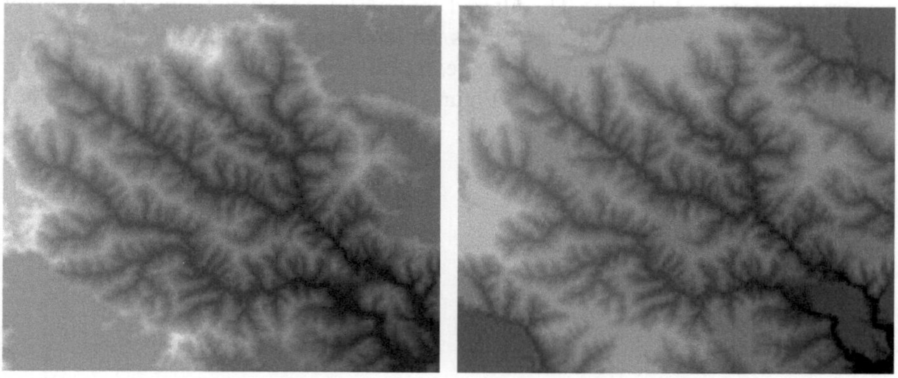

(A) Reconstruction by erosion of Figure 1A using as marker the bright point near the lower right corner of this latter image

(B) Reconstruction by dilation of Figure 1A using as marker the dark point near the upper left of this latter image

Fig. 2. Reconstruction by erosion (resp. dilation) of the DEM displayed in Figure 1a using a single point as marker

image used for the reconstruction by erosion (resp. dilation) consists of a constant image with maximum (resp. minimum) value except for the marker point which is set to a minimum (resp. maximum) value. Note that the reconstruction by erosion reconstructs exactly only those points of the DEM that can be reached by following an ascending path from the marker while the reverse applies to the reconstruction by dilation.

Finally, the concept of geodesic distance is heavily used in applications where the length of shortest path linking two points is not equal to the Euclidean distance owing to the presence of obstacles. In mathematical terms, the geodesic distance between two points p and q of a connected subset A of the Euclidean plane is the length of the shortest path, called the geodesic path, included in A and linking p to q. It is denoted by $d_A(p,q)$. Similarly, the geodesic distance between a pixel p of the geodesic mask A and a subset Y of A is defined as the smallest geodesic distance between p and any pixel q of Y: $d_A(p,Y) = \min_{q \in Y} d_A(p,q)$.

3. From Topographic Maps to DEMs

If elevation data are available only through topographic paper maps, DEMs can be created by scanning the maps and interpolating the elevation values of all pixels located in between the contour lines. For example, a sample of a digitised topographic map is shown in Figure 3a. The contour lines of this image can be automatically extracted using colour information. The resulting image is then cleaned using size criteria (area opening and closing). The filtered contour lines are then thinned to one pixel lines thanks to skeletonisation algorithms while disconnected lines are reconnected by analysing the distance separating their extremities as well as their orientation. The resulting contour lines are shown in Figure 3b. Once the connected elevation lines

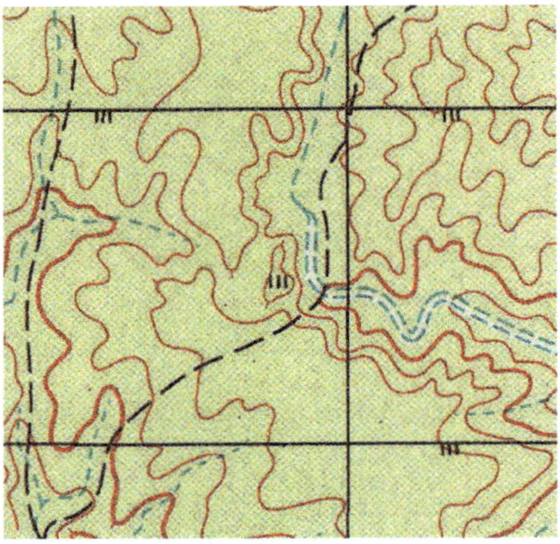

(A) Digitised topographic map

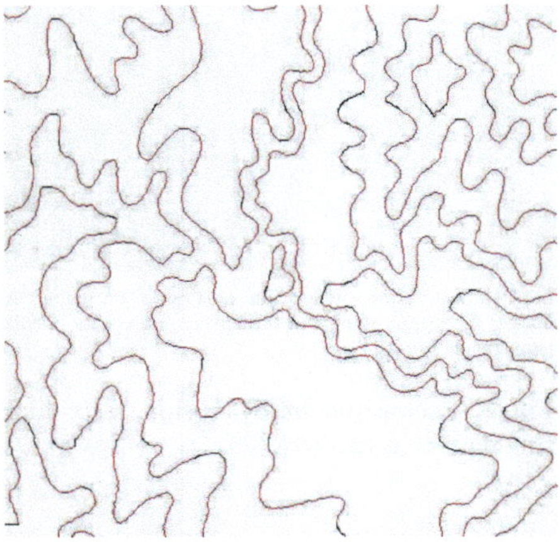

(B) Extracted contour lines with connections of initially disconnected lines shown in black

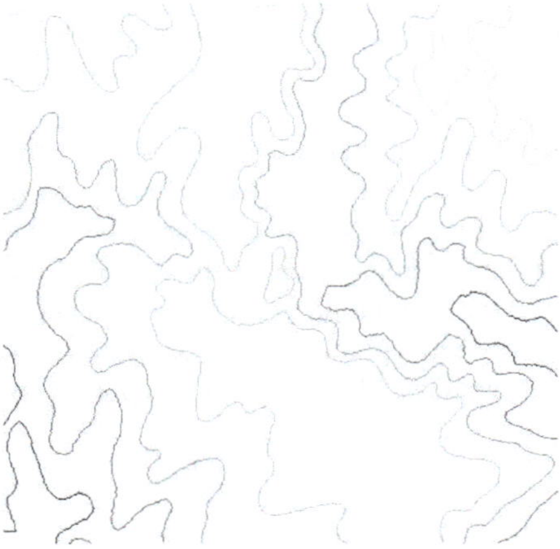

(C) Image CL of elevation contour lines with elevation values along the lines

Fig. 3. From a digitised topographic map to elevation contour lines: initially, the paper map is scanned *(A)*, then the contour lines are extracted while bridging the gaps caused by the superimposed layers *(B)*, and finally each line is set to the elevation value it refers to *(C)*

have been extracted from a topographic map the value of each pixel belonging to a contour line is set to the value corresponding to its elevation. This leads to the image of valued contour lines denoted by CL and displayed in Figure 3c. All these preliminary steps are described in (Arrighi and Soille 1999). In Figure 3c, the elevation values of the pixels belonging to the white regions in between two successive contour lines need to be interpolated to generate a grid DEM from this image. A wide variety of interpolation methods may be used. We briefly describe hereafter a method based on geodesic transformations and taking advantage of the topological and morphological properties of the contour lines (Soille 1991). The method is subdivided into two steps: (i) generation of two plateau images indicating the lower and upper limits of the interpolation and (ii) interpolation along the steepest slope lines passing through the considered point.

3.1. Plateau Image Generation

Those pixels that need to be interpolated are located in the regions delineated by two elevation lines with different elevation values. Given a pixel p of such a region, $c_l(p)$ refers to the lower contour line associated with p. Similarly, $c_u(p)$ refers to the upper contour line associated with p. The lower plateau image P_l holds for each pixel p the elevation of its lower contour line c_l as defined by the image of valued contour lines CL:

$$P_l(p) = \mathrm{CL}[c_l(p)]. \tag{8}$$

The upper plateau image P_u is defined accordingly. The computation of the lower and upper plateau images requires the definition of the following auxiliary image M:

$$M(p) = \begin{cases} \mathrm{CL}(p), & \text{if } p \text{ belongs to a contour line,} \\ h_{\max}, & \text{otherwise,} \end{cases} \tag{9}$$

where $h_{\max}$ denotes an arbitrary height higher than those of the considered map. The lower plateau image P_l is obtained by performing the morphological reconstruction by erosion of the image CL of contour lines using the auxiliary image M as marker image:

$$P_l = R^{\varepsilon}_{\mathrm{CL}}(M) \tag{10}$$

Similarly, the upper plateau image P_u is defined as the morphological reconstruction by dilation of the auxiliary image M using the image of contour lines CL as marker image:

$$P_u = R^{\delta}_{M}(\mathrm{CL}) \tag{11}$$

The methodology applied to the image of contour lines of Figure 3c outputs the plateau images displayed in Figure 4.

3.2. Interpolation Along Steepest Slope Lines

The two plateau images indicate the lower and upper bounds of the interpolated value of each pixel. In practice, we perform a linear interpolation along the shortest

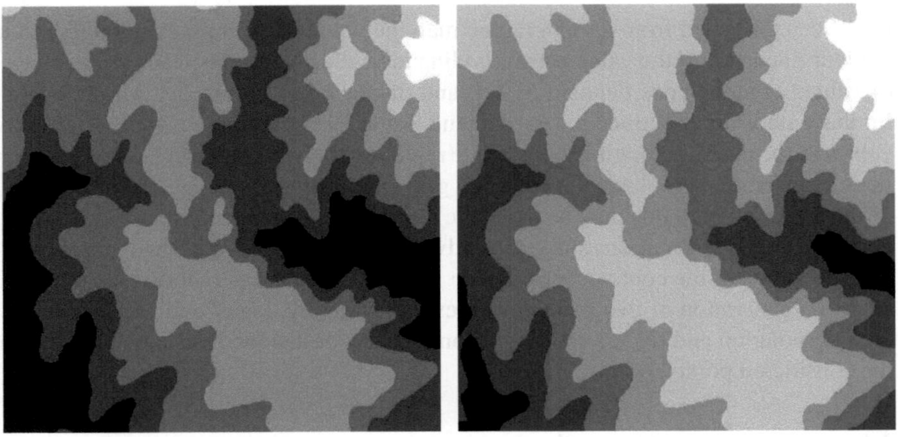

(A) Lower plateau image $P_l = R^{\varepsilon}_{\mathrm{CL}}(M)$ (B) Upper plateau image $P_u = R^{\delta}_{M}(\mathrm{CL})$

Fig. 4. Plateau images derived from the image of elevation contour lines displayed in Figure 3c

path linking the lower and upper elevation lines and going through the considered pixel. This shortest path is a geodesic path in the sense that it is constrained to remain within the region delineated by the lower and upper contour lines. Hence, the interpolated value $H(p)$ of p equals the weighted mean of $P_u(p)$ and $P_l(p)$ by the geodesic distances from p to $C_u(p)$ and p to $C_l(p)$ respectively:

$$H(p) = \frac{P_l(p)\, d_X[p, C_u(p)] + P_u(p)\, d_X[p, C_l(p)]}{d_X[p, C_l(p)] + d_X[p, C_u(p)]}, \tag{12}$$

where the geodesic mask X refers to the set of pixels that do not belong to any elevation line. This interpolation function is illustrated in Figure 5. Flat peaks and troughs can be avoided provided that the elevation and location of the peaks and troughs is mapped in the image CL of elevation lines (i.e., elevation lines reduced to 1 point and with only one such point per peak or trough). In the displayed example, only the elevation of the peak located in the largest closed contour line was available so that all smaller peaks as well as valley floors connected to the image border appear as flat regions in the interpolated DEM.

4. From DEMs to River Networks

Existing algorithms for delineating river networks from DEMs can be partitioned into two main groups depending on whether a geomorphological or hydrological outlook is adopted. Reviews of both approaches can be found in (Moore et al. 1991; Tribe 1992; Bertolo 2000).

From the geomorphological outlook, points belonging to the drainage network are defined from local morphologies (Johnston and Rosenfeld 1975; Peucker and

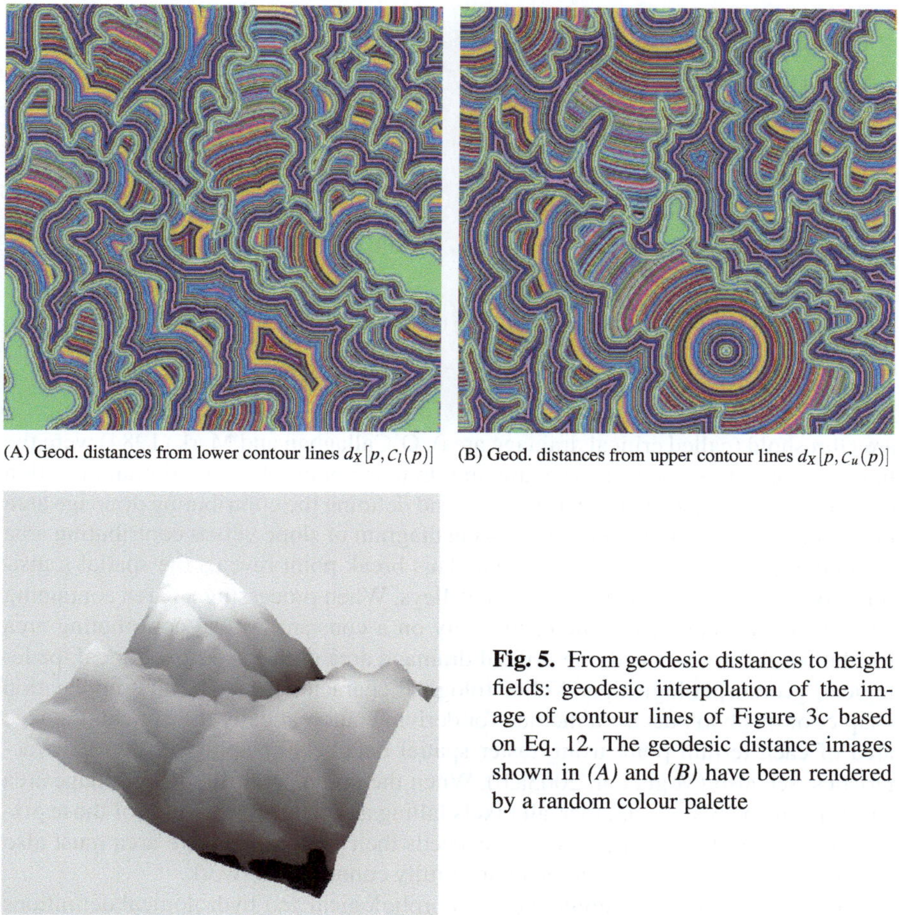

(A) Geod. distances from lower contour lines $d_X[p, C_l(p)]$ (B) Geod. distances from upper contour lines $d_X[p, C_u(p)]$

(C) Interpolated height field $H(p)$

Fig. 5. From geodesic distances to height fields: geodesic interpolation of the image of contour lines of Figure 3c based on Eq. 12. The geodesic distance images shown in *(A)* and *(B)* have been rendered by a random colour palette

Douglas 1975; Toriwaki and Fukumara 1978; Haralick 1983; Douglas 1986; Band 1986; Seemuller 1989; Zhang et al. 1990; Smith et al. 1990; Qian et al. 1990). For example, Johnston and Rosenfeld (1975) considered as drainage pixels all terrain points having a concave curvature coefficient higher than a given threshold. However, there exists no absolute threshold level to discriminate drainage pixels from other pixels. As a consequence, the resulting drainage networks are usually disconnected. Non-trivial postprocessing steps are therefore needed for connecting extracted pixels and removing irrelevant ones. This also applies to local morphologies defined in terms of morphological transformations. For instance, valleys narrower than a given width value n can be obtained by subtracting the input DEM from the closing of this DEM by a disk structuring element of diameter n (see Figure 6).

The hydrological approach is based on the assumption that channels form where the area drained by a given point (called contributing drainage area) exceeds some

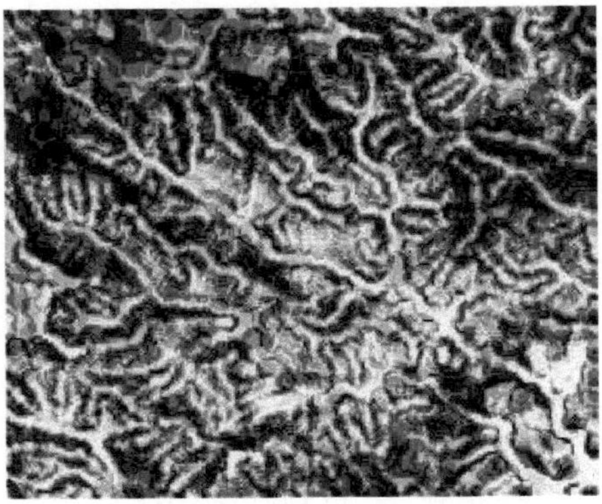

Fig. 6. Valley network enhancement by computing the difference between the closing of a DEM (see Figure 1f) and the initial DEM (see Figure 1a)

area threshold (called critical drainage area). O'Callaghan and Mark (1984) were the first to exploit this concept for extracting drainage networks from digital elevation data. Tarboton et al. (1989, 1991) recommend defining the contributing drainage area threshold as the break point of the log–log diagram of slope versus contributing area for individual cells within a catchment. This break point reveals the spatial transition from convex hill slopes to concave valleys. When processing an area containing heterogeneous landscapes, one cannot rely on a constant critical contributing area for the whole area. Indeed, the critical drainage area depends on geological, pedological, geomorphological, and climatological characteristics. The first application chapter of this book details a method for deriving automatically the critical drainage area of each terrain point using raster spatial data sets mapping all these characteristics, see also (Vogt et al. 2003a,b). When the critical contributing drainage area varies from one pixel to another, all pixels falling along the downstream of those pixels whose contributing drainage area exceeds their critical drainage area must also be considered as river pixels to generate a fully connected network.

Hybrid approaches combine the geomorphological and hydrological definitions in such a way that the sources of the rivers are defined from local morphologies while all downstream pixels of these sources define the river network (Chorowicz et al. 1992; Tribe 1992; Howard 1994; Rinaldo et al. 1995).

Contrary to the geomorphological approach, both the hydrological and hybrid approaches generate connected networks provided that the flow simulation algorithm handles the problem of spurious pits and flat areas which could cause channel discontinuities. These two issues are discussed hereafter.

4.1. Spurious Pits and their Suppression

The downstream of a point is defined as the steepest flow path linking this point to the catchment outlet. It follows that spurious pits must be removed before applying flow path detection algorithms to avoid flow paths being trapped in local minima. The need for this essential preprocessing step has been pointed out by (Mark 1984) and several solutions have been proposed since then. Essentially, three approaches may be considered as illustrated in Figure 7 on a 1-D signal: (i) pit filling whereby pits are progressively filled (by increasing their elevation values) until the elevation of their lowest outflow point is reached (Figure 7b), (ii) carving whereby a descending path is created from the bottom of the pit until the nearest point of lower elevation value is reached (Figure 7c), and (iii) an optimal approach combining the two first in such a way that the sum of the elevation differences between the input and output pitless DEM is minimised (Figure 7d). The optimal method requiring the understanding of the two first, all three methods are detailed hereafter.

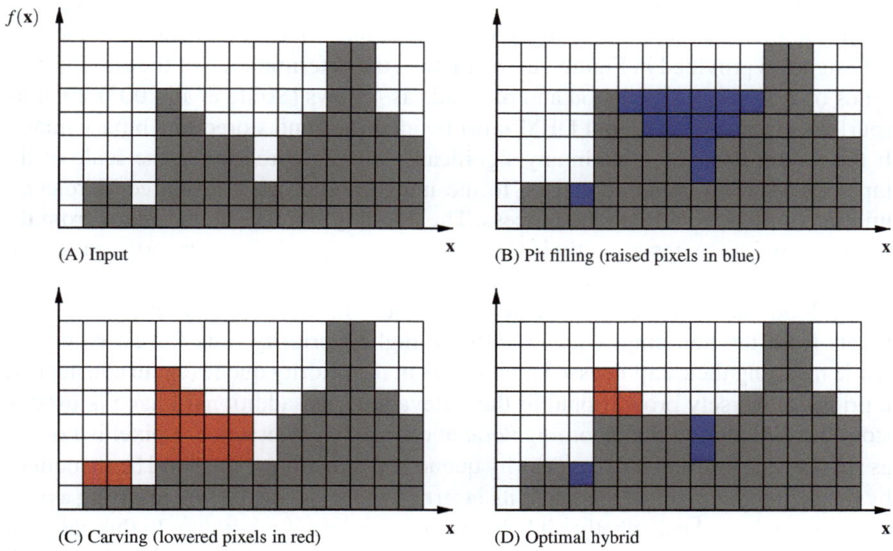

Fig. 7. Removal of spurious pits of a 1-D signal. The spurious pits are the two inner minima of this signal. See text for details

Pit Filling

Pit filling algorithms have been proposed in (Jenson and Domingue 1988; Martz and de Jong 1988). Simultaneously, it has been shown in (Soille 1988; Soille and Ansoult 1990) that mathematical morphology offers a suitable framework for the development of efficient pit filling procedures even in the presence of composite pits and natural depressions. The morphological pit filling algorithm is denoted by FILL and is defined as the reconstruction by erosion R^{ε} of the input digital elevation model f using a marker image f_m which is set to the maximum height $h_{\max}$ of the digital elevation model except along its borders and at the bottom of natural depressions where it inherits the values of the input digital elevation model:

$$\mathrm{FILL}(f) = R^{\varepsilon}_{f}(f_m), \tag{13}$$

where

$$f_m(\mathbf{x}) = \begin{cases} f(\mathbf{x}), & \text{if } \mathbf{x} \text{ lies either on the border of } f \text{ or} \\ & \text{at the bottom of a natural depression,} \\ h_{\max}, & \text{otherwise.} \end{cases}$$

This procedure is illustrated in Figure 7b on a 1-D signal.

Carving

Rather than suppressing pits with the fillhole transformation, the carving procedure (Soille 2002; Soille et al. 2003) suppresses each pit by creating a descending path from it to the nearest point having a lower elevation value. A didactic example on a 1-D signal is provided in Figure 7c. In practice, the determination of the carving path relies on a *flooding* simulation and proceeds as follows (Soille et al. 2003). First all spurious minima of the input DEM must be identified and stored in a binary image. If the terrain does not contain any significant natural depression at the scale of the input DEM, all minima connected to the image border are considered as relevant minima (outlets) to initiate the process. The flooding simulation then starts from the relevant minima by inserting their external boundary pixels into a priority queue, the priority being inversely proportional to the elevation value of the considered pixel. A rising flood that advances into the domain is then simulated by iteratively retrieving pixels from the non-empty queue with the highest priority (i.e., lowest elevation) while inserting their unprocessed neighbours in the priority queue (again considering a priority inversely proportional to their elevation). An additional image is used to store the direction of the incoming *flood* at each pixel. In practice, a pixel is flooded as soon as it is inserted into the priority queue, the direction of the flood being defined by the neighbour pixel which led to its insertion in the queue. Before inserting a pixel in the queue, we check whether it belongs to an irrelevant minimum. In this case, the stored directions enable us to backtrack the flooding path until a pixel of elevation less than or equal to that of the considered irrelevant minimum is reached. We then set all pixels along the detected path to this latter elevation. The reached minimum is then discarded from the binary mask of irrelevant minima while inserting all its unprocessed external boundary pixels in the priority queue. The process terminates when the priority queue is empty. By construction, all irrelevant minima are removed by the proposed carving procedure even in the presence of nested minima at arbitrary elevation levels and whatever the length of the carving paths.

Note that carving *decreases* the elevation of pixels occurring along a *path* while pit filling *increases* the elevation of the entire *region* falling below the elevation of the pour point of the considered pit. The sink unblocking (Morris and Heerdegen 1988) and the breaching (Martz and Garbrecht 1998, 1999) procedures for removing spurious pits are related to the carving procedure in the sense that they also proceed by decreasing the elevation values of some specific pixels. However, in contrast to carving, these latter methods fail to suppress complex pit configurations such as embedded pits.

Carving is also at the basis of adaptive drainage enforcement whereby river networks coming from other sources than the input DEM are imposed to the DEM only in places where the automatic river network deviates substantially from the given networks, see details in (Soille et al. 2003).

Optimal Hybrid

Rather than suppressing a pit by either pit filling or carving, an optimal algorithm combining both procedures has been proposed in (Soille 2004). That is, pits are filled up to a certain level and carving proceeds from this level. A didactic example on a 1-D signal is provided in Figure 7d. The method is optimal in the sense that the level at which pit filling stops and carving takes over is defined so as to minimise the sum of the elevation differences between the input and output pitless DEM. It can be shown that, when suppressing a single pit, the sum of the carving and pit filling elevation differences always displays a unique minimum (Soille 2004).

In the example of Figure 7, the 'cost' of the optimal pit suppression (Figure 7d) equals 6m while it amounts to 13m for both pit filling (Figure 7b) and carving (Figure 7c) procedures. If required by the application, the number of modified pixels can be minimised instead of the sum of the elevation differences. A comparison between all three pit removal procedures is shown on a sample of a real DEM in Figure 8.

Table 1 summarises the effect of all three pit filling methods for the DEM sample displayed in Figure 8.

Table 1. Sum of elevation differences (which is proportional to a volume) and number of modified pixels (which is proportional to an area) for pit filling, carving, and optimal hybrid minimising either the volume or area. Measurements were performed on the sample DEM displayed in Figure 8

	Sum of elevation differences (m)	Number of modified pixels
Pit filling	625640	34683
Carving	28833	4332
Optimal (volume based)	**17543**	4521
Filling part	10165	2854
Carving part	7378	1667
Optimal (area based)	20276	**4058**
Filling part	9378	2381
Carving part	10898	1677

4.2. Flow Directions

The flow direction of a given point is often defined as the direction of the 8-neighbour of this point producing the steepest downward slope. This simple algorithm is referred to as the D8 flow direction by O'Callaghan and Mark (1984). Adaptations

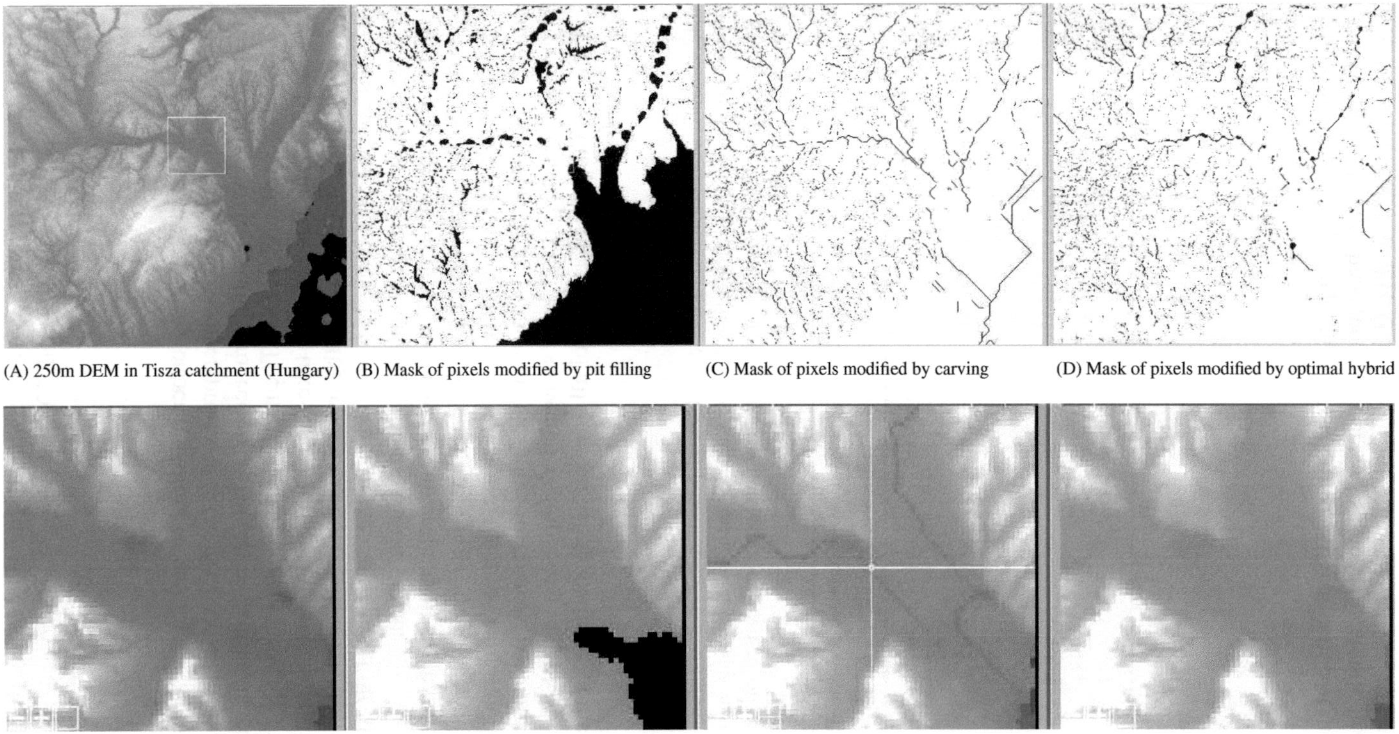

(A) 250m DEM in Tisza catchment (Hungary) (B) Mask of pixels modified by pit filling (C) Mask of pixels modified by carving (D) Mask of pixels modified by optimal hybrid

(E) Zoom on the Sajó-Bodva confluence (F) After pit filling (G) After carving (H) After optimal hybrid

Fig. 8. Pit removal procedures: filling, carving, and optimal hybrid on a 370×400 sample of a 250m DEM in the Tisza catchment. Top row: DEM and mask of modified pixels. Bottom row: zoom on input and filtered DEMs have been proposed to generate multiple flow directions (Tarboton 1997) to reflect flow dispersion occurring on hill slopes although flow dispersion is not desired for valley network demarcation. Tarboton (1997) represents the flow direction as a single angle taken as the steepest downward slope on the eight triangular facets centred at each grid point. Flow is then dispersed between the two grid cells adjacent to the resulting flow vector, proportionally to the angle between the flow vector and the vectors pointing to these grid cells.

Whether simple or multiple flow directions are required, flow directions on plateaus (regions of constant elevation) are undefined. However, they may be calculated by proceeding upwards from the descending border of the plateau (Jenson and Domingue 1988). The same effect can be achieved using concepts of mathematical morphology. The flow direction of each plateau point corresponds to that of the terrain defined by the geodesic distance function calculated from the descending border of the plateau and using the plateau itself as geodesic mask (Soille and Gratin 1994). However, these procedures tend to generate parallel flow lines. Garbrecht and Martz (1997) showed that better flow convergence can be obtained by adding the inverse of the geodesic distance computed from the ascending border of the plateau to the geodesic distance computed from the descending border of the plateau but the resulting relief may itself contain plateaus so that the procedure must be iterated. Recently, (Soille 2002; Soille et al. 2003) proposed a solution to this problem by defining the topography on the plateau as the geodesic time function using the descending border of the plateau as marker set and the inverse of the geodesic distance from the ascending border as geodesic mask. This procedure is illustrated in Figure 9 together with that with simple geodesic distance computations for comparison purposes.

4.3. Contributing Drainage Areas

The contributing drainage area (CDA) of a pixel is equal to its area (i.e., 1 if it is computed in number of pixels) plus the sum of the CDAs of its neighbouring pixels draining through it. The image of CDAs is therefore initialised to 1. We then start to simulate the flow from the pixels at the highest elevation and let them drain along its flow direction. When a pixel drains towards another, the CDA of the latter is incremented by that of the former. The process is then repeated for the second highest elevation and so forth until the lowest elevation is reached. A detailed implementation of this algorithm is described in (Soille and Gratin 1994). Figure 10 illustrates the main steps of the flow simulation process leading to a connected river network for a fixed critical contributing drainage area of 1 square kilometre.

5. Dividing Lines

Dividing lines are key morphological features. They are extensively used in mathematical morphology to extract meaningful edges and object boundaries of an image. Since objects of a grey tone image often correspond to regions of homogeneous intensity, the difference between the maximum and minimum value within a given

	<	1	2	3	4	5	6							
	<	1	2	3	4	5	6							
	<	1	2	3	4	5	6	7	8	9	10	11		
				3	4	5	6	7	8	9	10	11	12	
				4	4	5	6	7	8	9	10	11	12	
						5	6	7	8	9	10	11	12	
								7	8	9	10	11	12	
								8	8	9	10	11	12	
								9	9	9	10	11	12	
	18	17	16	15	14	13			10	10	10	11	12	
	18	17	16	15	14	13	12	11	11	11	11	11	12	
	18	17	16	15	14	13	12	12	12	12	12	12		
	18	17	16	15	14	13	13	13	13	13	13			
	18	17	16	15	14	14	14	14	14	14				

(A) Geodesic distance from descending border

		←	←	←	←	←	←							
		←	←	←	←	←	←							
		←	←	←	←	←	←	←	←	←	←	←		
				↖	←	←	←	←	←	←	←	←	←	
				↑	↖	←	←	←	←	←	←	←	←	
						↖	←	←	←	←	←	←	←	
								↖	←	←	←	←	←	
								↑	↖	←	←	←	←	
								↑	↑	↖	←	←	←	
	→	→	→	→	→	↘			↑	↑	↖	←	←	
	→	→	→	→	→	→	→	↗	↑	↑	↑	↖	←	
	→	→	→	→	→	→	↗	↑	↑	↑	↑	↑		
	→	→	→	→	→	↗	↑	↑	↑	↑	↑			
	→	→	→	→	↗	↑	↑	↑	↑	↑				

(B) Resulting D8 flow directions

	<	3	5	7	9	11	13							
	<	2	4	6	8	10	13							
	<	3	5	7	8	10	13	15	17	19	21	22		
				8	9	10	12	14	16	18	19	22	23	
				11	11	12	12	14	16	17	19	20	23	
						14	15	15	16	17	18	20	23	
								18	17	17	18	20	23	
								20	19	19	18	20	23	
								22	22	20	19	20	23	
	39	38	37	35	34	32			23	21	21	21	23	
	39	36	35	34	32	31	29	26	24	23	23	24	24	
	39	36	34	33	32	30	28	26	25	25	26	26		
	39	36	35	34	32	30	28	27	27	28	28			
	39	38	37	35	33	31	30	30	30	30				

(C) Geodesic time from descending border

		←	↙	↙	↙	↙	↙							
		←	←	←	←	←	←							
		←	↖	↖	↖	←	←	↙	↙	↙	↙	↙		
				↖	↖	↖	←	←	←	←	↙	←	↙	
				↑	↖	↖	↖	←	←	←	←	↙	←	
						↖	↑	↖	↖	←	←	←	←	
								↑	↖	↖	←	←	←	
								↗	↑	↑	↖	←	←	
								↗	↑	↗	↑	↖	←	
	↘	↘	↘	↘	↘	↘			↗	↗	↑	↖	↖	
	→	↘	↘	→	↘	↘	→	↗	↗	↑	↑	↑	↖	
	→	→	→	→	→	→	→	↗	↗	↑	↑	↖		
	→	↗	↗	→	→	→	↗	↗	↑	↑	↖			
	↗	↗	↗	↗	↗	↗	↗	↑	↑	↖				

(D) Resulting D8 flow directions

Fig. 9. On the computation of flow directions on plateaus. Top row: based on the geodesic distance function from the descending border of the plateau using the plateau as geodesic mask. Bottom row: based on the geodesic time function using the descending border of the plateau as marker set and the inverse of the geodesic distance from the ascending border as geodesic mask. The descending border of the plateau consists of the 3 pixels marked by a < sign in (A) and (C)

(A) Shaded pitless DEM

(B) Colour coded D8 flow directions

(C) Contributing drainage areas

(D) Terrain points draining at least 1 km^2

Fig. 10. From DEM to river networks using flow simulation on a DEM pre-filtered for suppressing all pits and define flow directions on plateaus. Example based on a 400×400 SRTM sample gridded at 100m in the ETRS-89 LAEA projection (Annoni et al. 2003) and centred on the Gerlachovsky peak (highest point of the High Tatras mountain range, near the border between Slovakia and Poland). The colour codes for the flow directions are as follows: red for West, green for East, blue for North, white for South, magenta for North-West, cyan for North-East, pink for South-West, and light green for South-East neighbourhood (i.e., a morphological gradient, see section 2) is usually computed beforehand to generate dividing lines along the object boundaries and other salient edges. This preprocessing step is not needed when processing DEMs owing to their very nature. Mathematical morphology provides a formal framework for the definition of watersheds and crest lines.

5.1. Watersheds

Surprisingly, the watershed paradigm is the cornerstone of the mathematical morphology approach to the segmentation of grey scale images (Beucher 1982; Meyer and Beucher 1990). Indeed, provided that the input image has been transformed so as to output an image whose minima mark relevant image objects and whose crest lines correspond to image object boundaries, the so-called watershed transformation partitions the image into meaningful regions, each image object corresponding to a catchment basin. An efficient implementation of a formal definition of the watershed transformation in terms of morphological operators (erosions and dilations) can be found in Vincent and Soille (1991). Interestingly, this definition of watersheds is based on an immersion rather than a flow simulation. Other morphological approaches to the computation of watersheds using topological concepts and also leading to efficient algorithms have been recently developed (Couprie et al. 2005).

5.2. Crest Lines

Crest lines of a DEM can be extracted using the notion of skeletonisation. The idea is that the DEM is iteratively lowered but while making sure that the local topology is not modified. At the end of this iterative process, each catchment is set to the elevation of its outlet while crest lines remain unaltered to preserve the homotopy of the input DEM. In practice, the iterative process is realised through an operation called thinning and that can be defined in terms of morphological erosions with a pair of non intersecting structuring elements (Ranwez and Soille 2002). Figure 11 illustrates river and crest line networks extracted from a DEM near Santiago in Spain.

6. Concluding Remarks

Because the topographic representation of a grey tone image is at the basis of the set representation of grey tone images, mathematical morphology is ideally suited for the processing of digital elevation models. This is further highlighted by the morphological terms used for describing an image: crest lines, watersheds, plateaus, etc. In addition, active research on the development of efficient algorithms allows for the processing of data sets of ever increasing size owing to the ever increasing resolution and area coverage of DEMs. Although all concepts presented in this chapter have been illustrated on DEMs defined over a square lattice of points, they all apply to triangulated irregular networks where the terrain is sampled at surface specific points.

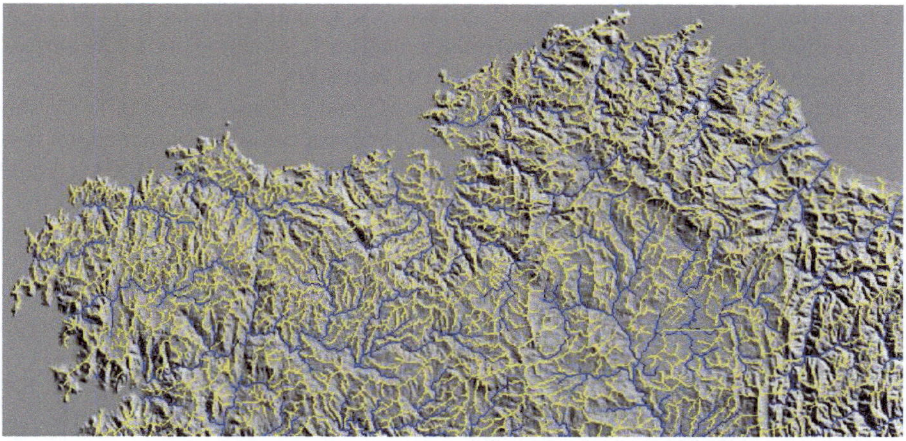

Fig. 11. Crest lines (yellow) and valley networks (blue) on a 1km DEM around Santiago de Compostela and La Coruña in Spain

Indeed, mathematical morphology applies to graphs such as those defined by the Delaunay triangulation of an arbitrary set of points. For example, an algorithm for watersheds on graphs is detailed in (Vincent and Soille 1991).

The river network extraction methods described in this chapter are at the basis of the pan-European river and catchment database presented in this volume (Vogt et al. 2007), see also (Colombo et al. 2007).

References

Annoni, A. (2005). *European Reference Grids*, volume EUR 21494 EN. European Commission, Joint Research Centre. URL http://sdi.jrc.it/publist/annoni2005eurgrids.pdf

Annoni, A., Luzet, C., Gubler, E., and Ihde, J., editors (2003). *Map Projections for Europe*, volume EUR 20120 EN. European Commission, Joint Research Centre. URL http://sdi.jrc.it/publist/annoni-etal2003eur.pdf

Arrighi, P. and Soille, P. (1999). From scanned topographic maps to digital elevation models. In Jongmans, D., Pirard, E., and Trefois, P., editors, *Proc. of Geovision'99: International Symposium on Imaging Applications in Geology*, pages 1–4. Université de Liège, Belgium

Band, L. (1986). Topographic partition of watersheds with digital elevation models. *Water Resources Research*, 22(1):15–24

Bertolo, F. (2000). Catchment delineation and characterization: a review. Technical Report EUR 19563 EN, European Commission, Joint Research Centre

Beucher, S. (1982). Watersheds of functions and picture segmentation. In *IEEE Int. Conf. on Acoustics, Speech and Signal Processing*, pages 1928–1931, Paris

Chorowicz, J., Ichoku, C., Riazanoff, S., Kim, Y.-J., and Cervelle, B. (1992). A combined algorithm for automated drainage network extraction. *Water Resources Research*, 28(5):1293–1302. doi:10.1029/91WR03098

Colombo, R., Vogt, J., Soille, P., Paracchini, M., and de Jager, A. (2007). On the derivation of river networks and catchments at European scale from medium resolution digital elevation data. *Catena*. doi:10.1016/j.catena.2006.10.001. [Available online 22 November 2006]

Couprie, M., Najman, L., and Bertrand, G. (2005). Quasi-linear algorithms for the topological watershed. *Journal of Mathematical Imaging and Vision*, 22(2–3):231–249. doi:10.1007/s10851-005-4892-4

Douglas, D. (1986). Experiments to locate ridges and channels to create a new type of digital elevation model. *Cartographica*, 23(4):29–61

Garbrecht, J. and Martz, L. (1997). The assignment of drainage direction over flat surfaces in raster digital elevation models. *Journal of Hydrology*, 193:204–213. doi:10.1016/S0022-1694(96)03138-1

Haralick, R. (1983). Ridges and valleys on digital images. *Computer Vision, Graphics, and Image Processing*, 22:28–38

Howard, A. (1994). A detachment-limited model of drainage basin evolution. *Water Resources Research*, 30(7):2261–2285. doi:10.1029/94WR00757

Jenson, S. and Domingue, J. (1988). Extracting topographic structure from digital elevation data for geographic information system analysis. *Photogrammetric Engineering and Remote Sensing*, 54(11):1593–1600

Johnston, E. and Rosenfeld, A. (1975). Digital detection of pits, peaks, ridges, and ravines. *IEEE Transactions on Systems, Man, and Cybernetics*, 5:472–480

Mark, D. (1984). Automated detection of drainage networks from digital elevation models. *Cartographica*, 21:168–178

Martz, L. and de Jong, E. (1988). CATCH: a FORTRAN program for measuring catchment area from digital elevation models. *Computers and Geosciences*, 14(5):627–640. doi:10.1016/0098-3004(88)90018-0

Martz, L. and Garbrecht, J. (1998). The treatment of flat areas and depressions in automated drainage analysis of raster digital elevation models. *Hydrological Processes*, 12:843–855. doi:10.1002/(SICI)1099-1085(199805)12:6¡843::AID-HYP658¿3.0.CO;2-R

Martz, L. and Garbrecht, J. (1999). An outlet breaching algorithm for the treatment of closed depressions in a raster DEM. *Computers and Geosciences*, 25(7):835–844. doi:10.1016/S0098-3004(99)00018-7

Matheron, G. (1967). *Eléments pour une théorie des milieux poreux*. Masson, Paris

Matheron, G. (1975). *Random sets and integral geometry*. Wiley, New York

Meyer, F. and Beucher, S. (1990). Morphological segmentation. *Journal of Visual Communication and Image Representation*, 1(1):21–46. doi:10.1016/1047-3203(90)90014-M

Moore, I., Grayson, R., and Ladson, A. (1991). Digital terrain modelling: a review of hydrological, geomorphological, and biological applications. *Hydrological Processes*, 5:3–30

Morris, D. and Heerdegen, R. (1988). Automatically derived catchment boundaries and channel networks and their hydrological applications. *Geomorphology*, 1:131–141. doi:10.1016/0169-555X(88)90011-6

O'Callaghan, J. and Mark, D. (1984). The extraction of drainage networks from digital elevation data. *Computer Vision, Graphics, and Image Processing*, 28:323–344

Peucker, T. and Douglas, D. (1975). Detection of surface-specific points by local parallel processing of discrete terrain elevation data. *Computer Graphics and Image Processing*, 4:375–387

Qian, J., Ehrich, R., and Campbell, J. (1990). DNESYS—An expert system for automatic extraction of drainage networks from digital elevation data. *IEEE Transactions on Geoscience and Remote Sensing*, 28(1):29–45. doi:10.1109/36.45743

Ranwez, V. and Soille, P. (2002). Order independent homotopic thinning for binary and grey tone anchored skeletons. *Pattern Recognition Letters*, 23(6):687–702. doi:10.1016/S0167-8655(01)00146-5

Rinaldo, A., Vogel, G., Rigon, R., and Rodriguez-Iturbe, I. (1995). Can one gauge the shape of a basin? *Water Resources Research*, 31(4):1119–1127. doi:10.1029/94WR03290

Seemuller, W. (1989). The extraction of ordered vector drainage networks from elevation data. *Computer Vision, Graphics, and Image Processing*, 47(1):45–58. doi:10.1016/0734-189X(89)90053-4

Serra, J. (1982). *Image Analysis and Mathematical Morphology*. Academic Press, London

Serra, J., editor (1988). *Image Analysis and Mathematical Morphology. Volume 2: Theoretical Advances*. Academic Press, London

Smith, T., Zhan, C., and Gao, P. (1990). A knowledge-based, two-step based procedure for extracting channel networks from noisy DEM data. *Computers and Geosciences*, 16(6):777–786. doi:10.1016/0098-3004(90)90003-C

Soille, P. (1988). Modèles numériques de terrain et morphologie mathématique : Délimitation automatique de bassins versants. Master's thesis, Université catholique de Louvain, Louvain-la-Neuve, Belgique

Soille, P. (1991). Spatial distributions from contour lines: an efficient methodology based on distance transformations. *Journal of Visual Communication and Image Representation*, 2(2):138–150. doi:10.1016/1047-3203(91)90004-Y

Soille, P. (1992). *Morphologie Mathématique : du Relief à la Dimensionalité — Algorithmes et Méthodes—*. PhD thesis, Université catholique de Louvain ; en collaboration avec l'Ecole des Mines de Paris

Soille, P. (2002). Advances in the analysis of topographic features on discrete images. *Lecture Notes in Computer Science*, 2301:175–186. URL http://springerlink.metapress.com/link.asp?id=hkufqkd2ehbe43hl

Soille, P. (2003). *Morphological Image Analysis: Principles and Applications*. Springer-Verlag, Berlin Heidelberg New York, 2nd edition

Soille, P. (2004). Optimal removal of spurious pits in grid digital elevation models. *Water Resources Research*, 40(12):W12509. doi:10.1029/2004WR003060

Soille, P. and Ansoult, M. (1990). Automated basin delineation from digital elevation models using mathematical morphology. *Signal Processing*, 20:171–182. doi:10.1016/0165-1684(90)90127-K

Soille, P. and Gratin, C. (1994). An efficient algorithm for drainage networks extraction on DEMs. *Journal of Visual Communication and Image Representation*, 5(2):181–189. doi:10.1006/jvci.1994.1017

Soille, P., Vogt, J., and Colombo, R. (2003). Carving and adpative drainage enforcement of grid digital elevation models. *Water Resources Research*, 39(12):1366. doi:10.1029/2002WR001879

Tarboton, D. (1997). A new method for the determination of flow directions and upslope areas in grid digital elevation models. *Water Resources Research*, 33(2):309–319. doi:10.1029/96WR03137

Tarboton, D., Bras, R., and Rodriguez-Iturbe, I. (1989). Scaling and elevation in river networks. *Water Resources Research*, 25:2037–2051

Tarboton, D., Bras, R., and Rodriguez-Iturbe, I. (1991). On the extraction of channel networks from digital elevation data. *Hydrological Processes*, 5:81–100

Toriwaki, J.-I. and Fukumara, T. (1978). Extraction of structural information from grey pictures. *Computer Graphics and Image Processing*, 7:30–51

Tribe, A. (1992). Automated recognition of valley lines and drainage networks from digital grid elevation models: a review and a new method. *Journal of Hydrology*, 139:263–293. doi:10.1016/0022-1694(92)90206-B

Vincent, L. and Soille, P. (1991). Watersheds in digital spaces: an efficient algorithm based on immersion simulations. *IEEE Transactions on Pattern Analysis and Machine Intelligence*, 13(6):583–598. doi:10.1109/34.87344

Vogt, J., Colombo, R., and Bertolo, F. (2003a). Deriving drainage networks and catchment boundaries: a new methodology combining digital elevation data and environmental characteristics. *Geomorphology*, 53(3-4):281–298. doi:10.1016/S0169-555X(02)00319-7

Vogt, J., Colombo, R., Paracchini, M., Soille, P., de Jager, A., and Folving, S. (2003b). A European landscape stratification reflecting drainage density. In Helming, K. and Wiggering, H., editors, *Sustainable Development of Multifunctional Landscapes*, pages 95–110. Springer-Verlag, Berlin Heidelberg New York

Vogt, J., Soille, P., Colombo, R., Paracchini, M., and de Jager, A. (2007). Development of a pan-European river and catchment database. In Peckham, R. and Jordan, G., editors, *Digital Terrain Modelling*, pp. 121–144. Springer-Verlag

Zhang, M., Campbell, J., and Haralick, R. (1990). Automatic delineation of drainage basins within digital elevation data using the topographic primal sketch. *Mathematical Geology*, 22(2):189–209

Optimisation of Interpolation Parameters Using Cross-validation

Jaroslav Hofierka, Tomáš Cebecauer and Marcel Šúri

1. Introduction

Quality of data used in the GIS-support tools is a critical issue, as the decisions that affect locations or areas are to be made effectively, in time and with adequate accuracy. At present, a number of European Union and national policy strategies rely on the use of quality digital elevation model (DEM) data. The accuracy, smoothness and representativeness are properties of DEM determining outputs of the support systems that are designed for assessment of renewable energy resources, flood forecasting, disaster and security management. Similarly, suitability analysis, calculating of environmental indicators and water quality monitoring within the catchments are based on the use of DEM and the decisions taken have financial and legal implications. Thus, a prerequisite for full exploitation of the potential of DEMs is to make them available for the community at sufficient accuracy and detail for a variety of applications.

Spatial complexity of terrain features in different landscapes (e.g. floodplains, mountains) imposes a challenge to automated digital terrain mapping. New high-resolution data sources, such as SRTM and LIDAR are now being used, but their processing is not straightforward and requires more empirical experience and in-depth understanding. Diverse approaches have to be adopted to meet specific demands of users and their applications. For example, requirements for a DEM in hydrological and sediment-flow applications put stress on the hydrological connectivity and topological links among terrain features and are different from those in resource mapping where the height precision of specific terrain features may be crucial regardless of the topological connectivity.

There are many interpolation methods that can be used in digital terrain modelling. The interpolation is generally controlled by a set of parameters that enable the modification of the behaviour of the mathematical function so that a resulting surface meets the criteria of specific applications. Often, proper setting of these parameters is difficult and not clear to those who are responsible for the DEM data processing.

The interpolation methods implemented in geographic information systems (GIS) often have control parameters set up as default constants or their functionality is poorly explained and their influence on results is unclear. Instead of a clearly controlled procedure, the spatial interpolation becomes a black box for the user. In such cases it is difficult to provide the end user with information on the uncertainty associated with the estimates generated during the interpolation. Therefore robust interpolation methods are needed that offer comprehensible parameterisation.

There are various approaches to finding optimal interpolation parameters. In practice, a trial-and-error method is often used, but much effort is needed, and less-experienced users can get quite poor results. There are also user-independent computational approaches. The best known is cross-validation (CV), widely used in the interpolation applications (e.g. Hancock and Hutchinson 2003). The main advantage of the method is a clearly defined and user-independent algorithm that can be easily implemented in the software. The method is less reliable for surfaces with insufficient number of representative input points and resulting error tends to be lower in areas of high density of input points compared to areas were points are sparsely distributed (Jeffrey et al. 2001).

The goal of this chapter is to test the reliability and applicability of CV in an automatic optimization of interpolation control parameters on an example of the Regularized Spline with Tension (RST) and elevation data. The input elevation points are derived by random sampling with various densities from grid-based digital elevation models with two different spatial resolutions (25m and 100m). The CV is applied on each sampled set of points to identify the optimum RST parameters. These parameters are subsequently applied in the interpolation of new surfaces, and their deviations from the original DEMs are analysed.

2. Interpolation by Regularised Spline with Tension

The RST method has been described by Mitasova and Mitas (1993) and Mitasova et al. (1995). We briefly recall the basic principles of the method and focus on issues related to parameters that control the interpolation result.

The RST function $S(\mathbf{x})$ is a radial basis spline based on the condition of minimising the deviation from the measured points and, at the same time, its smoothness seminorm $I(S)$ (e.g. Wahba 1990):

$$\sum_{j=1}^{N} \left| p^{[j]} - S\left(\mathbf{x}^{[j]}\right) \right|^2 w_j + w_0 I(S) = minimum \tag{1}$$

where $p^{[j]}$ are the values measured at discrete points $\mathbf{x}^{[j]} = \left(x_1^{[j]}, x_2^{[j]}\right)$, $j = 1,\dots,N$ within a region of a 2-dimensional space, w_j, w_0 are positive weighting factors and $I(S)$ is the measure of smoothness (smooth seminorm or roughness penalty). For $w_j / w_0 = 0$ the function $S(\mathbf{x})$ passes exactly through the data. The general solution of the minimisation problem given by equation (1) can be expressed as a sum of two components (Talmi and Gilat 1977):

$$S(\mathbf{x}) = T(\mathbf{x}) + \sum_{j=1}^{N} \lambda_j R\left(\mathbf{x}, \mathbf{x}^{[j]}\right) \tag{2}$$

where $T(\mathbf{x})$ is a 'trend' function and $R(\mathbf{x}, \mathbf{x}^{[j]})$ is a *radial basis function* with an explicit form depending on the choice of the $I(S)$. The smoothness seminorm $I(S)$ for the RST method has been designed to synthesise in a single function properties of several previously known splines, such as Thin Plate Spline (Duchon 1976), Thin Plate Spline with Tension (Franke 1985; Mitas and Mitasova 1988), Regularised Thin Plate Spline (Mitas and Mitasova 1988). These desired properties include an explicit form, multivariate formulation, smooth derivatives of higher orders, variational freedom through tension, and anisotropy. In 2-D formulation, the RST function generally defined by (2) has the following explicit form (Mitasova et al. 1995):

$$S(\mathbf{x}) = a_1 + \sum_{j=1}^{N} \lambda_j \left\{ -\left[E_1(\rho) + \ln \rho + C_E \right] \right\}, \tag{3}$$

where $\rho = (\varphi r/2)^2$, $r^2 = \sum_{i=1}^{2} \left(x_i - x_i^{[j]}\right)^2$ is the squared distance, C_E=0.577215 is the Euler constant, $E_1(.)$ is the exponential integral function (Abramowitz and Stegun 1964), and φ is a generalised tension parameter which provides the control over the influence of derivatives of certain order on the resulting function. The tension parameter controls the behaviour of the resulting surface from a thin membrane to a stiff steel plate.

3. The RST Control Parameters

One of the advantages of the RST interpolation is its flexibility within the single radial basis function, as opposed to often subjective selection of a suitable variogram in kriging. To properly consider the variability of a modelled phenomenon, the function can be tuned by a set of the following parameters:

- tension φ,
- smoothing w,
- anisotropy: rotation and scale (θ, s),
- minimum and maximum distances between points.

The tension φ, smoothing w, anisotropy (θ, s) are internal RST parameters and they control the character of the resulting surface (Mitas and Mitasova 1999). The parameters can be selected empirically, based on the knowledge of the modelled phenomenon, or automatically, by minimisation of the predictive error estimated by a CV procedure (Mitasova and Mitas 1993). In this study, we have tested only the tension and smoothing, as these are the main parameters that control the interpolation result.

The tension parameter φ plays a key role in areas with a steep change of modelled phenomenon where overshoots and undershoots of the interpolated surface occur. The RST method is scale dependent and the tension works as a rescaling parameter (Neteler and Mitasova 2002). A high tension 'increases the distances between the points' and reduces the range of impact of each point, low tension 'decreases the distance' and the points influence each other over longer range. A surface with tension set too high behaves like a membrane (rubber sheet stretched over the data points) with peak or pit ('crater') in each given point and everywhere else the surface goes rapidly to trend. A surface with tension set too low behaves like a stiff steel plate and overshoots can appear in areas with rapid change of gradient (Hofierka 2005).

Using the smoothing parameter w, the RST behaves like an approximation function, i.e. the resulting surface does not pass through the given points, but approximates the input values. This parameter is useful in modelling of noisy data, where higher smoothing can filter out the noise, or alternatively, when a phenomenon is to be modelled at a lower level of detail.

The anisotropy parameters (θ, s) can be used in the interpolation of anisotropic data. The orientation of the perpendicular axes characterising the anisotropy is defined by the rotation parameter θ and the scaling ratio of the perpendicular axes (a ratio of axes sizes) is defined by the scale parameter s. These parameters scale distances (i.e. the value of tension as well) in 2 perpendicular directions that should fit the spatial pattern of the anisotropic phenomenon.

Minimum and maximum distances between points control the number of points that are actually used in the interpolation after reading the input data. The maximum distance can be used only for vector data in the GRASS GIS version 5.7 and above. However, this parameter internally influences the effect of the tension, because the tension works as a distance-scaling factor. Therefore, the tension can be set with or without normalisation. The data density does not affect the normalised tension parameter. In this study we use only the normalised tension.

The RST interpolation has been implemented as *s.surf.rst* and *v.surf.rst* commands and is widely used by the GRASS GIS users community (Neteler and Mitasova 2002). Very recently, the RST method in GRASS GIS has been enhanced by a CV procedure. This option enables automation of the interpolation in the sense of choosing optimised interpolation parameters, but the applicability to different datasets is still not fully investigated. Also, the computational demands are rather high, and therefore it is currently applicable only to small datasets (up to 10^3-10^4 input points). Large datasets must be divided into smaller, representative subsets.

4. Evaluation of Interpolation Accuracy

The interpolation accuracy can be measured by different methods. The most straightforward is to evaluate deviations between interpolated surface and the input points. The overall error, measured e.g. by Root Mean Squared Error (RMSE) then characterises the interpolation accuracy in the given points. However, this approach does not provide information about the accuracy in areas between these points. The interpolation accuracy in areas between the input points is actually a predictive accuracy of the interpolation method. This contrasts with approximation methods that do not necessarily pass through the given input points.

One of the options to evaluate the predictive accuracy between points is to use an evaluation dataset that contains data not used in the interpolation.

For each evaluation point the deviation between actual and interpolated value are calculated and the overall accuracy is tested. However, in many applications due to a limited number of input points, such an independent evaluation dataset is difficult to select. Moreover, the accuracy information is available only for these independent points and they rarely cover the entire area of interest with a sufficient density.

The accuracy assessment by a cross-validation is based on removing one input data point at a time, performing the interpolation for the location of the removed point using the remaining samples and calculating the residual between the actual value of the removed data point and its estimate. The procedure is repeated until every sample has been, in turn, removed. This form of CV is also known as the 'leave-one-out' method (Tomczak 1998).

The CV of this particular form is especially suitable for relatively dense data sets, since removing points from already under-sampled areas can lead to misrepresentation of the surface to be interpolated. Despite the wide use of this technique for assessing interpolation schemes, one should be aware of its shortcomings. Specifically, CV will usually overestimate the interpolation error because the estimate is being computed at a location where data are genuinely available. In addition, the computed surface and hence the cross-validated estimate may be altered by the removal of the point being cross-validated. In practice these issues are unavoidable but with increasing number of input data points they have less impact (Jeffrey et al. 2001).

The overall performance of the interpolator is then evaluated by statistical means such as the Root Mean of Squared Residuals, Mean Absolute Error (MAE) or Mean Error (Tomczak 1998). Low Root Mean Squared Error (RMSE) indicates an interpolator that is likely to give most reliable estimates in the areas with no data. The minimum RMSE calculated by CV can be used to find the optimum interpolation control parameters (Mitasova and Mitas 1993). However, Hutchinson (1998) has found that CV does not always represent a reliable estimate of the model error, especially when a short-range correlation in data is present.

Therefore it is appropriate to use additional evaluation methods to ensure the reliability and consistency of the predictions. For example, in the interpolation of precipitation, a comparison with expert hand-drawn precipitation maps (e.g. Custer et al. 1996; Hofierka et al. 2002) and maps of derived hydrological processes can provide some insight into the capabilities of the interpolation algorithms to adequately represent the behaviour of the modelled phenomena, including the features which may not be directly incorporated in the data.

By CV that has been implemented in the GRASS GIS, the user can calculate prediction errors that are saved in the database as a point data set. As the computational demands are high, currently only a few thousand points can be analysed in a reasonable time. However, the improvement of the algorithm is currently under way by simplifying the matrix calculations and by removing unused points that are located outside the inner interpolation segment.

5. Application to Digital Terrain Modelling

The suitability of CV for optimisation of interpolation control parameters was tested on samples of digital elevation data from Slovakia. The study area is located in the Central Slovakia at the contact of Turiec River floodplain with Veľká Fatra Mountains (Figure 1). The region consists of two parts: a relatively flat area in the North-West with gentle slopes and a floodplain containing terrain microforms and a mountainous area in the South-East with steep slopes and narrow valleys. The size of the area is 5.2 x 5.2 km, with the elevation ranging from 465 to 1380 m a.s.l. (Figure 2).

Fig. 1. Location of the test site

The spatial resolution of the DEM determines its application properties and its ability to represent the terrain features in a desired detail (Weibel and Heller 1991). Therefore samples were chosen from two original raster DEMs with resolution 25 and 100 meters, respectively (Figure 2). These two reference DEMs were generated by the GRASS *s.surf.rst* method

(GRASS 2000) from contour lines with 10m interval, elevation points and a river network, all representing a scale 1:50 000.

The 25m DEM represents the original map variability of surface with high conformity. The resulting surface is smooth, detailed features fit well with the original contour lines, small differences are visible in very narrow valleys and ridges. The surface represented by the 100m DEM does not represent all the details of the original elevation structure, especially in the areas of high-frequency variability at slopes, but the overall trend of the terrain fits well. On the other hand, the 100m DEM holds as much as possible of the original variability, and its smoothness and spatial autocorrelation is lower than in the 25m DEM.

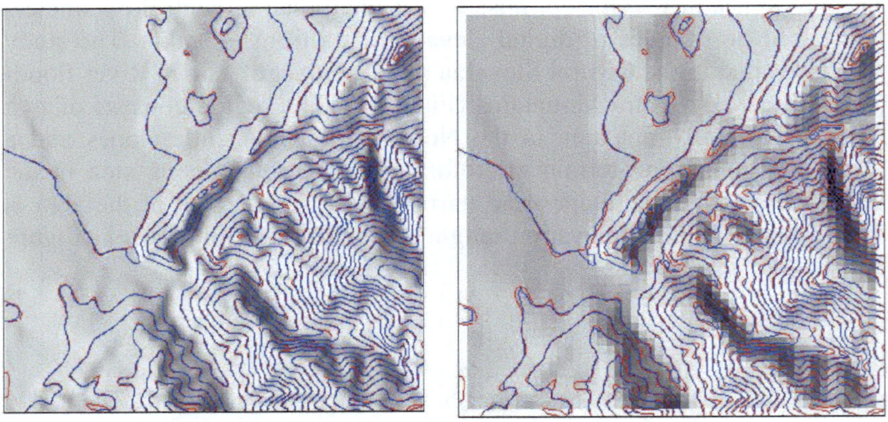

Fig. 2. Comparison of the two reference DEMs. Left: DEM with spatial resolution of 25 m. Right: DEM with spatial resolution of 100 m. The red lines represent original contours from topographic maps that were used for generation of both reference DEMs. Blue lines represent contours that were derived from the created DEMs (contour interval 50 m)

To assess CV suitability, the samples at different density were randomly taken and evaluated for the DEMs of both resolutions. The sample sets consist of 20, 10, 5, 2, 1, 0.5 % of the total number of cells (43,264) for 25m DEM and 40, 20, 10, 5% of the total number of cells for 100m DEM (2 704 cells).

Each random sample set was the subject of an iterative interpolation with tension and smoothing parameters varying from 10 to 90 and from 0.1 to 0.9, respectively. For each iteration, the CV procedure was realised and an overall prediction error calculated using minimum, maximum error, mean error, standard deviation and RMSE. The predictive error for each individual sample point was stored in a GIS database. The overall results were compared in order to find the minimum statistical errors.

Generally, the lowest RMSE or MAE (mean absolute error) indicates the optimal parameter combination that should provide interpolation with the lowest prediction error. For each iteration, the prediction errors for individual points were analysed in order to see spatial differences in errors and sensitivity of parameters. The results of CV were also analysed by a visual inspection of the interpolated elevation and other topographic parameters (slope, aspect, profile and tangential curvatures). More information on topographic parameters used in the analysis can be found in Mitasova and Hofierka (1993).

For each combination of tension and smoothing, the CV deviations were compared with deviations between the reference surfaces and surfaces interpolated form random samples (elevation, slope and curvatures) on a per-pixel basis. This approach supplies detailed information about errors for the entire study area.

6. Results and Discussion

For the 25m DEM, the results become consistent and reliable when using the 8,653 points (20% of total grid cells). As the optimum parameters, tension = 20 and smoothing = 0.1 were selected. This combination gives the overall RMS error of 1.45 m, which is fairly acceptable for the scale of the original data (Table 1, Figure 3).

Table 1. The RST parameters optimised by the cross-validation and by the statistical comparison of surfaces for different samplings and the reference 25m DEM. The last row presents optimum parameters and RMSE for vertices of contours used to interpolate a reference surface

Percentage of 25-m DEM cells	Number of points	CV			Surface comparison (reference vs. interpolated)		
		Tension	Smoothing	RMSE	Tension	Smoothing	Standard dev.
20%	8653	20	0.1	1.45	20	0.1	1.26
10%	4326	30	0.1	3.36	30	0.1	3.35
5%	2136	30	0.1	6.69	30	0.1	5.95
2%	865	30	0.1	13.92	30	0.1	13.68
1%	433	45	0.1	26.54	30	0.1	20.08
0.5%	216	25	0.1	26.63	35	0.1	31.96
contours	10216	30	0.2	3.77			

While sampling a lower number of points (i.e. in our case 2-10% of grid cells), CV still offers optimum set-up of the interpolation control parameters (tension = 30 and smoothing = 0.1), however the interpolation error becomes worse. This reflects the fact that the lower number of sampling points does not represent sufficiently the high variability of the reference surface. The decrease in the accuracy with lower sample density is clearly seen in the Figure 3 where the differences between reference and interpolated surface rapidly increase both in amplitude and extent.

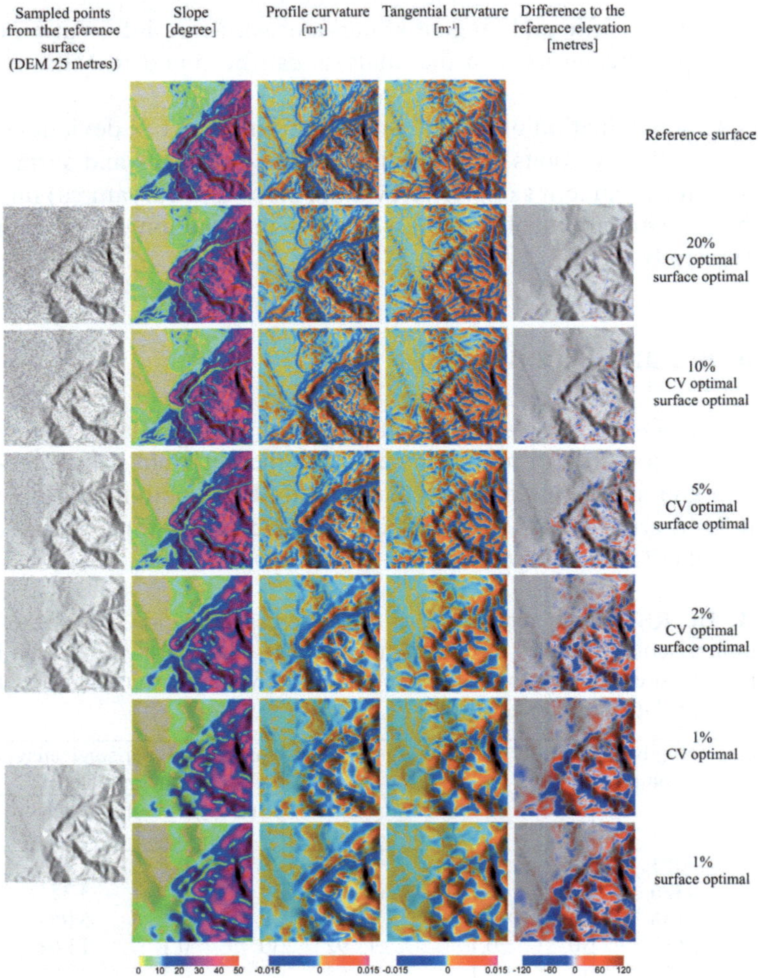

Fig. 3. Comparison of the reference and interpolated surfaces (for random samplings taken from the DEM with 25 m resolution) using the optimised parameters from CV and from the comparison of interpolated surfaces with the reference. The values of optimal parameters are presented in the Table 1

While both 10% and 5% samplings fit the overall surface trend well, the samplings of 2% and less do not represent correctly important features of the reference surface. This is also evident from the derived topographic parameters. As regards the 2% sampling, it is important to note that despite the deviations in the vertical dimension, CV still offers the optimal interpolation parameters in the sense of minimum surface differences.

The CV procedure does not work properly when selecting less than 1% of the cells (Figure 2). A number of points in this sampling does not capture representatively the terrain features, therefore the prediction provides results with unacceptably high RMSE. Visual inspection of the results confirms these statements. Many smaller terrain features (e.g. ridges, valleys) are missing or improperly captured by the resulting surface. Visualised slope and curvatures reveal that decreasing the number of sampling points increases the surface smoothness and generalisation. The evident leap in the surface smoothness is between the 5% and 2% samplings.

Table 2. The RST parameters optimised by the cross-validation and by the statistical comparison of surfaces for different samplings and the reference 100-m DEM

Percentage of 100-m DEM cells	Number of points	CV			Surface comparison (reference vs. interpolated)		
		Tension	Smoothing	RMSE	Tension	Smoothing	Standard dev.
40%	1082	30	0.1	12.61	30	0.1	9.51
20%	541	30	0.1	19.20	30	0.1	15.66
10%	270	25	0.1	28.63	40	0.1	26.62
5%	135	20	0.1	43.91	55	0.1	40.78

Similar results were obtained from analysing the random samples from the 100m DEM. While CV procedure gives optimum parameters (tension = 30 and smoothing = 0.1) and acceptable results for input data representing 40% and 20% of the total cells (that is around 2% level in the 25m DEM), the lower point densities make CV unusable for finding optimal interpolation parameters (Table 2, Figure 4).

Despite the high point density in the 40% sample, the RMSE predictive error is rather high (12.61 m) (Table 2). This fact relates to the lower smoothness of the reference 100m DEM that decreases the prediction accuracy of values at missing points. The visual inspection of the results (Figure 4) reveals that even the 40% sampling density significantly smoothes out the resulting surface. In analogy to the 25m DEM case, the decreasing sampling density smoothes and generalises the original terrain structures.

While a surface interpolated from the 20% sampling fits the main structure of the reference surface well, the 10% and less samples fail to represent the terrain forms sufficiently (Figure 3).

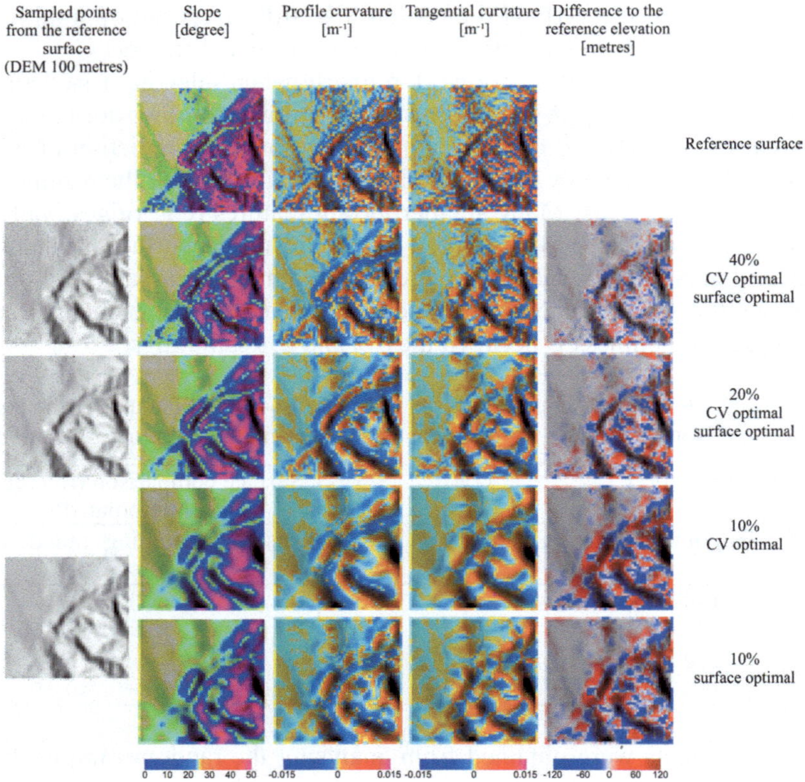

Fig. 4. Comparison of the reference and interpolated surfaces (for random samples taken from the DEM with 100m resolution) using the optimised parameters from CV and from the comparison of interpolated surfaces with the reference. The values of optimal parameters are presented in the Table 2

The analysis of the spatial distribution of the predictive error is done from the CV deviations that were calculated for each sample point. The Figure 5 shows an apparent error dependency on the vertical differentiation of the terrain. While the error amplitude is low in the flat area these values become quite high in the mountainous area. This trend is obvious in all samples, and the range of errors increases with decreasing sampling density. In high-density samples, a significant number of points with low deviations can be found in the mountainous area, while the low-density samples contain only few points with acceptable deviation.

The presence of high deviations (more than +/- 10 m) in relatively dense samples (even in 10% sample) indicates that there are local maxima of absolute prediction error that should be treated carefully when evaluating the resulting surface. It is apparent that the frequency of local maxima is high in the area of ridges and valleys, which generally represent the local extremes, typically with high curvature. This imposes a question whether the optimisation of interpolation parameters based on the averaging of CV deviations from all points (such as RMSE) is suitable for each application. In some cases, other criteria such as minimisation of local maximum deviations could be considered.

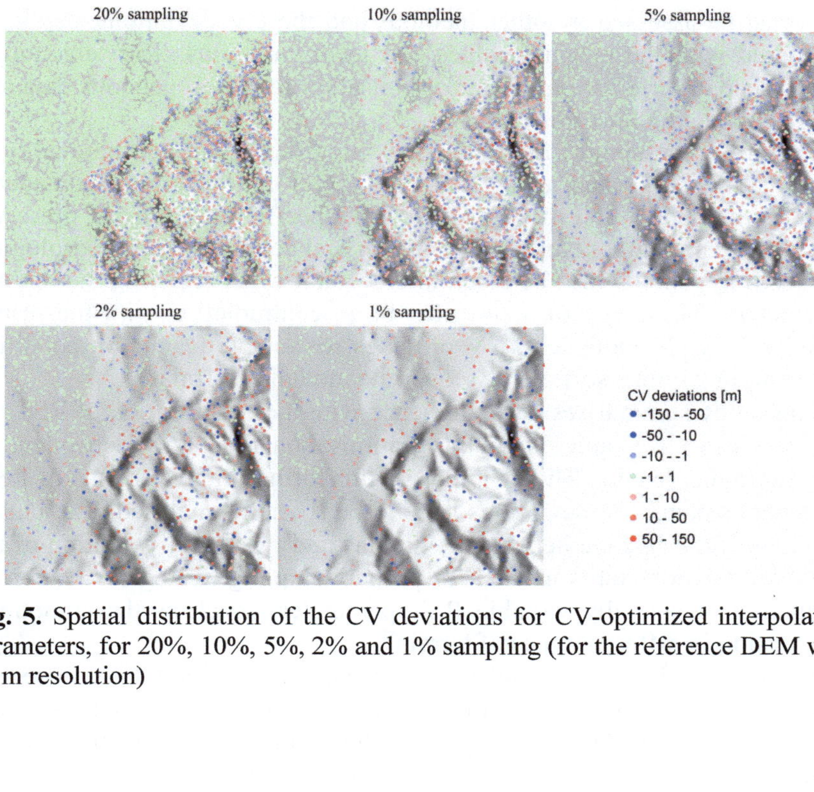

Fig. 5. Spatial distribution of the CV deviations for CV-optimized interpolation parameters, for 20%, 10%, 5%, 2% and 1% sampling (for the reference DEM with 25 m resolution)

7. Conclusions

This study has demonstrated that the Regularised Spline with Tension and cross-validation are appropriate and effective tools in digital terrain modelling if density of the input elevation data is sufficient. With optimisation of parameters by CV, the RST method can be fully automated and it can be applicable in various mapping tasks, not only for generation of the DEM, but also for interpolation of other spatially-distributed phenomena. The optimisation of RST control parameters by CV helps to minimise the prediction error and maximise the exploitation of the method. Considering the fact, supported by many studies (e.g. Mitas and Mitasova 1999; Hofierka et al. 2002), that the RST is one of the most accurate interpolation methods, this provides the user a combination of tools that are valuable in many applications. This approach makes the use of advanced interpolation methods such as RST easier to use for less experienced users.

The main drawback of CV is that it does not guarantee optimum parameters in all situations. The reliability of CV depends on density and homogeneity of input points. For lower data densities, the applicability of CV must be assessed by other methods and the CV deviations should be compared with the original values and local variations. The selection of appropriate criteria for parameter optimisation for heterogeneous data still poses a challenge for further study.

The pre-processing of input data may significantly affect the quality of interpolation (Cebecauer et al. 2002) and therefore it should be considered prior to the application of CV. Other task-specific criteria (e.g. preservation of extreme values, surface smoothing and generalisation, hydrological connectivity) should also be taken into account when selecting the optimal parameters. These criteria may often be in contradiction with the results found by CV. Nevertheless, CV provides reasonable results on condition that sampling points sufficiently cover the modelled terrain.

This study summarises our experience with the optimisation of RST parameters using CV and shows that optimisation is heavily dependent on data sampling density. This is important for further improvements of RST. In the next stage, CV deviations have to be spatially analysed. The understanding of the linkage between deviations and the spatial variation of the modelled phenomena is needed for proper sampling strategies and for introduction of spatially variable RST parameters (such as the smoothing and tension in different types of terrain).

The digital elevation model is one of the most important data used by hydrological, erosion and pollution models. The cross-validation assists in improvement of the DEM quality and in the error assessment and thus contributes to the improved decision support applications.

Acknowledgement

This work was supported by the EC JRC Action No. 2324 'SOLAREC' and by the VEGA Grant Agency of the Slovak Republic within the scientific project No. 1/3049/06.

References

Abramowitz M and Stegun IA (1964) Handbook of mathematical functions. Dover, New York, pp 297-300, 228-231

Cebecauer T, Hofierka J, Šúri M (2002) Processing digital terrain models by regularized spline with tension: tuning interpolation parameters for different input datasets. In: Ciolli M, Zatelli P (eds) Proceedings of the Open Source Free Software GIS - GRASS users conference 2002, Trento, Italy, 11-13 September 2002

Custer GS, Farnes P, Wilson JP, Snyder R D (1996) A comparison of hand- and spline-drawn precipitation maps for Mountainous Montana. Water Resources Bulletin 32:393-405

Duchon J (1976) Interpolation des fonctions de deux variables suivant le principe de la flexion des plaques minces: R.A.I.R.O. Analysis Numerique 10:5-12

Franke R (1985) Thin plate spline with tension. Computer Aided Geometrical Design 2:87-95

GRASS 2000 Geographic Resources Analysis Support System. http://grass.itc.it/

Hancock PA and Hutchinson MF (2003) An iterative procedure for calculating minimum generalised cross validation smoothing splines. Australian and New Zealand Industrial and Applied Mathematics Journal 44:290-312

Hofierka J (2005) Interpolation of radioactivity data using regularized spline with tension. Applied GIS, 1:16/01–16/13

Hofierka J, Parajka J, Mitasova H, Mitas L (2002) Multivariate interpolation of precipitation using regularized spline with tension. Transactions in GIS 6:135-150

Hutchinson MF (1998) Interpolation of rainfall data with thin plate smoothing splines: I. Two dimensional smoothing of data with short range correlation. Journal of Geographic Information and Decision Analysis 2:152-167

Jeffrey SJ, Carter JO, Moodie KB, Beswick AR (2001) Using spatial interpolation to construct a comprehensive archive of Australian climate data. Environmental Modelling and Software 16:309–330

Mitas L and Mitasova H (1988) General variational approach to the interpolation problem. Computers and Mathematics with Applications 16:983-992

Mitas L and Mitasova H (1999) Spatial interpolation. In: Longley P, Goodchild MF, Maguire DJ, Rhind DW (eds) Geographical information systems: principles, techniques, management and applications. GeoInformation International, Wiley, pp. 481-492

Mitasova H and Hofierka J (1993) Interpolation by regularized spline with tension: II. Application to terrain modeling and surface geometry analysis. Mathematical Geology 25:657-671

Mitasova H and Mitas L (1993) Interpolation by regularized spline with tension: I. Theory and implementation. Mathematical Geology 25:641-655

Mitasova H, Mitas L, Brown BM, Gerdes DP and Kosinovsky I (1995) Modeling spatially and temporally distributed phenomena: new methods and tools for GRASS GIS. International Journal of GIS, Special Issue on Integration of Environmental Modeling and GIS 9: 443-446

Neteler M and Mitasova H (2002) Open source GIS: A GRASS GIS approach. Kluwer Academic Publishers, Boston.

Talmi A and Gilat G (1977) Method for smooth approximation of data. Journal of Computational Physics 23:93-123

Tomczak M (1998) Spatial interpolation and its uncertainty using automated anisotropic inverse distance weighting (IDW) - cross-validation/jackknife approach. Journal of Geographic Information and Decision Analysis 2:18-33

Wahba G (1990) Spline models for observational data. CNMS-NSF Regional Conference Series in Applied Mathematics 59, SIAM, Philadelphia, Pennsylvania

Weibel R and Heller M (1991) Digital Terrain modelling. In: Maguire DJ, Goodchild MF, Rhind DW (eds) Geographical information systems: principles and applications. Longman, Essex, 269-297

Scale-Dependent Effect of Input Data Design on DEM Accuracy

Radoslav Bonk

1 Introduction

Terrain geomorphometric properties are major input parameters for majority of GIS-supported topographic models and applications such as erosion-deposition model, solar radiation model (Suri and Hofierka 2005), soil erosion model or river basin modelling (De Roo et al. 2000).

However, the process of DEM utilisation is not trivial, but it involves numerous factors which influence the DEM quality. Such as, input data design, terrain complexity and interpolation method (Figure 1). Moreover, the DEM quality is not influenced only by these factors, but also by the internal fluctuation of each factor itself. For example, the distribution of input data for DEM calculation does not vary only by its spatial pattern, but it also varies by the number of points; the interpolation procedure used to produce elevation field does not vary only by choice of the interpolation function, but it also varies by the tuning parameters of the interpolation function itself. In systematic investigation in order to study the effect of one factor on DEM accuracy, all the other factors have to remain constant. Also it has to be noted, that from the practical and methodological point of view it is desired to understand the DEM *quality* as a combination of DEM *accuracy* and DEM *suitability*. While DEM accuracy refers to the fact how representative the DEM is compared to the real landscape (e.g.: from statistical point of view), DEM suitability refers to how suitable the DEM is for certain application or model (e.g.: hydrologically correct DEM, DEM for engineering applications, preprocessed DEM where high frequency variation was filtered out etc. Weather driven natural hazards action – WDNH(1) within the Land management unit – LMU(2) in JRC Ispra is carrying out scientific tasks and research where DEM uncertainty and quality is crucial. The tasks involve: development and testing of early flood warning system – EFAS(3), flood risk mapping, and mitigation and forecasting case studies for Elbe and Danube catchments. These tasks provide policy support for DG Environment, which is preparing the European Action Programme on flood risk management(4). Also, the tasks are benefiting from the methodological and practical outcomes of the research presented here, and from the scientific knowledge acquired during the test study. Moreover, new global DEM datasets, such as the

SRTM(5) DEM, are now available and first accuracy assessments were already carried out. Further research involve utilisation of the global SRTM DEM with already existing European spatial datasets(6).

Research objectives of this investigation was to determine the effect of input data

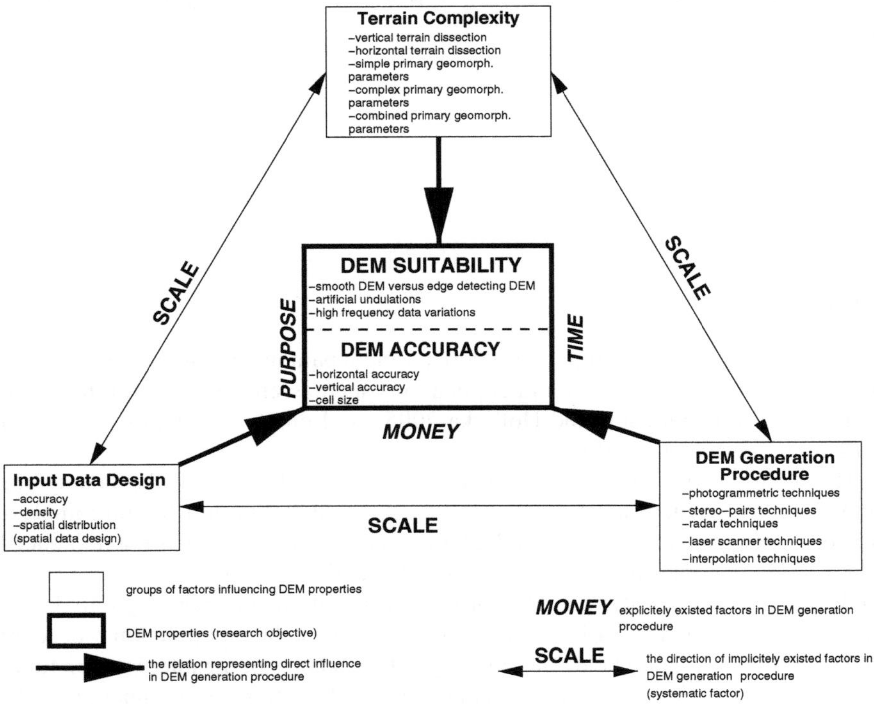

Fig. 1. Group of factors influencing DEM quality and their adjacent interactions. DEM quality consists of DEM accuracy and DEM suitability

points design on magnitude and spatial distribution of elevation residuals (elevation residuals are pixel values representing the difference of computed DEM and the reference DEM, they were produced by subtracting the computed DEM from the reference DEM). Specifically, compare random sampling design and grid sampling design, internally varying in number of input data points.

Input data design can be defined as a spatial distribution and as a number of elevation input data used by interpolation technique to produce continuous elevation surface. There are various types of distributions and probably the best known is the one where input data is located along the contours (Bonk 2003). Other distributions cover that of rectangular grid, random spatial distribution, or points distributed along critical topographic lines (Figure 2). Each distribution, coupled with the number of data points, affects the quality of the resulted elevation surface. Specifically, it has direct connection with hierarchical order of topographic features it represents (Minar

1998; Schmidt and Dikau 1999). Low density sampling produces only trend terrain features, and does not cover topographic details, or fine terrain undulations. While high density sampling can cover fine topographic features, it can often overestimate the number of input data really needed for certain terrain types. Also, higher number of input data points does not automatically means the better DEM accuracy (Li 1992, 1994). Systematic study of direct factors on reliability and accuracy of DEM was

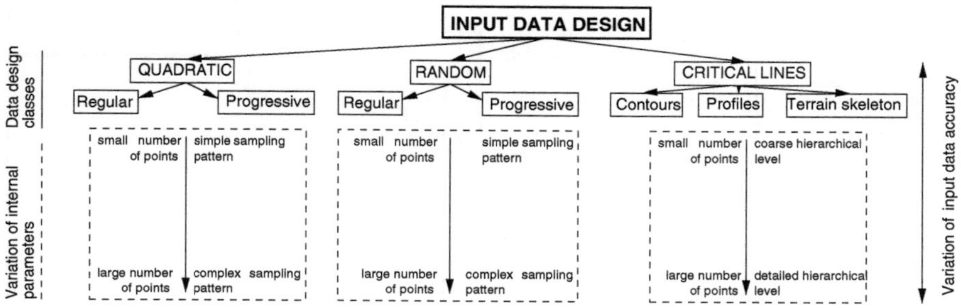

Fig. 2. Various types of input data designs (sampling designs) based on spatial distribution of input data and sampling density

presented by Li (1990, 1991, 1992, 1994). In his first work Li (1990) identified the set of parameters or factors which affect the DEM accuracy: three attributes of the source data (accuracy, density, and distribution), characteristics of the terrain itself, method used for the construction of the DEM, characteristics of the DEM surface which is constructed from the source data.

Also it was recognised by several other authors, for example Makarovic (1972), and Ackerman (1980), that sampling interval has the most significant effect on DEM accuracy. Li (1991) presented theoretical rules on reliability of the DEM accuracy estimates in sense of required accuracy, number and distribution of check points used in DEM accuracy assessment. His main objective was to determine how each of these three attributes of check points affect the reliability of the DEM accuracy estimates in context of experimental tests. Li (1994) deals with the comparative study on accuracy of DEM derived from four different data models. However, the problem of DEM accuracy was not answered in its full complexity, and that it is worth to continue the research in both theoretical and experimental directions. The author proposed to investigate the accuracy of DEMs derived from various data designs to establish some relationships between data designs and DEM accuracy (Li 1994) and this is an example of such a research research. More in depth overview of literature regarding the DEM accuracy and methods is covered in Bonk (2003).

2 Study Area

The research region of Toporec is located in northern Slovakia in the High Tatras Mts. upland (Figure 3). The study area has a rectangular shape 1230 m wide, and 1920 m

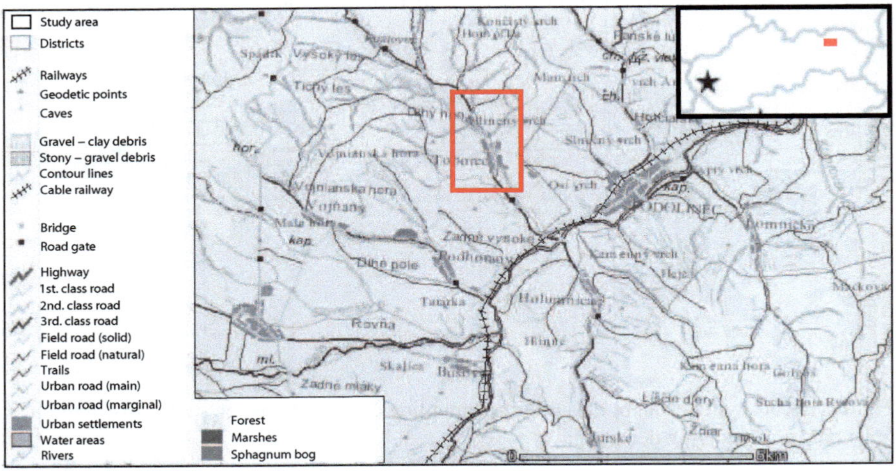

Fig. 3. Location of Toporec study area within the High Tatras Mts. region at northern Slovakia

long. While relatively flat topography covers southern part and south-western part of the area (slope angle < 4°). The steepest slopes are located at the slopes separating river valley bottom and adjacent topography (slope angle > 18°) and at the highest point located near the Hlinený peak at the north-eastern part of the area (Figure 4). Minimum elevation in the study area is 591.6m and maximum elevation is 678m. Minimum slope angle is 0.1° and maximum is 24°.

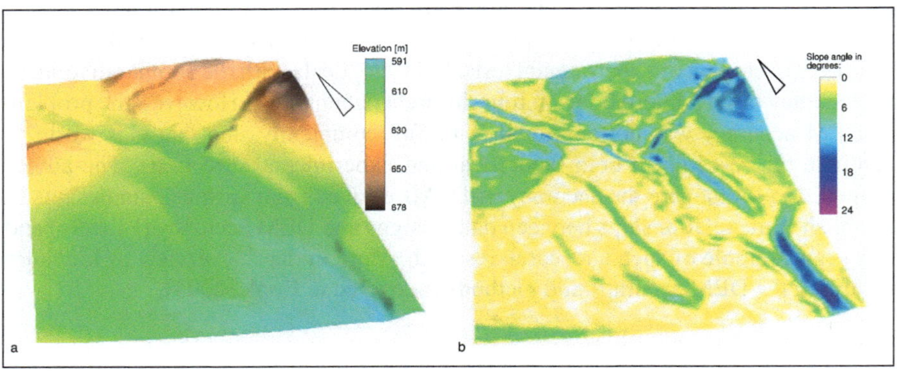

Fig. 4. a. elevation (2 times exaggerated), b. slope angle (2 times exaggerated) maps at Toporec in 3D view

Two sampling designs were chosen: (1) random sampling design (its related datasets are indicated by the lower index "$_R$"), and (2) grid sampling design (its related datasets are indicated by the lower index "$_G$"). Datasets for both sampling designs were extracted from the reference elevation surface and used in interpolation to produce elevation.

3 Methodology

Reference DEM for the Toporec region was computed based on grid of regularly spaced discrete data points, obtained from a photogrammetric surveying (Bonk 2003). Points were equally spaced in 20 × 20m grid.

To obtain realistic topographic surface, the following parameters were chosen for regularised spline with tension (Mitasova and Hofierka 1993) interpolation procedure: $B_a = 0.5$, $\varphi = 20$, $dmin = 15$ (B_a is smoothing, φ is tension, $dmin$ controls the distance between the input data points). Here the number of input data points in original dataset was 7218; 217 points were excluded after $dmin$ parameter was specified, and 1479 were outside of the region. Thus actual number of points used for interpolation was 5522. The resolution for the reference surface was set to 10m. Root mean squared deviations, of the newly computed reference surface (Z_E^T), from the photogrammetric input data points was 0.38m, maximum elevation was 678 m, and minimum elevation was 592m (Figure 4).

The core of this research is the effect of the varying sampling design on resulted DEM, while all other factors, such as: terrain complexity, DEM resolution, interpolation function, and surface representation remain constant. The only changing factor is the design of the input data points. DEM related factors represented in this research are as follows:

- Terrain complexity (constant) - the same region is used for analysis, thus the terrain complexity remains constant.
- DEM resolution (constant) - the resolution of the resulted elevation surfaces is 10m × 10m.
- Interpolation method (constant) - RST interpolation method (Mitasova and Hofierka 1993) with constant parameters ($B_a = 40$, $\varphi = 0$, $dmin = 0$) is used in each interpolation.
- Surface representation (constant) - regular grid of continuous points (commonly known as "raster") is used for resulted surface representation.
- Sampling design (varying) - two sampling designs are used, random sampling design, and grid sampling design. Each of them varies in number of data points. The number of points was chosen as a percentage from the total number of pixels covering Toporec region with 10m resolution (192 x 123 = 23 616 pixels). The percentages and actual number of pixels were as follows: 30.6%/(7218 points), 10.2%/(2406 points), 3.4%/(802 points), 1.1%/(267 points), 0.4%/(90 points). The number of data points increases approximately by 33% from the previous number, such that the highest number of points (7218) is significantly lower than the number of pixels covering Toporec research region.

Input data were extracted from the reference surface for Toporec region (Z_E^T). Random sampling was performed by using random number algorithm implemented in GRASS GIS (Neteler and Mitasova 2002). Grid-like spatial sampling was also performed within the GRASS GIS. This sampling has resulted in five datasets of randomly distributed data and five datasets of grid-like spatial data distribution.

Then, each of these datasets were used in interpolation producing a particular elevation surface. To ensure that the resulted surface passes through the input data, the following parameters were used in RST interpolation function: $B_a = 0$, $\varphi = 40$. The resulted elevations based on the random sampling, and grid sampling data points were labelled as follows. Random sampling elevations $(Z_R)_{30.6}, (Z_R)_{10.2}, (Z_R)_{3.4}, (Z_R)_{1.1}, (Z_R)_{0.4}$ and grid sampling elevations $(Z_G)_{30.6}, (Z_G)_{10.2}, (Z_G)_{3.4}, (Z_G)_{1.1}, (Z_G)_{0.4}$. The datasets shared the lower indexes based on the percentage form the total number of pixels in Toporec region. Then, the resulted surface ($(Z_R)_{0.4}$) was subtracted from the reference surface (Z_E^T) to produce surface elevation residuals ($(Z_R^R)_{0.4}$). Grid-like input points were used in the same manner.

3.1 Statistical Analysis of Interpolated Surfaces

The aim of this analysis was to find out if the pixels with the same value have a tendency to cluster together. In other words to find out if there is any systematic spatial pattern in the distribution of residuals. If there is a systematic spatial pattern present, it might suggests some relation between the pattern and the reference DEM. In order to find out measurements of spatial autocorrelation were applied. Several methods exist for the measurement of spatial autocorrelation. Two related measures that are often referred to in geographical literature are Geary's Coefficient, and Morain Coefficient (MC) (Clif and Ord 1981; Goodchild 1986). Spatial autocorrelation refers to the spatial ordering of a single variable, and to the relationship between pairs of observations of this variable. The orderings of n observed values of some variable X is usually described with the aid of connectivity matrix C. Non-zero c_{ij} entries in $n \times n$ matrix indicate that the corresponding row, and columns units are juxtaposed. To study the presence of autocorrelation between the elevation residuals, spatial statistics measures - Morain's Coefficient (MC) was applied.

$$MC = \frac{n \sum_{i=1}^{n} \sum_{j=1}^{n} c_{ij}(x_i - \bar{x})(x_j - \bar{x})}{\sum_{i=1}^{n} \sum_{j=1}^{n} c_{ij} \sum_{i=1}^{n} (x_i - \bar{x})^2} \tag{1}$$

Where $\bar{x}$ is the mean of the respective attribute values x_i, x_j, at the pair of zones i, j of which there are n. Morain Coefficient (MC) deals with covariations, and the expected value of MC is $\frac{-1}{n-1}$. For similar values in juxtaposition values, MC $\rightarrow 1$. For dissimilar values in juxtaposition, MC $\rightarrow -1$.

Statistics, including root mean squared error (RMSE), Morain Coefficient (MC), and others are presented at Table 1 for random data sampling design and at Table 2 for grid-like sampling design. Exact locations of extracted input data points together with the spatial pattern of residuals are presented at Figure 6 and 7 respectively.

$$RMSE = \sqrt{\frac{\sum_{i=1}^{n}(t_1 - x_1)^2}{n-1}} \quad (2)$$

RMSE represents overall estimation performance, and is calculated as the square root of the mean of the squared differences between measured (t_i), and estimated (x_i) morphometric parameters (here refereed to as residuals) for n sample locations RMSE parameter is an important information since it describes which method, or in this case which sampling design, has better prediction ability. The lower the RMSE, the better the interpolation method, or sampling design.

4 Results

4.1 Statistical Analysis

Statistical analysis of residuals representing deviation of the Z derived using random sampling design from the reference surface (Z_E^T), is presented in Table 1, and using grid sampling design in Table 2. It was revealed, that there are higher extremes at the residuals resulting from random sampling design, than those of grid sampling design. However at the 30.6% of the input data points, extremes for the random sampling design are lower that those for the grid sampling design. Also, at the 10.2% of input data points extremes are almost similar. Overall, at the 30.6% of the input data lower extremes are those for the random sampling design, ranging from -0.77 to 2.36m. Further analysis of extreme residuals might not provide useful results, since they might have character of outliers (Bonk 2003). Thus further analysis is focused on the RMSE, as it is more representative parameter. The RMSE values for

Table 1. Statistics of residuals for random sampling data design at Toporec region (MC - Morain coefficient)

Dataset	Min. [m]	Max. [m]	RMSE [m]	MC	< 0 [%]	> 0 [%]	= 0 [%]
0.4%	-10.57	37.95	4.85	0.226	47.72	53.17	0.39
1.1%	-9.64	18.16	1.95	0.258	47.25	52.87	1.15
3.4%	-6.21	14.77	1.04	0.256	51.1	46.73	3.45
10.2%	-2.55	2.87	0.26	0.213	46.88	44.02	10.39
30.6%	-0.77	2.36	0.08	0.165	35.05	35.19	31.05

random sampling design are higher in general, and have higher standard deviation ($\sigma = 1.94$m), while the RMSE values for the grid sampling design are lower, as well as its standard deviation ($\sigma = 0.9$m). However, the RMSE for the random sampling design computed at the 30% of input data (RMSE=0.08m), is significantly lower that that for the grid sampling design (RMSE=0.26m). It suggests, that the random sampling design gives better results when the number of input data is relatively high (around 30% of total pixels or more), while the grid sampling design performs better when the number of input data is lower (up to around 10% of total pixels or less). To study the correlation between the number of input data and RMSE value for both

Table 2. Statistics of residuals for grid sampling data design at Toporec region (MC - Morain coefficient)

Dataset	Min. [m]	Max. [m]	RMSE [m]	MC	< 0 [%]	> 0 [%]	= 0 [%]
0.4%	-12.64	16.5	2.38	0.254	46.6	53.4	0
1.1%	-5.43	8.88	1.11	0.21	55.1	44.9	0
3.4%	-3.8	3.96	0.48	0.217	53.2	46.8	0
10.2%	-1.77	2.9	0.28	0.238	52.3	47.7	0
30.6%	-2.56	1.66	0.26	0.211	50.6	49.3	0

sampling designs, simple line graph was produced (Figure 5). Line graphs showed, that there is certain degree of dependency between the RMSE value and number of input data points for both samplings designs. That dependency can be characterised by a curve similar to that of hyperbolic function. Correlation coefficient for random sampling design was -0.63, and that for grid sampling design was -0.59. However, when the same graph was plotted with *log* values, curves seemed to be almost straight and correlation coefficients has increased to -0.97 for random sampling design and to -0.99 for grid sampling design respectively. This analysis suggest, that there might be some mathematical representation of relationship between the RMSE values and number of input data points. However, that statistics is strongly dependent on pixel-size and terrain complexity.

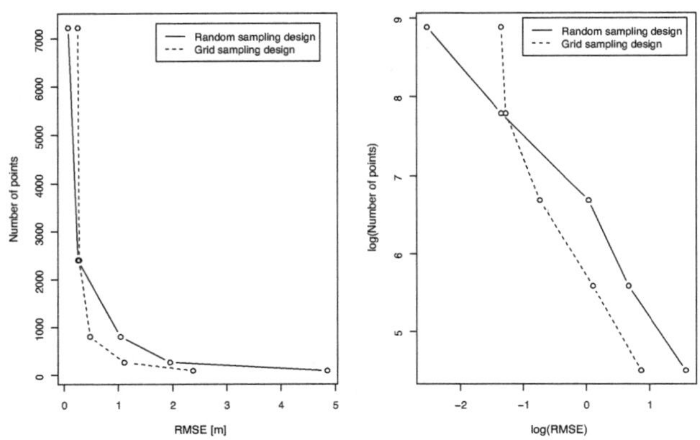

Fig. 5. The relationship between the RMSE and number of input data points for both samplings designs (left) and log values (right) respectively for the region of Toporec

4.2 Spatial Autocorrelation

To obtain information on spatial autocorrelation of the residuals, spatial distribution of the residuals was visualised on maps (Figure 6, and 7), and Morain Coefficient (MC) was computed for each dataset.

Maps of residuals for the random sampling design showed, that the lowest extremes have tendency to locate along the valley bottoms (for 0.4%, 1.1%, and partialy also for 3.4% of total pixel number), while the highest residuals are located at the slopes with the highest slope angle. This, however, is not valid for the datasets computed using the number of points equivalent to 10.2%, and 30.6% of total pixel size. Moreover, the spatial pattern of residuals with increasing number of input points is super fine, and there is an abrupt change of residuals values over the small area. On the other hand, spatial pattern of residuals derived using low number of input data is more homogeneous (as for spatial changes of positive and negative residuals). Also, this pattern proves, that there is a tendency for negative and positive residuals to clump together and to form clumps rather large in area. Maps of residuals also revealed the fact, that the spatial occurrence of extreme residuals (both positive and negative) can be explained by the absence of input data points on those locations coupled with the terrain complexity (see Figure 4).

Maps of residuals derived by the grid sampling design showed the same spatial pattern as those for the random sampling design (applicable for 0.4%, and 1.1% of total number of pixels). However, spatial pattern for 3.4%, 10.2%, and 30.6% of data points are distorted by the internal segmentation procedure implemented in RST interpolation function. In other words, the grid sampling design is evidently visible in the spatial pattern of these residuals. This might show an evidence that the pattern is directly influenced by the RST interpolation function. That pattern has a mesh-like shape, and orthogonal artifacts are visible (Figure 7). Generally, the effect of systematic error increases with the number of input data points (however, there might be a systematic error due to the random sampling as well, unfortunately with the research tools used in the research it was not possible to discover or describe it).

The Morain coefficient (MC) shows low spatial autocorrelation between the residuals and no systematic variation was found between its values. Overall, the lowest MC was found at the datasets with highest number of input points (MC=0.165 for the random sampling design, and MC=0.211 for the grid sampling design respectively). However as it was mentioned in above paragraph, and suggested by maps of residuals for grid sampling design (Figure 6 and 7), the effect of systematic errors increases with higher number of input data points. Thus, there is a strong evidence, that those two maps (and related MC as well) should be interpreted with caution. Also, this analysis proved that RST interpolation function, with its default parameters, might not be directly applicable for the grid sampling design containing relatively high number of input data points (number of points equivalent to more than 30% of total number of pixels).

5 Discussion

The results at Toporec region confirmed that data design, and number of input data points are important factors of DEM accuracy.

Often the nature of the studied phenomena is so complicated (urban areas, mountainous areas, flat areas etc.), that irregular sampling design is required. It means that certain sampling density is required for areas with low slopes, certain for peaks, and certain density is required to depict linear objects (such as mounds) as precise as possible. Even certain geodetic standards for number of surveying points exist, but so far, there is no generally applicable methodology on number of points, and its sampling design for specific categories of terrain complexity (Cebecauer et al. 2002). Thus, it would be very helpful from the methodological and practical point of view, to study terrain complexity – sampling design relationship, by analysing results with theoretically defined accuracy parameters.

Such a study might consists of the following procedures: (1) selection of the terrain morphometric properties to describe terrain complexity (2) definition of accuracy parameters for certain terrain type, (3) suggestion of sampling design based on previous two steps, including minimal number of points required for interpolation fulfilling the criteria of accuracy. However, such a methodology does not exist so far and DEM generation except the geomatic standards, still rely on professional skills of GIS operator. Thus, progressive sampling, usually discovered by empirical studies, and long-term experiences is necessary, especially in real world application, and in complicated terrain (Cebecauer et al. 2002).

The aim of this research was to reveal the effect number of input data points can have on DEM accuracy. The aim from the technical point of view was to depict enough datasets with different number of points, so that spatial variation of residuals, and also RMSE–number of points correlation could be studied. This problem might be divided into two parts: (1) to chose the highest and the lowest number of points, (2) to chose the step between the datasets. Highest number of points was limited by the number of total pixels in Toporec region which was 23 616 and its resolution which was 10×10m. If the highest number of input points was to high, RST function could not have "enough freedom" to interpolate surface within the data points. Our empirical studies proved, that the number of points corresponding to 50% of total number of pixels is too high (Bonk 2003). Also, this assumption proved to be true when the number of points equivalent to 30.6% was chosen as the highest number. In other words, increasing number of data points (30.6% of total pixel size, or more) would not correlate with the quality of information obtained from such a dataset.

Another fact that supports the selection of highest and lowest number of input data points, is the shape of curve representing the RMSE–number of points correlation (Figure 5). It shows that the highest effect the number of data points have on the RMSE value is when number of data points is represented by 1.1%, 3.4%, and 10.2% of total pixel number are chosen (Figure 5). In other words, it would not make a sense to lower the lowest number of points and increase the highest number of points since those parts of the curve approximate to 0, or ∞ respectively (under the assumption, that the curve is a hyperbolic function). The interpretation of the curve, has another practical application. It tells, what should be the number of points used for the interpolation, to end up with certain acceptable value of the RMSE at the region with similar terrain complexity and the same resolution. It also tells, that starting from certain percentage of the input data points, it does not make a sense to increase this percentage, since the RMSE value decreases only in a limited fashion (Figure 5). The step between the two datasets was defined such that lower number of points at particular dataset was around 3 times lower, than the number of points in adjacent dataset with higher number of points (number of points equivalent to 30.6% of total number of pixels was defined, as the highest one). Selection of such a step allowed even low number of points to be studied into certain degree of rationality.

6 Conclusions

The accuracy of the DEM generated from random sampling design is highly correlated with the sampling density. The accuracy of the DEM based on grid sampling design is also highly correlated with sampling density, however standard deviation (σ) of its RMSE is significantly lower ($\sigma = 0.9$), than that for random sampling design ($\sigma = 1.9$). The best accuracy in the sense of the lowest RMSE however, was observed at random sampling design (RMSE = 0.08m) using 7218 input points (30.6% of total number of pixels in the region). The correlation between the sampling density, and RMSE value however is not linear, and it is similar to a hyperbolic function (Figure 5).

From the practical point of view it means, that if dense enough random sampling design is possible (sampling points reaching theoretically around 30% of total number of pixels) it should be preferred to grid sampling design. On the other hand, if there there is not the possibility for high density random sampling design grid sampling design should be preferred, since, if number of grid sampling points is lower than approximately 10% of total pixels number, it produces lower RMSE.

Also the revealed function representing the sampling density–RMSE value relationship (Figure 5) might help identify the number of sampling points, and the type of sampling design required to obtain the DEM with the specified RMSE.

Generally, extreme values of positive residuals have tendency to be higher that those of negative residuals for both sampling designs. However, the spatial distribution of positive and negative residuals is approximately similar. At random sampling design, the percentage of residuals equal zero is the same as the percentage of input data points used for the particular DEM production. It is assumed (and histogram analysis supports this assumption) that, there is a similar percentage grid of sampling residuals oscillating around zero value, as those of random sampling design.

The analysis revealed that, if grid sampling design is used special attention should be paid to adjustment of internal RST parameters. Especially those which control minimum distance between the sampling points at interpolation (`dmin`), maximum number of points within segment (`segmax`), and minimum number of points used for interpolation (`npmin`). If it is desired the resulted surface to pass through the sampling points, and adjustment by using the internal RST parameters does not help, non-grid sampling design might be considered.

Spatial statistics, based on Morain coefficient, showed that there is no significant autocorrelation between the spatial locations of residuals (random sampling design: $MC \subset \langle 0.165, 0.258 \rangle$, grid sampling design: $MC \subset \langle 0.21, 0.254 \rangle$). Visual analysis of maps representing spatial location of residuals confirmed, that for surfaces generated from 0.4%, and 1.1% of total number of pixels, there is a significant tendency for negative, and positive residuals to be clumped together. Spatial distribution of those residuals is highly homogeneous, forming isolating islands of negative or positive residuals. That homogeneity is disrupted with increasing number of input points used for interpolation (3.4%, 10.2%, and 30.6% of total number of pixels), and it becomes coarse, and heterogeneous.

As for the spatial location of the extreme residuals, it was confirmed that those are located at the places where sampling points were missing, and also at the places with high terrain complexity such as river terraces, steep slopes and hills.

Results presented here are valuable because they are being used in JRC tasks where DEM utilisation is fundamental. These tasks include high quality DEM for flood risk mapping (Baredo et al. 2005), 2D flood inundation modelling (Bates and De Roo 2000). Although the demand for DEM quality for EFAS is not crucial (De Roo et al. 2003), other activities such as soil erosion modelling (De Roo 2000) and potential flood hazard and risk mapping (De Roo et al. 2005) definitely are. These are the core activities at WDNH where DEM utilisation is crucial to support the European Commission policy (announced in the communication (COM(2002)-481) - the development of a European Flood Alert System - EFAS(3).

DEM quality should be always tied to its application, the purpose, the DEM was developed for. Thus the particular DEM should not labelled as a "good" one, or "bad" one based on results of statistical analysis purely. Its purpose should be considered as well.

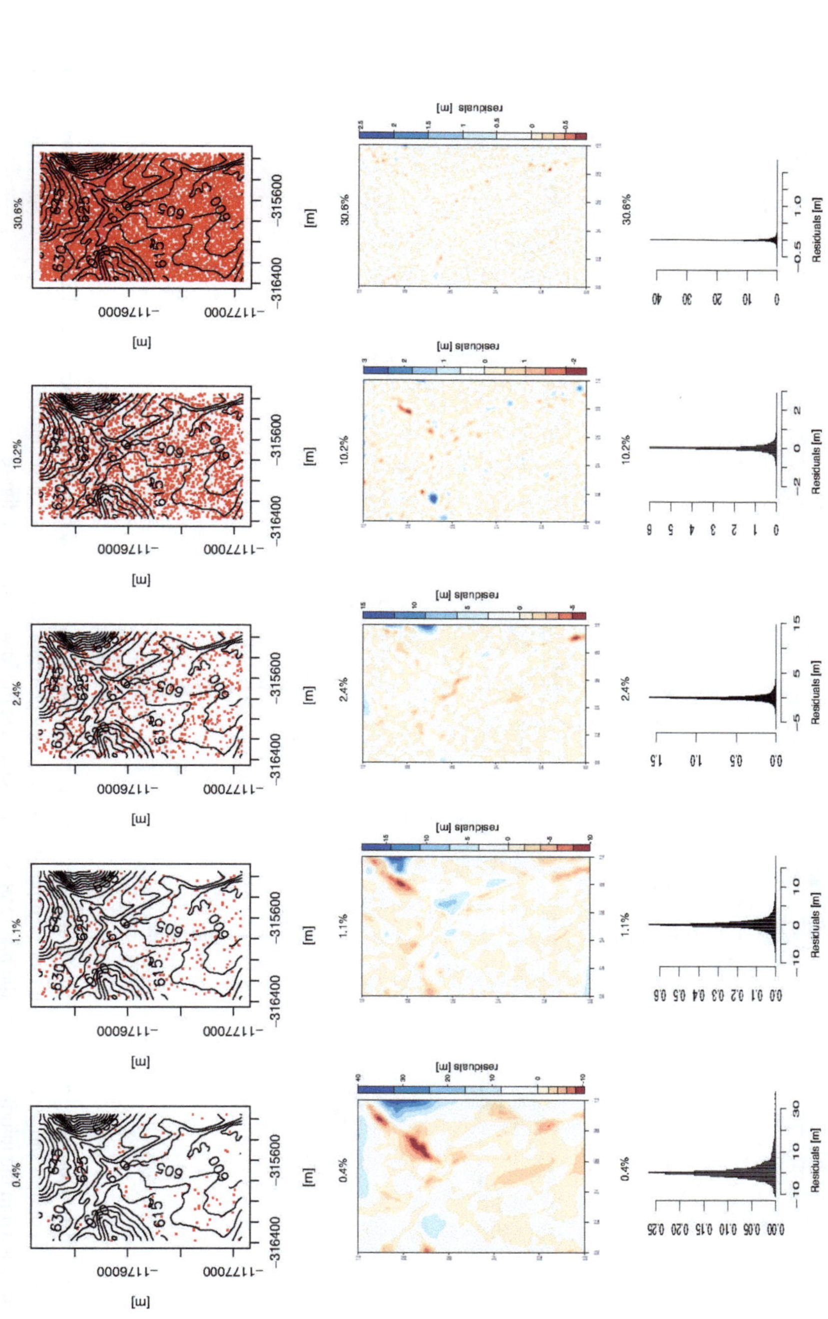

Fig. 6. Spatial distribution of random sampling input data points at the region of Toporec. There are actual locations of input data points shown at the upper part. Each column shows varying number of input data points as a percentage of the total pixel number in the map. The spatial locations of the residuals are shown the middle part and the histograms are presented at the lower part of the figure

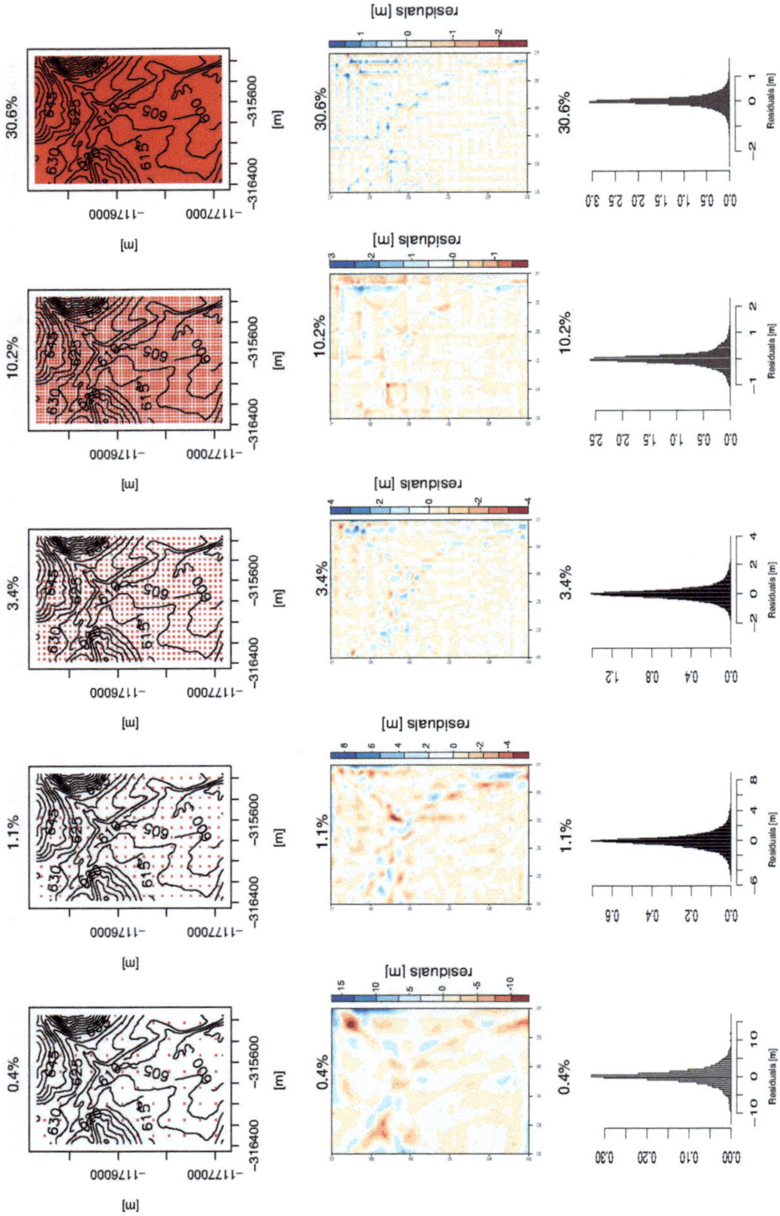

Fig. 7. Spatial distribution of grid sampling input data points at the region of Toporec. There are actual locations of input data points shown at the upper part. Each column shows varying number of input data points as a percentage of the total pixel number in the map. The spatial locations of the residuals are shown in the middle part and the histograms are presented at the lower part of the figure

References

Ackerman, F. (1980). The accuracy of digital terrain models. In *Proceedings of the 37th Photogrammetric Week*, pages 113–143, University of Stuttgart

Baredo, J., Lavalle, C., and De Roo, A. P. J. (2005). European flood risk mapping. *European Commission Joint Research Centre. Special Publication.* S.P.I.05.151.EN

Bates, P. and De Roo, A. P. J. (2000). A simple raster-based model for flood inundation simulation. *Journal of Hydrology*, 236:54–77

Bonk, R. (2003). *Scale-dependent impact of selected factors on morphometric parameters accuracy and automated geomorphological mapping*. PhD thesis, Comenius University Bratislava, Slovakia, Faculty of Natural Sciences

Cebecauer, T., Hofierka, J., and Suri, M. (2002). Processing digital terrain models by regularized spline with tension: tuning interpolation parameters for different input datasets. In Ciolli, M. and Zatelli, P., editors, *Proceedings of the Open Source GIS - GRASS users conference 2002, Trento, Italy, 11-13 September*

Clif, A. D. and Ord, J. K. (1981). *Spatial Processes: Models and Applications*. London: Pion

De Roo, A. P. J. (2000). Modelling runoff and sediment transport in catchments using GIS. In Gurnell, A. M. and Montgomery, D. R., editors, *Hydrological Applications of GIS*, Advances in Hydrological Processes, page 184. John Wiley

De Roo, A. P. J., Barredo, J., Lavalle, C., Bodis, K., and Bonk, R. (2005). Potential flood hazard and risk mapping at pan-European scale. In Peckham, R. and Jordan, G., editors, *Digital elevation modelling. Development and applications in a policy support environment*. Joint Research Centre, European Commission, Ispra. in press

De Roo, A. P. J., Gouweleeuw, B., Thielen, J., Bates, P., and Hollingsworth, A. (2003). Development of a European flood forecasting. *International Journal of River Basin Management*, 1(1):49–59

De Roo, A. P. J., Wesseling, C., and Van Deursen, W. P. A. (2000). Physically-based river basin modeling within a gis: The LISFLOOD model. *Hydrological Processes*, (14):1981–1992

Goodchild, M. (1986). Spatial autocorrelation. In *Concepts and Techniques in Modern Geography*. Norwich: Geo Books

Li, Z. (1990). *Sampling strategy and accuracy assessment for digital terrain modeling*. PhD thesis, University of Glasgow, Glasgow. 299 pp.

Li, Z. (1991). Effects of check points on the reliability of dtm accuracy estimates obtained from experimental tests. *Photogrammetric Enineering & Remote Sensing*, 57(10):1333–1340

Li, Z. (1992). Variation of the accuracy of digital terrain models with sampling interval. *Photogrammetric Records*, 14(79):113–128

Li, Z. (1994). A comparative study of the accuracy of digital terrain models (DTMs) based on various data models. *ISPRS Journal of Photogrammetry and Remote Sensing*, 49(1):2–11

Makarovic, B. (1972). Information transfer in construction of data from sampled data. *Photogrammetria*, 28(4):111–130

Minar, J. (1998). *Georelief and geoecological mapping in large scales.* PhD thesis, Dept. of Physical Geography and Geoecology, Faculty of Natural sciences, Comenius University, Bratislava. in Slovak

Mitasova, H. and Hofierka, J. (1993). Interpolation by regularized spline with tension: II. Application to terrain modeling and surface geometry analysis. *Mathematical Geology*, 25(6):657–669

Neteler, M. and Mitasova, H. (2002). *Open Source GIS: A GRASS GIS Approach.* Kluver Academic Publishers. 1st.edition

Schmidt, J. and Dikau, R. (1999). Extracting geomorphometric attributes and objects from digital elevation models - semantics, methods, future needs. In R. Dikau and H. Saurer, editor, *GIS for Earth Surface Systems*, pages 154–172. Gebrüder Borntraeger, D-14129 Berlin D-70176 Stuttgart

Suri, M. and Hofierka, J. (2005). Application of DEM in solar radiation modelling (natural resources: solar energy). In Peckham, R. and Jordan, G., editors, *Digital elevation modelling. Development and applications in a policy support environment.* Joint Research Centre, European Commission, Ispra. in press

Referenced Web Sites

[1] `http://natural-hazards.jrc.it`

[2] `http://ies.jrc.cec.eu.int`

[3] `http://efas.jrc.it`

[4] `http://www.europa.eu.int/comm/environment/water/flood_risk`

[5] `http://srtm.usgs.gov`

[6] `http://dataservice.eea.eu.int/dataservice`

SRTM as a Possible Source of Elevation Information for Soil-landscape Modelling

Borut Vrščaj, Joël Daroussin and Luca Montanarella

1. Introduction

"The Shuttle Radar Topography Mission (SRTM) obtained elevation data on a near-global scale to generate the most complete high-resolution digital topographic database of Earth' [1]. Availability to the public, uniform cover of the Earth's surface combined with a cell size acceptable for medium-scale soil modelling (90m on average) introduced the SRTM digital elevation model (DEM) as promising datasets for small and medium scale soil-landscape modelling and applications.

Soil is an important non-renewable natural resource and a great majority of the terrestrial ecosystems are defined above all by the soil properties. Recently a broad range of policy instruments of the European Community directly comprehends soil protection, which depends to a great extent on the availability of data on soils (ECC 2002; EPR 2003). At present the European Soil Information System (EUSIS) is the only available homogenous soil information covering the European continent (Kirkby et al. 2003; Daroussin et al. 2004). The EUSIS basic component dataset is the Soil Geographical Database of Eurasia at scale 1:1,000,000. This geographically seamless soil database is accompanied with interpretative models and is used in continental and regional modelling and analyses (Kirkby et al. 2003). The EUSIS data has several limitations inherited from national databases (Daroussin et al. 2004), but one of the most important is the coarse scale. The spatial accuracy of the European soil databases can be improved by applying different methods. An accurate DEM of proper resolution and DEM derived data represent the key input datasets into many GIS-based models for soil-landscape modelling (MacMillan et al. 2000; Dobos et al. 2000; Bui and Moran 2001; MacMillan 2001; MacMillan et al. 2004; Langanke et al. 2005) or for the improvement of coarse soil information (Zhu and Mackay 2001; Bock et al. 2005). The DEM of better resolution represents the topography and slope processes and movements more accurately and, consequently, can serve as a more reliable source of information for soil-landscape relationships modelling. The SRTM elevation data, if accurate enough, could represent a valuable uniform source of DEM for the improvement of the 1:1,000,000 soil maps of Europe.

2. Gaps and Noise in SRTM Data

Altitudes of water surfaces are generally badly detected by the SRTM sensor. This produces either gaps (i.e. absence of altitude values, also called NODATA values in the used/applied ArcInfo® GRID terminology) or wrong, noisy altitudes in water body areas. Figure 1 shows an area of the Greek island Lefkada. The thin white line shows the coastline dividing the land from the sea. 'A' zones are examples of NODATA values whereas 'B' zones show examples of wrong, noisy altitude values in the water surface areas ranging from –11 to +7 m in these examples.

A possible solution for this problem would be to mask out water bodies by simply replacing all concerned cell values by a NODATA value. This raises the problem of obtaining a good quality water bodies and coastlines database (difficult/expensive to obtain for the continental scale). Another solution would be to replace the cell values lying within these identified water bodies by a value computed from surrounding altitudes. We have not investigated a methodology for achieving this task.

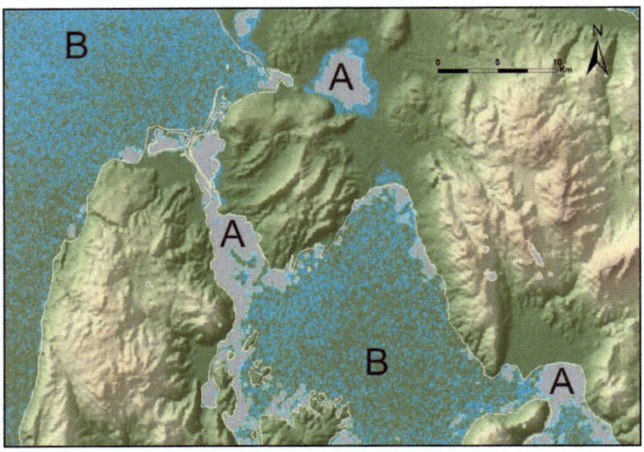

Fig. 1. Examples of (A) NODATA values and (B) wrong, noisy altitude values in water bodies around Lefkada island (Greece) from raw SRTM data

Figure 2 shows examples of gaps shows examples of gaps (NODATA values) on either (A) very steep slopes or (B) lakes and glaciers in mountainous area of Slovenian Julian Alps. For steep slopes this is due to their lying in the 'shadow' of the SRTM sensor (i.e. hidden from it). Lakes and glaciers have the same reflectance problem (B) as mentioned above for the sea.

In the context of our objectives, these problems have to be solved before further use of the SRTM data for soil modelling. The missing values and/or wrong altitude values would prevent modelling or radically affect the quality of output results. Possible solutions for 'filling' these gaps have been investigated and are presented hereafter.

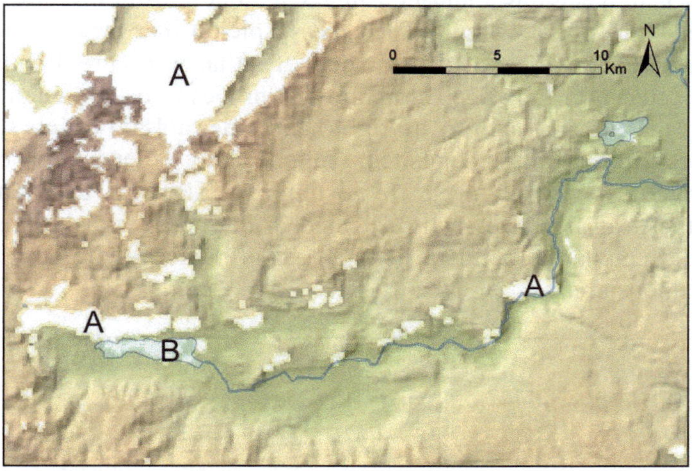

Fig. 2. Examples of NODATA values in (A) steep slopes and (B) lakes and glaciers in the Julian Alps mountainous area (Slovenia) from raw SRTM data

3. Digital Surface Model vs. Digital Elevation Model

Raw DEM data generally contains erroneous altitude values originating from either measurement errors or interpolation methods or even computational errors (e.g. rounding of numbers). These various errors often produce cell or groups of cell values that are artificially lower/higher in altitude than their surrounding cells. When using such a DEM in hydrological modelling, this may produce unrealistic or erroneous results (e.g. wrong flow directions, disrupted drainage lines, etc.). Correcting a DEM is therefore necessary in order to build a so-called 'hydrologically correct DEM' before it can be used in a number of applications such as erosion or soil-landscape modelling. In particular, a hydrologically correct DEM represents the basic requirement in soil-landscape modelling especially at large (MacMillan et al. 2004) or medium (Zhu and Mackay 2001; Langanke et al. 2005) scales.

Visual inspection of SRTM DEM in several places of Europe has shown another type of error that affect its hydrological correctness. Two examples located in the plane of southern Poland illustrate these. Figure 3 clearly shows patches of angular shaped areas (A) that can be detected west of a lake. The differences in altitudes between these areas (A) and their neighbouring areas (B) are mainly (lying) in the range of 15 to 20 m. Confrontation with a satellite Landsat 5 image available from the Internet -[9] (51°25′26″N; 19°48′13″E) confirmed the suspicion that these angular uplifted structures are areas of dense woods. Similarly, Figure 4 shows that the DEM detects a 190 m deep structure that proves to be a mine pit (in the middle of the picture) and, to its right-hand side, a 180 m high structure that proves to be a hill of deposited gangue. Many other natural or man-made structures can be 'seen' by examining SRTM DEM data. But whereas they can be quite easily detected in flat areas with low relief intensity, they can hardly be identified in areas of higher relief intensity such as mountainous areas.

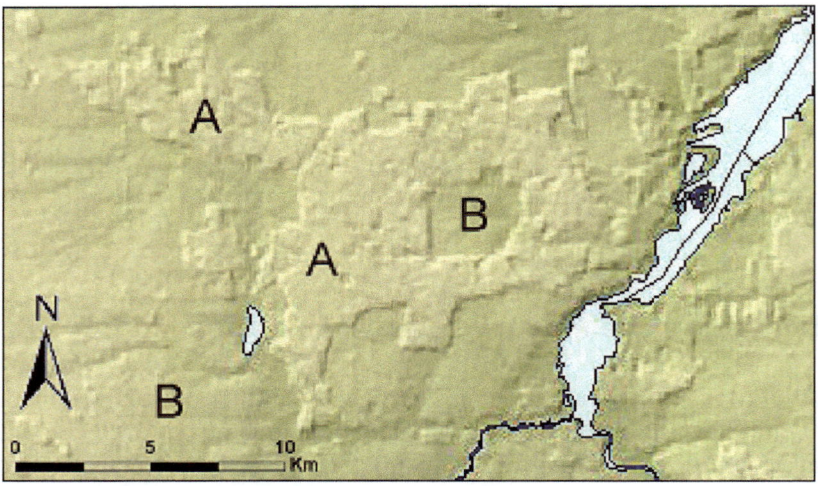

Fig. 3. Areas covered with woods (A) or without (B) as detected from a raw SRTM DEM in the southern plain of Poland (51°25′26″N; 19°48′13″E). Approximate scale

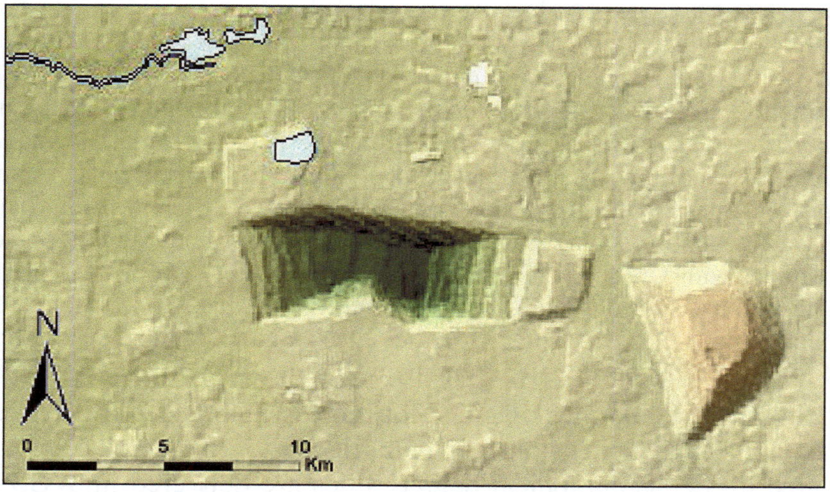

Fig. 4. A human-made structure, open coal mine pit and deposit, as detected from a raw SRTM DEM near the Kamiensk in Southern Poland (51°14′40″N; 19°18′58″E). Approximate scale

The goals of this research were (1) to examine the accuracy and the suitability of SRTM data for medium to small scale soil-landscape modelling and (2) consequently its possible use for the improvement of the existing European soil databases.

4. Materials and Methods

4.1 Description of the Test Area

The test area covers a 110km^2 rectangular zone located at the southern Alpine mountain range in the Sava River alluvial plain of central Slovenia. Its altitude ranges from 323 to 421m above sea level with a mean of 374m. The flat area is intersected in the west by the Sava River gorge. The land use within the area is mixed, mainly agricultural with patches of predominantly coniferous woods and small villages. These village settlements are mainly with independent family houses and gardens except for the city of Kranj lying near the northwestern border of the area and the Brnik airport located in its eastern part. The city of Kranj is a mixed composition of two-floor and up to five-floor buildings.

4.2 SRTM Elevation Data

The SRTM data is available on request via the USGS Seamless Data Distribution System following the instructions at [2] web page or can be instantly downloaded as separate files from the [3] FTP site. The reader should be aware that these web pages are subject to frequent changes and improvements. SRTM DEM data is available in two different resolutions: the 1 arc second (~30 m) resolution is available only under conditions of agreement with US authorities, and the 3 arc seconds (~90 m) resolution, which was used for our study, which is freely available. Data is delivered in 'hgt' format files covering 1x1 degree tiles. A more detailed description of SRTM data can be accessed from the web pages listed in the Appendix.

Raw SRTM data files obtained for this research were downloaded in separate '.hgt' extension files compressed to '.zip' files covering an area of 1x1 geographical degree. Files are delivered in WGS94 unprojected geographic latitude-longitude coordinates.

4.2.1 Converting and Joining SRTM Tiles to an ArcInfo® GRID Raster Layer

Each raw individual SRTM '.hgt' file covers a 1x1 degree tile. Before conversion, the file was renamed to '.bil' extension and an associated header file is created according to the information provided in the file name. Tiles can then be converted to ArcInfo® GRID raster format using the IMAGEGRID command or by using several scripts available on the Internet to achieve this. We used the SRTMGRID.AML ArcInfo® routine developed at USGS available from the SRTM web page [2]. The syntax of the [c1] command from the Arc: prompt is available in Annex of this paper. Assembling separate SRTM tiles can be done by mosaicking with the MOSAIC tool available in ArcInfo® GRID using the [c2] command syntax from the ArcInfo® GRID command prompt. This MOSAIC function creates one single grid from two or more adjacent grids with a smooth transition of output grid values in the overlapping areas of neighbouring grids. The concept is well explained in ArcInfo® Help documentation.

4.2.2 Filling Gaps in SRTM Data

Raw SRTM DEMs contain gaps (zones with no data) for areas in the shadow of the radar sensor. These have to be filled to fulfil the need for a complete cover of the study area. Several methods were tested to achieve this for our test sites. A number of software applications emerged in the years 2003-2004 for SRTM correction, mostly developed for filling gaps [s1; s2; s3; s4] (see Appendix – Software applications). The applications are generally poorly documented and behave as 'black boxes'. Control over the DEM correcting process is either limited or not enabled at all. They interpolate from available values surrounding the zones without values. Two of the used software packages, 3dem and SRTMfill applications, were tested and they achieved similar but unsatisfactory results. ArcInfo® suite of tools also offers several possibilities to patch the SRTM gaps. The interpolation procedure is more controlled and better documented. Two ArcInfo® methods for filling SRTM were compared: TOPOGRID and FOCALMEAN.

Similarly to the above-mentioned methods, the FOCALMEAN command in ArcInfo® GRID module interpolates inward from the edge of the void in order to fill it with estimated values. The syntax of the [c3] command line issued from the ArcInfo® GRID module prompt.

The TOPOGRID command is based on the Hutchinson procedure (Hutchinson 1989). It can generate hydrologically correct grids of elevation from point, line and polygon elevation coverages. At the first stage the SRTM grid cells are converted to a contour lines coverage using the CONTOUR command [c4] from the ArcInfo® GRID module.

In the second phase, the resulting contour lines coverage is used as an input to the TOPOGRID command to produce an intermediate gridded DEM. The command [c5] has its own dialog prompts and was used from the ArcInfo® prompt.

The tool is used to generate an elevation grid covering the same extent and resolution as the SRTM grid but without gaps. This intermediate grid is then used to patch the gaps found in the original SRTM DEM using the [c6] syntax.

The resulting grid is identical to the raw SRTM DEM except that any zone of no data is filled with values interpolated by the TOPOGRID tool which has the advantage of taking into consideration not only the immediate surroundings of the gaps but also the general structure of the terrain. Moreover, if any other elevation data sources can be found that falls into gaps, these can be used as inputs to TOPOGRID to improve the result of the inward interpolation.

4.2.3 Effects of Projecting SRTM Data

The 1x1 degree SRTM tiles are provided in WGS84 unprojected geographic coordinates and have to be projected to the local projection system when needed by applications. For the European continent, several projections are suggested and well documented [4]. We projected SRTM grids to Lambert Azimuthal (LA) and to Slovenian Gauss-Krueger projections.

During the projection process, the pixels are resampled and new altitudes are calculated based on a resampling algorithm. We used the Arc Info® PROJECT command [c7].

A north to south column of cells taken from an SRTM tile was converted to a polygon of the same size and position as the raster cells. Both datasets were then projected to the LA projection and the position of projected polygons and projected raster cells was visually compared (see Figure 7) to illustrate the effects of projecting on cell positions.

4.2.4 Comparing SRTM Data to DEMs Available in Slovenia

A visual inspection of SRTM for several European regions revealed some problems in SRTM altitudes related to land use. Two DEM datasets, with well-known quality characteristics available from the Slovenian authorities, were used to carry out a more rigorous evaluation assessment of the SRTM DEM quality. Both are available as raster datasets in ArcInfo® GRID format in the Gauss-Krueger projection. DEM25 is a 25 m cell size DEM originally generated from 1:5,000 and 1:10,000 Slovenian topographic maps. DEM100 is a 100 m cell size DEM generated from point data selected from Slovenian military 1:25,000 maps. For this comparison exercise, these two DEMs were used as they were delivered, without any further processing or corrections.

SRTM data was projected to the Gauss-Krueger projection and visually compared to the Slovenian datasets. The altitude differences from SRTM to DEM25 and DEM100 grids were calculated on a cell-by-cell basis using simple ArcInfo® GRID map algebra [c8; c9].

The results were visually inspected using ArcScene® (Figure 9), and the grid value attribute tables were transferred to MS Excell® for further statistical evaluation.

Additionally, Corine Land Cover data (CLC) and Landsat TM images were used both for selecting evaluation zones based on their land use type and to provide help in the identification and understanding of problems. Corine Land Cover data was used to identify three main different land use types within the test zone: agricultural areas (where ploughed fields predominate), forested areas (predominantly coniferous forests), and urban areas. DEM25 altitude values were considered as reference values (best available altitude information) and were used to evaluate the consistency of DEM100, which in turn could be used to evaluate the SRTM DEM.

4.2.5 Software

To achieve the goals specified above, a set of software applications was used. We used ESRI ArcGIS® version 8.3 geographical information system software package. ArcMap® and ArcScene® were used for data query and visualisation, as well as for map production. DEM processing was done using ArcInfo® Workstation along with the GRID module or with ArcGIS® Spatial Analyst. Specific routines for automating processing were developed with the Arc Macro Language (AML) programming facility. Tabular data management and analysis was done with the combined use of Microsoft Access® 2000 and Microsoft Excel® 2000. Processing gaps in SRTM data were tested using 3dem software, SRTMfill application, and the ArcInfo® TOPOGRID tool.

5. Results and Discussion

5.1 Possible Solutions for Data Improvement of SRTM

The 90 m resolution of the tested SRTM dataset is expected to make it useful for processing small to medium scale soil maps. (Bui and Moran 2001; MacMillan 2001]

Yet the inspected SRTM datasets showed several deficiencies. Some of them strongly affect and influence the use of SRTM in soil-landscape modelling. The main deficiencies found are the following:

- SRTM data represent the dense vegetation canopy surface instead of ground surface,
- SRTM altitude is that of the top of human-made structures,
- Gaps (no value areas; or areas with undefined value) in areas of very steep slopes and some parts of the water bodies,
- The reflectance from other parts of the water bodies caused wrong altitude values in SRTM dataset,
- Adjacent SRTM tiles have an overlapping line of cells with a different altitude.

Each of these problems is illustrated hereafter.

5.1.1 Testing Methods for Filling Gaps

Four different methods for filling gaps (NODATA areas) were used and compared: the TOPOGRID tool available from ArcInfo®, the SRTMFill application [s3], the FOCALMEAN command available from ArcInfo® GRID prompt, and the 3dem application [s1]. These methods were tested over a set of different situations where gaps occurred. The results were visually inspected on maps and topographic transects. Figure 5 shows an example of this kind of analysis.

It appeared that the more credible results are obtained from the TOPOGRID method. It produces a smooth surface and avoids 'noisy' results. Other advantages of the TOPOGRID method over the others include:

- Developed for the specific purpose of processing topographic data and therefore well adapted in this context,
- Well documented (Annoni 2001),
- It offers the possibility to include additional spatial datasets (e.g. measured altitude points, landscape and water feature vector datasets such as ridges, streams, lakes, etc.), which act as structuring and corrective information to improve the final results,
- Better control and parameterisation of the process,
- It includes behaviour for producing hydrologically correct results.

Other methods are either undocumented 'black boxes' (SRTMFill, 3dem) or rather rough and not well adapted to the context of topographic data processing (FOCALMEAN). The comparison of the altitude transect and the differences of the output grids generated by four different method used are presented by Figure 5.

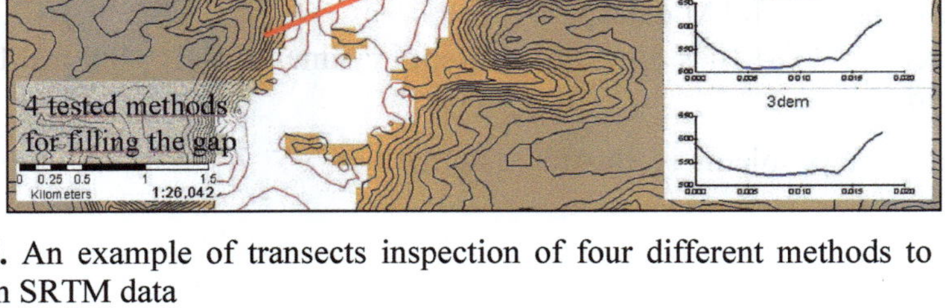

Fig. 5. An example of transects inspection of four different methods to fill the gaps in SRTM data

A more objective, quantitative investigation could be carried out for evaluating the quality of the gap-filling methods. By ‘creating’ artificial gaps in areas with actual data, i.e. artificially ‘blanking’ areas, then filling them, then comparing the result with the original actual data, one could compare not only the methods, but also their behaviour in various topographic conditions (slope, relief variability, slope aspect, etc.) as well as in various gap sizes and shapes.

5.1.2 Mosaicking SRTM Tiles

Mosaicking of 1x1 degree SRTM tiles caused a significant problem in this study. As opposed to the statement of the SRTM documentation, the tiles show an overlapping line of cells on their borders. Moreover, differences in altitude values of these overlapping cells were identified in the order of a magnitude of ±10 m (Figure 6). We were not able to find an explanation for these differences.

The MOSAIC command used for assembling tiles handles overlapping cells by averaging overlapping cell values in the corresponding output grid cells. This default behaviour is acceptable as no knowledge is available to allow privileging one value or the other. Thus, this method was used in the Slovenian case study.

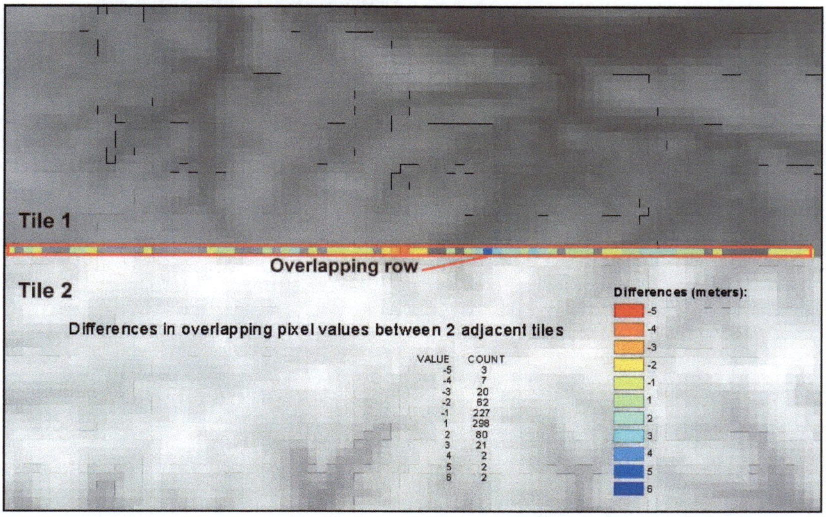

Fig. 6. Altitude differences on the row of overlapping cells of two adjacent SRTM files

5.1.3 The Effect of Projecting SRTM Data

SRTM data is provided in WGS84 unprojected geographic coordinates. These data often have to be projected to other systems such as Lambert Azimuthal used for Europe or Gauss-Krueger used in Slovenia. It is important to be aware that during the projection process, cell values (altitudes) can change. While being projected to Lambert Azimuthal, a 1-degree SRTM tile is not only rotated and translated (Figure 7), but also distorted. It is both 'stretched' in the latitude direction and 'contracted' in the longitude direction. Moreover, because the new, projected, grid is still made of vertical columns and horizontal rows of cells, new cells have to be inserted in the vertical direction while the number of cells in the horizontal direction is reduced. During the process, altitude values are interpolated from neighbouring values.

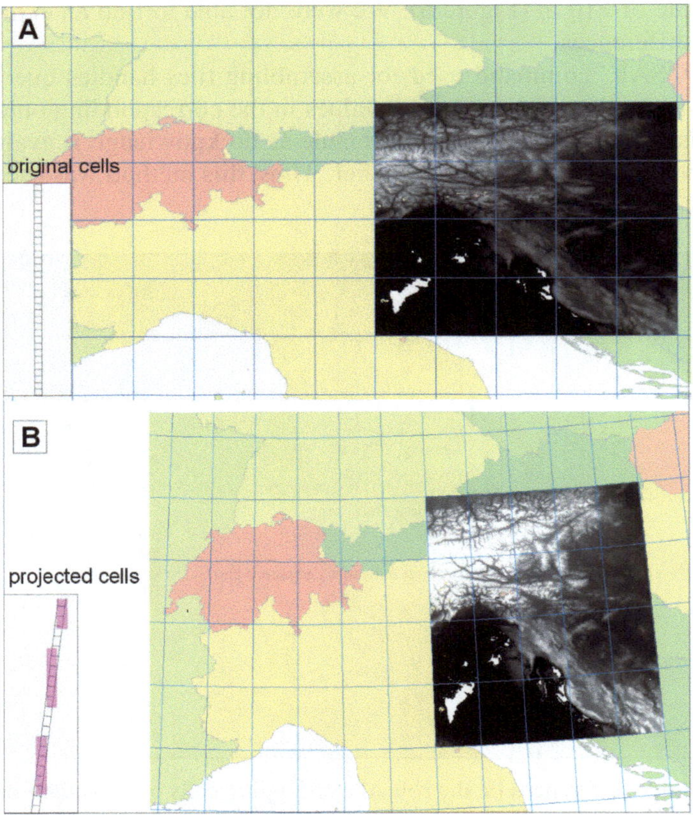

Fig. 7. Illustrating distortion due to projecting a grid. A. A column of cells from SRTM90 before projecting. B. What it looks like after projecting to Lambert Azimuthal co-ordinate system

To illustrate the distortion effect, a column of raster cells was selected from an SRTM tile and converted to vector polygons, each polygon corresponding to one cell (Figure 7a). Both the raster and vector representation of that column of cells were projected from WGS84 to Lambert Azimuthal (Figure 7b). In Figure 7b white squares show the position of cells as coming from the projected polygons. These are assumed to be the 'good' position and serve as a reference to evaluate the distortion of the pink squares showing the position of corresponding cells but coming from the projected grid.

Such a distortion can have important consequences for further use of the DEM in spatial analysis. For instance, if a DEM has been corrected for ensuring its hydrologic continuity (see the FILL command in ArcInfo GRID), projecting the DEM will most probably alter and invalidate this property. As a consequence, a user of DEM data in general should be aware of this problem and carefully choose the right moment and procedure to change the projection system in the analysis process.

6. Case Study

6.1 Quantitative Comparison of the SRTM DEM to National DEMs

To push the analysis further, the SRTM dataset was visually and quantitatively compared to the Slovenian 100m DEM (DEM100) in the area of the Sava River alluvial plain in central Slovenia. To make sure that the 100 m Slovenian DEM was appropriate for this comparison task, it was compared to the high-accuracy and independently derived 25m DEM of the same area (DEM25). Our interpretation hypothesis of the differences between SRTM and DEM100 were again checked against Landsat TM images.

Figure 8a shows an example for the result of computing the elevation difference, on a cell-by-cell basis, between SRTM projected to Slovenian Gauss-Krueger projection and the Slovenian DEM100 (difference = SRTM - DEM100). Pink to red areas are uplifted areas in SRTM (where SRTM > DEM100) in the range of 5 to 18 m. They mostly correspond to the tree canopy. This closely matches the size, location and shape of forests identified in brown colours on the corresponding superimposed 3D view of the Landsat TM image in Figure 9. Green areas are only small patches where SRTM is in accordance with DEM 100 (where SRTM = DEM100) and correspond mainly to agricultural areas. Blue areas are the places where SRTM altitudes are below DEM100 (where SRTM < DEM100). These differences remain within the acceptable range of 5m and their visual inspection shows no spatial pattern. The differences in altitude values within 5m were expected considering the different methodologies used to build the compared DEMs.

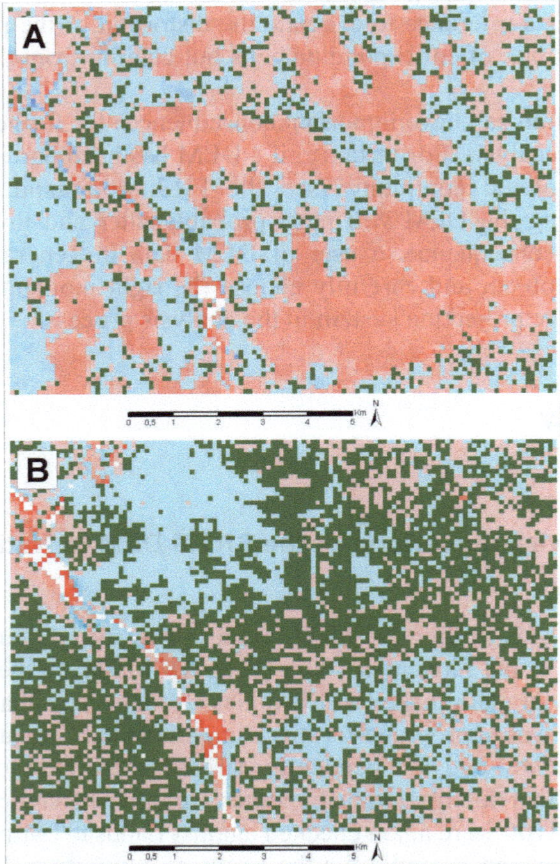

Fig. 8. Slovenian DEM100 in the Sava River gorge. A. SRTM DEM. B. DEM25 taken as a reference. See text for details

Making the same comparison between DEM100 and DEM25 (difference = DEM100 - DEM25), one can see from Figure 8b that their concordance is much higher. The green areas (where DEM100 = DEM25) are much larger. Even where they are different, the differences remain within a few meters and are not correlated to vegetation patterns (compare to Figure 9). The most significant red areas (where DEM100 > DEM25) lie in the Sava River gorge and are due to the effect of resolution differences: a higher resolution better accounts for small, narrow relief patterns whereas a lower resolution averages altitudes in the gorge to higher surrounding altitudes.

Fig. 9. 3D stack views in the Sava River gorge of the Landsat TM image (bottom), and the Slovenian DEM100 (middle) compared to SRTM DEM and DEM25 (top) taken as a reference

The differences between SRTM DEM and DEM100 were further analysed. Corine Land Cover 1995 level 2 classes were used to separate analysis for the three classes that are sensitive to the measurement of surface altitudes as opposed to ground elevation: urban, agriculture and forest. The difference grid (SRTM - DEM100) was split into three separate subset grids: one for urban areas (CLC classes 100 to 199), one for agricultural areas (CLC classes 200 to 299) and one for forested areas (CLC classes 300 to 399). Figure 10 and Table 1 illustrate the distribution of the differences in altitudes in the compared DEMs for each of these land use classes.

In agricultural areas, both DEMs are highly correlated as shown by Figure 9. Most differences lie between –1m and +1m. This demonstrates that the SRTM accuracy is comparable to that of the national elevation grid in the ploughed fields or grassland areas. In urban areas, the distribution of altitude differences is mostly between +1m and +3m which roughly corresponds to the mixed structure of rooftop heights in settlements in that region of Slovenia, whereas in the forest land use class, most of the differences lie between +10m to +16m corresponding to a less variable structure of forest treetops.

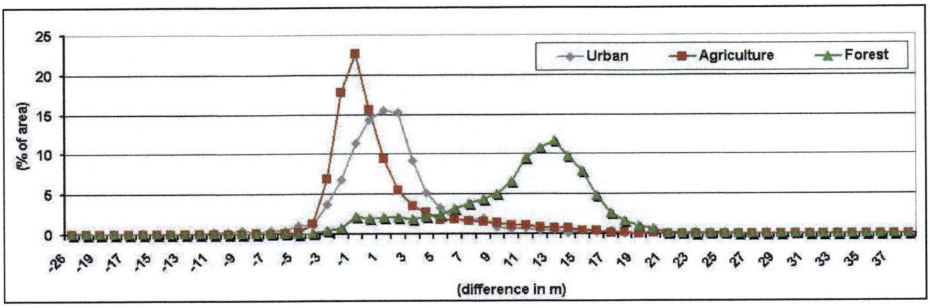

Fig. 10. SRTM-DEM100 differences in altitude in % of area for three different land use classes

It is worth noting that DEM comparison and altitude error detection was possible in regions of low relief intensity (plains, plateaus). But in more or less hilly regions these artefacts are difficult to discriminate from the background relief variability and thus they are undetectable directly from the DEM data.

Table 1. Interpretation of the differences in altitudes between grids SRTM and DEM100 for three land use classes

	Urban	Agriculture	Forest
The most common differences lie between	+1m to +3m	-1m to +1m	+10m to +16m
→ % of the total area	39,96%	47,64%	56,26%
The majority of differences lie between	-3m and +9m	-3m and +3m	+6m and +19m
→ % of the area	90,90%	77, 50%	82,62%

In this study we did not investigate techniques to correct these problems illustrated above. A possible method would be to use accurate land use information as an additional GIS data layer to locate the areas of potential occurrence of a problem such as forests and anthropic features.

From this analysis it can be concluded that SRTM DEM data can be used in soil-landscape modelling at a small or small to medium scale. The scale at which altitude differences significantly influence the results still needs to be ascertained and depends also on the accuracy and aggregation of modelled soil information. SRTM DEM data can be used for critical and demanding small-scale soil modelling only if the dataset is rigorously corrected.

7. Two Possible SRTM Processing Workflows

Despite the limitations identified above there is a strong interest for building a Pan-European SRTM dataset because the source data is easy to access and at no cost. As the area to cover is large (from 35 to 60° latitude North and from -15 to 30° longitude), a workflow needs to be set up or procedures implemented to achieve these pre-processing tasks. The two proposed methods described below aim at homogeneously producing a seamless ArcInfo® grid covering Europe from raw SRTM DEM tiles usable in non-critical applications, mainly for 3D visualisation, as a relief background for thematic maps and low resolution / continental-wide relief modelling.

7.1 Workflow for Building a Seamless Pan-European DEM for Non-critical Applications

The following workflow relies thoroughly on graphic user interface based software packages. Thus the process is not automatic but the user can keep control over each step.

1. Download [2] and move SRTM files in subdirectories, one by 40x40 degrees region that all together cover the whole area of interest. This can be done using the AML script SRTM_list.aml available from (Hutchinson 1989).
2. Automatically unzip all zip files contained in each directory.
3. Run one of the black-box freeware software that fills gaps by interpolation [5].
4. Convert the resulting tiles to GeoTiff files using the 3dem freeware software [6].
5. Reduce GeoTiff files from 32 to 16 bits and convert them to ArcInfo® GRID format using the Raster Calculator tool available in ArcGIS® Spatial Analyst.
6. Mosaic tiles into 10x10 degrees blocks with the MOSAIC function in ArcGIS® Spatial Analyst clipped to the GISCO NUTS0 [7] administrative vector data layer used for coastlines identification.
7. Mosaic separate grid tiles into one single seamless grid.
8. Generate a hill shade grid with the HILLSHADE function in ArcGIS® Spatial Analyst. This grid can be used in ArcMap® as an overlay layer with a 50% transparency for enhancing visualization impact.
9. Project the resulting data to Lambert Azimuthal projection at 90 m resolution.

7.2 Procedure for Building a Seamless DEM

A fully automated procedure was built upon AML scripting language to build a seamless DEM; corrected for gaps; starting from a defined area of interest (AOI). The procedure makes use of the USGS SRTMGRID.AML ArcInfo® routine [c1] which is based on the ArcInfo® TOPOGRID tool for gap correction. Figure 11 gives an overview of the goals achieved by the script and the data input to it. If lakes and seashore data were available as ancillary data, it could also be an input to the procedure to improve the result. As mentioned above, the methodology to do this remains to be defined.

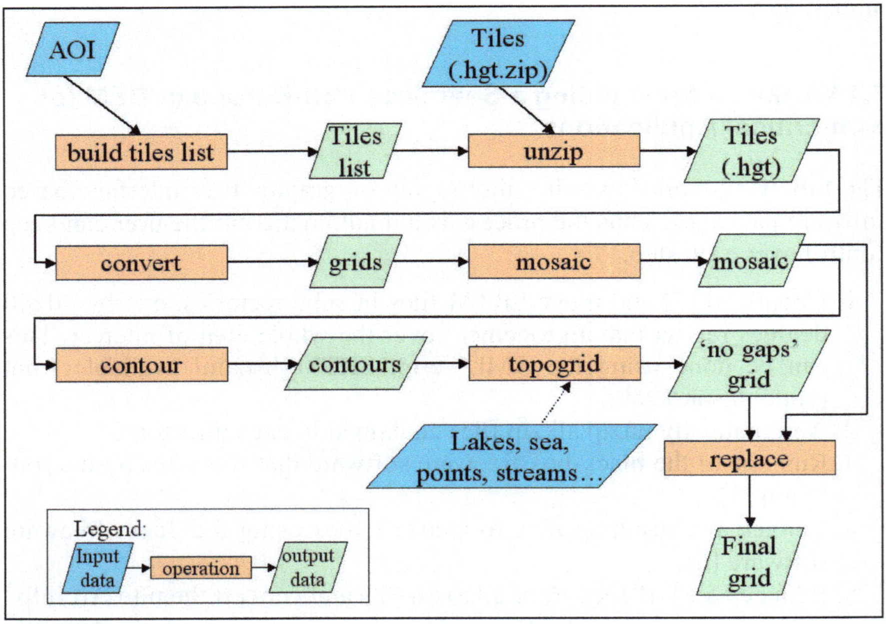

Fig. 11. Flow diagram in a procedure built to automate DEM production from SRTM raw data files

8. Conclusions

Availability of the quasi-global 3 arc seconds resolution (~90 m) SRTM is seen as a great opportunity for use in continental-wide soil-landscape modelling. Nevertheless, investigation of the quality and problems inherent to this data source has shown some of its limitations to fulfill the needs on small to medium scale soil-relief relationships modelling unless the quality of raw SRTM data is improved. Areas of missing or noisy wrong altitude values occur in some zones of steep slope and in water bodies and sea. Methods for correcting these which seem quite satisfactory can be found. But the main disadvantage of the SRTM data is the fact that the surface of the vegetation and human-made structures is detected instead of the ground surface.

The investigated examples confirm that the SRTM DEM is actually a digital model of the surface (digital surface model; DSM) rather than a digital model of the true ground elevation (DEM). This problem is crucial in soil-landscape and all environmental modelling where hydrologically correct DEMs constitute basic input information. In case the 'surface' is not corrected to 'elevation', extruded areas such as forests or buildings can play the role of barriers whereas excavations behave as spurious pits. Such uncorrected datasets cannot be used at detailed scales or in high accuracy demanding applications such as erosion modelling.

No final solution has yet been set up in time of writing this article to overcome this problem that represents a serious obstacle for the use of the SRTM in soil-landscape modelling at medium scales.

SRTM DEM data still can prove to be useful in small or small to medium-scale applications such as visualisation and improvement of small scale soil maps where minor altitude inaccuracies do not seriously affect the final results. After proper pre-processing and correction it can represent a suitable source of uniform and continental-wide elevation data for improving of soil databases such as the Geographical Soil Database of Europe at scale 1:1,000,000 or for elaboration of 1:250.000 scale European soil databases.

Acknowledgements

The research was partially funded by and performed at the European Soil Bureau, Institute for Environment and Sustainability, Joint Research Centre of the European Commission, Ispra, Italy, through a visiting scientist and a detached national expert programme within the 5th Framework Programme, SPIN project (Spatial Indicators for European Nature Conservation), under contract N° EVG1-CT-2000-019.

References

Annoni A, Luzet C, Gubler E, Ihde J (2001) Map projections for Europe. Institute for Environment and Sustainability, Joint Research Centre, Ispra, EUR 20120 EN. 131 p.

Bock M, Rossner G, Wissen M, Remm K, Langanke T, Lang S, Klug H, Blaschke T, Vrščaj B (2005) Spatial indicators for nature conservation from European to local scale. Ecological Indicators 5:267-374

Bui EN, Moran CJ (2001) Desegregation of polygons of surficial geology and soil maps using spatial modelling and legacy data. Geoderma 103:89-94

Daroussin J, King D, Le Bas C, Vrščaj B, Dobos E, Montanarella L (2004) Methods to improve the Soil Geographical Database of Eurasia at scale 1:1,000,000. Abstracts of the Global Workshop on Digital Soil Mapping, Montpellier, France, 2004

Dobos E, Micheli E, Baumgardner MF, Biehl L, Helt T (2000) Use of combined digital elevation model and satellite radiometric data for regional soil mapping. Geoderma XX:367-391

ECC (European Commission Communication) (2002) Towards Thematic Soil Strategy.

EPR (European Parliament Resolution on the Commission communication) (2003) Towards a Thematic Strategy for Soil Protection. COM (2002) 179 - C5-0328/2002 - 2002/2172 (COS)

Hutchinson MF (1989) A new procedure for gridding elevation and stream line data with automatic removal of spurious pits. Journal of Hydrology 106:211-232

Kirkby MJ, Jones RJA, Irvine B, Gobin A, Govers G, Cerdan O, Van Rompaey AJJ, Le Bissonnais Y, Daroussin J, King D, Montanarella L, Grimm M, Vieillefont V, Puigdefabregas J, Boer M, Kosmas C, Yassoglou N, Tsara M, Mantel S, Van Lynden GJ, Huting J (2003) Pan-European soil erosion risk assessment: the PESERA map, Version 1 October 2003. Explanation of Special Publication Ispra, 2004, No.73 (S.P.I.04.73). European Soil Bureau Research Report No.16, EUR 21176

Langanke T, Rossner G, Vrščaj B, Lang S, Mitchley J (2005) Selection and application of spatial indicators for nature conservation at different institutional levels. Journal for Nature Conservation 13:101-114

MacMillan RA (2001) A vision for automated landform classification: a literature review and synthesis of concepts and techniques for automated classification of a multi-level hierarchy of landform types and landform elements from digital elevation data. LandMapper Environmental Solutions, Edmonton, Alberta

MacMillan RA, Pettapiece WW, Nolan SC, Goddard TW (2000) A generic procedure for automatically segmenting landforms using DEMS. Fuzzy set and systems 113:81-109

MacMillan RA, Jones RK, McNabb DH (2004) Defining a hierarchy of spatial entities for environmental analysis and modelling using digital elevation models (DEMs). Computers, Environment and Urban Systems 28:175-200

Zhu AX, Mackay DS (2001) Effects of spatial detail of soil information on watershed modelling. Journal of Hydrology XX:54-77.

Referenced Web Sites

[1] http://www2.jpl.nasa.gov/srtm/
[2] http://edc.usgs.gov/srtm/data/obtainingdata.html
[3] http://edcftp.cr.usgs.gov/pub/data/srtm/
[4] http://eusoils.jrc.it/gisco_dbm/ad/nu/dbm/adnu.htm:GISCON NUTS vector data & descriptions
[5] http://gisweb.ciat.cgiar.org/sig/90m_data_tropics.htm
[6] http://www.3dnature.com/srtmfill.html.
[7] http://www.visualizationsoftware.com/3dem.html
[8] http://eusoils.jrc.it/gisco_dbm/ad/nu/dbm/adnu.htm
[9] http://www.mapmart.com

Appendix

1. Commands

[c1] Arc: &run srtmgrid <input HGT file> <output grid>
[c2]Grid: <output grid> = mosaic(<input grid>, …, <input grid>)
[c3] Grid: <ouput grid> = con(isnull(<input grid>), ~
focalmean(<input grid>, rectangle, 4, 4), <input grid>)
[c4] Grid: <output contours> = contour(<input grid>, interval, ~
<contour line interval>)
[c5] Arc: topogrid <intermediate output grid> <output cell size>
TopoGrid: datatype contour
TopoGrid: contour <input contour cover> <input elevation item>
TopoGrid: tolerances <sink tolerance> <horizontal std error> ~
<vertical std error>
TopoGrid: end
For more information on command [5], please refer to the software documentation, especially concerning the tolerances.

[c6] Grid: <output grid> = con(isnul(<original SRTM grid>), ~
<intermediate grid>, <original SRTM grid>)
[7] Grid: PROJECT GRID <input> <output> ~

{projection file} {NEAREST | BILINEAR | CUBIC} ~
{out cellsize} {x register} {y register}
[c8] Grid: D100SRTM = DEM100 - SRTM90
[c9] Grid: D25SRTM = DEM100 – DEM25

2. Software applications

The URL addresses accessed in October 2004:
[s1] http://www.visualizationsoftware.com/3dem/downloads.html: 3dem
[s2] http://www.geomantics.com/hgt.html: LandFormer Pro
[s3] http://www.3dnature.com/srtmfill.html: SRTMFill
[s4] http://www.dgadv.com/dgvk/: SRTM VoidKiller

3. Selected data sources

Availability and form of on-line data are often subject to changes. The list below was generated in October 2004. It can serve the reader as possible sources for data and software.

http://edcdaac.usgs.gov/gtopo30/gtopo30.asp: GTOPO30 is a global digital elevation model (DEM) with a horizontal grid spacing of 30 arc seconds (~1 km)

http://www.ngdc.noaa.gov/mgg/global/seltopo.html: ETOPO5 and TerrainBase 5 minute gridded DEM

http://www.ngdc.noaa.gov/mgg/fliers/01mgg04.html: ETOPO2 is a new available global elevation database replacing ETOPO5, gridded at 2-minute resolution

http://www2.jpl.nasa.gov/srtm/: Home page of the Shuttle Radar Topography Mission (SRTM)

http://www2.jpl.nasa.gov/srtm/dataprelimdescriptions.html: Description of SRTM data

http://www.arch.cam.ac.uk/comp/ac056/: Using the Shuttle Radar Topography Mission (SRTM) Digital Elevation Models in ArcGIS. University of Cambridge

http://edc.usgs.gov/products/elevation.html: Download page for SRTM data

http://edcftp.cr.usgs.gov/pub/data/srtm/: FTP download for SRTM data

http://seamless.usgs.gov/: USGS Seamless data server

http://gisdata.usgs.net/website/Seamless/: USGS Seamless data viewer

Development of a pan-European River and Catchment Database

Jürgen Vogt, Pierre Soille, Roberto Colombo, Maria Luisa Paracchini and Alfred de Jager

1. Introduction

The availability of a European-wide database of consistent and hierarchically structured river networks and associated drainage basins is a major asset for the implementation and follow-up of environmental policies in the European Union and an important basis for international modelling and monitoring activities. In fact, drainage basins are important reference units for many biophysical processes and rivers represent major transport networks not only for water, but also for sediments, nutrients and pollutants.

At the European scale, structured flow networks and related catchment boundaries are required for environmental monitoring and modelling both by international organisations and research networks. An example is the European Environment Agency (EEA), which manages the Eurowaternet station network, a representative sample of several thousand national monitoring stations throughout Europe. The correct location of these monitoring stations on the flow network, the delineation of their catchments, and the analysis of their topological relationships are a major asset for the analysis of pollution threads and environmental pressure indicators. Pressure indicators can, for example, be calculated on the basis of catchment characteristics such as area distribution of land cover types, altitude parameters, terrain form, prevailing soil and rock types, and climate characteristics. These pressure indicators can then help to judge the situation also in ungauged catchments. Other examples are the use of a coherent flow network for analysis and reporting tasks under the Water Framework Directive (DG Environment), the modelling of nutrient and pollutant transport in support of the implementation of the Nitrates Directive (DG Environment and DG Agriculture), or the use in flood and drought forecasting and monitoring activities at European level.

The required digital river networks and catchment limits are, however, not readily available. Existing data cover either limited areas, or are available on very small scales only. In addition, these datasets do not represent coherent flow networks but are rather digital cartographic products.

In response to this need, the Catchment Characterisation and Modelling (CCM) activity of JRC's Institute for Environment and Sustainability has developed a methodology and algorithms for producing a pan-European database of river networks and related catchment boundaries corresponding to a mapping scale of 1:250.000 to 1:500.000. The methodology was implemented with a grid digital elevation model (DEM) of 250m grid-cell size and is now repeated using a DEM of 100m grid-cell size, based on the Space Shuttle Radar Topography (SRTM) mission [1]. Specific algorithms have been developed for handling flat areas and for the so called carving, which overcomes the drawbacks of the well-known flooding for pit filling (Soille et al. 2003). Finally, a landscape stratification for drainage density was introduced, enabling the use of a variable drainage area threshold for channel initiation in order to reproduce the naturally varying drainage density across the continent (Vogt et al. 2003a; Vogt et al. 2003b).

This paper details the processing steps performed for the development of the CCM River and Catchment database version 1.0 and highlights the improvements to be expected from version 2.0, which will become available in late 2006.

The development of this database is considered a major step ahead since previously available river and catchment data do neither provide a consistent flow network, nor sufficient detail for the necessary assessments. It became possible following the increasing availability of extended DEMs and the development of massive processing capabilities for spatial data using specific software and GIS.

In section two we outline the study area and then briefly describe the data underlying the analysis in section three. In section four we detail the methodologies for deriving the landscape stratification, extracting the river network and delineating the drainage basins. In section five some validation results are presented and in section six we present the implemented Pfafstetter hydrological feature coding system. Finally, we discuss and summarise our findings in section seven.

2. Study Area

The area covered by version 1.0 of the CCM River and Catchment database, released in 2003, extends from the Mediterranean Sea to northern Scandinavia and from the Atlantic Sea to roughly 38 degrees Eastern longitude. It includes the Azores and Canary Islands to the West and follows the limits of major drainage basins in the East (Figure 1).

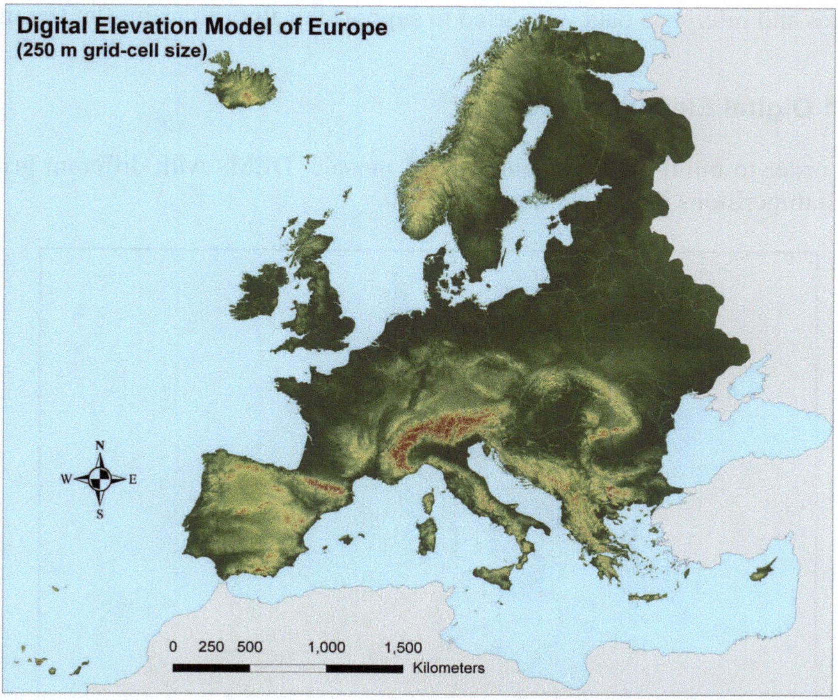

Fig. 1. Area covered by the CCM River and Catchment Database, version 1.0. The elevation data shown are based on the DEM mosaic with a 250m grid-cell

Version 2.0 of the database will be extended eastwards up to the Urals and will include the territory of Turkey and the Caucasian states in the south-East.

3. Input Data

An important constraint to large-area applications is the availability of basic input data. The methodology to be developed, therefore, had to be based on data readily available over all or at least most of the continent. Besides the DEM, data on climate, land cover, soils and geology were used for the development of the landscape stratification. A specific inland lake layer was further compiled in order to constrain the river extraction routines and reference data were used to support the algorithms on flat terrain.

3.1 Digital Elevation Data

In order to build the continental DEM mosaic, DEMs with different grid-cell dimensions had to be merged.

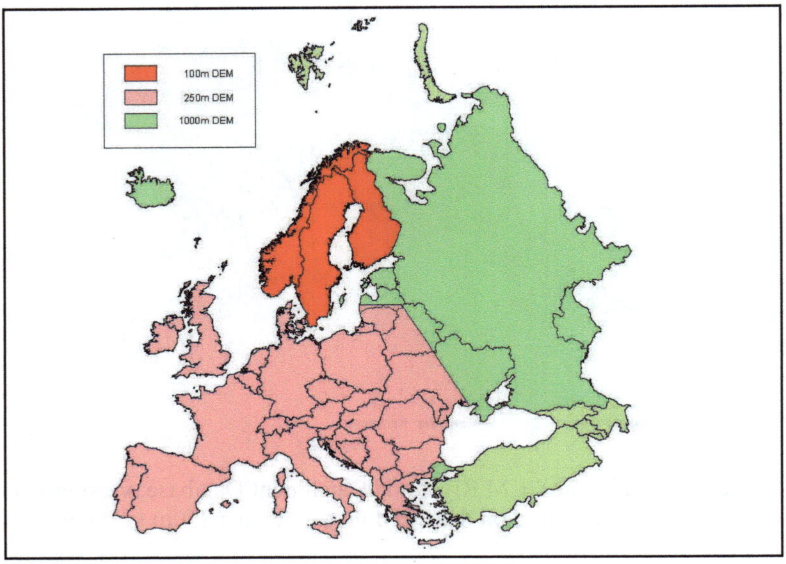

Fig. 2. Grid cell resolution of the original DEMs. Most of Turkey and the Caucasian States are not covered

Original grid-cell dimensions correspond to 100 meters and 250 meters for most of the area covered. For areas where DEMs at this resolution could not be acquired (e.g., Iceland, Russia, Atlantic Islands), data from the HYDRO1K global digital elevation dataset with 1000m grid-cell resolution were used. Figure 2 shows the area coverage of different DEM resolutions.

The quality of the input DEMs is varying, depending on the grid-cell resolution and source. While the positional accuracy of the grid is generally within one grid-cell, the altitude accuracy is varying depending on DEM and relief type. It ranges from a few meters RMSE for the Scandinavian DEMs to about 18 meters for the HYDRO1K DEM.

In order to merge the different datasets, they were initially sub-sampled to a grid with a resolution of 0.001 decimal degrees, using a nearest neighbour algorithm. This grid-cell size represents a conservative value and allowed maintaining the integrity of elevation values. It corresponds to about 20 meters at the highest latitudes and to about 80 meters for the Mediterranean countries. All DEMs were then merged, giving priority to the DEMs with best quality. In cases were the quality of two datasets was comparable, the country border defined the extent of the dataset within the final mosaic.

Finally, the DEM mosaic at a grid resolution of 0.001 decimal degrees was projected into a Lambert Equal Area Azimuthal projection with a 250m grid-cell size, using a bilinear resampling technique. This projection system was chosen for its equal area property (Annoni et al. 2003).

For the case of islands close to the coast or marine straights narrower than 250 meters, the projection to a coarser resolution resulted in a topological degradation. This effect was mainly visible in Scandinavia, where the original DEM resolution was 100 meters. Therefore, in this area the continent was separated from the islands. Islands were then processed separately through the whole processing chain.

3.2 Inland Water Body Layer

The preparation of an inland water body layer was necessary to ensure that the extracted river network is coherent with lakes and transitional waters. Transitional waters are defined as intertidal flats, coastal lagoons and estuaries. In addition, large internal drainage basins (depressions without outlet to the sea) were considered when extracting the river network.

From CORINE Land Cover (EEA 1997, 250m grid) and SWISS Land Cover (Nippel and Klingl 1998, 250m grid) the classes corresponding to inland marshes (code 35), peat bogs (code 36), salt marshes (code 37), salines (code 38), intertidal flats (code 39), water courses (code 40), water bodies (code 41), coastal lagoons (code 42), estuaries (code 43), and glaciers and perpetual snow (code 34) were selected. These datasets were further extended by merging them with the GISCO Lakes at 1:50,000 scale

(HY Lake) and with the PELCOM 1000 meter Land Cover grid [2]. CORINE Land Cover was the primary data set. Where CORINE Land Cover was not available, the other sources were used in order of decreasing quality (Figure 3).

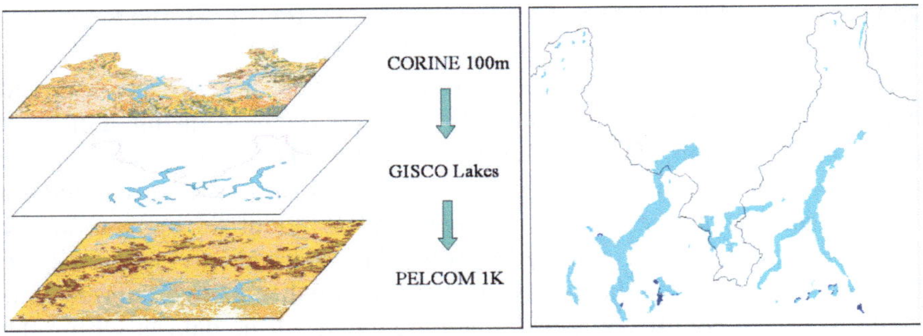

Fig. 3. Preparation of the inland lake layer. Left: Input data. Right: Resulting layer for the area of Lago Maggiore, Lago di Lugano and Lago di Como, northern Italy. Inland marshes are shown in dark blue, inland lakes in light blue. North-south extension of the lakes shown is about 50 km

Intertidal flats represent areas periodically submerged by seawater and they were considered as sea. In such cases the final river network stops at the landward edge of the intertidal flat. Lakes, inland marshes and coastal lagoons with no connection to the sea and larger than 8 grid-cells (0.5 km^2), were used to force the drainage channels to flow through the water body along its centre line. These water bodies are often situated in flat areas and without such constraint the extracted river network could cross the water surface many times causing incoherence between river and water body. When coastal lagoons have an outlet to the sea, they were considered as sea and the drainage channel stops at the edge of the lagoon.

3.3 Environmental Data Layers

The following data layers were used to derive a landscape stratification reflecting the terrain aptitude to develop a certain drainage density:

- Climate data from the MARS meteorological database, which is available on a 50 km grid for the whole of Europe [3],
- The grid digital elevation model with a 250m grid-cell resolution described above,
- CORINE and SWISS Land Cover data with a grid-cell resolution of 250m, complemented by IGBP land cover data (1km resolution) and

PELCOM data (1 km resolution) in areas where CORINE data were not available,

- The Soil Geographical Database of Europe (ESBSC 1998) at a scale of 1:1,000,000.

4. Methodology

Figure 4 provides an overview of the methodology implemented. The following main processing blocks can be identified: (1) modelling surface runoff as a basis for extracting the drainage network, including the pre-processing of the DEM, (2) preparation of a landscape stratification reflecting drainage density, (3) preparation of additional information on water bodies, used to constrain the final river network, (4) data processing (i.e. extracting the river network), (5) ordering and coding of rivers, (6) extraction and characterisation of catchments, and (7) data validation (not shown in the figure).

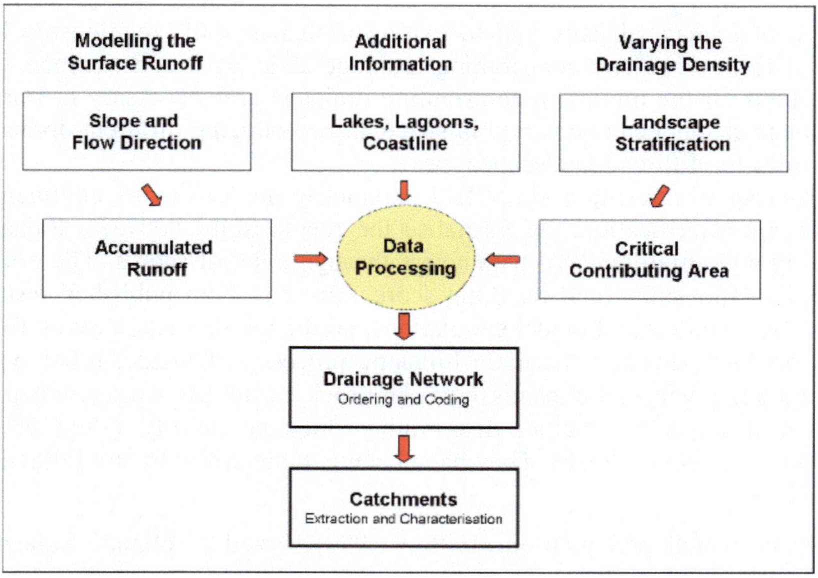

Fig. 4. Major Data Processing Steps (see text for details)

4.1 Landscape Stratification

Algorithms for the extraction of drainage channels from DEMs are now implemented in many standard GIS. Major limitations stem, however, from inadequate capabilities to handle large datasets, from the flooding routines used to handle artificial pits and from the limited capabilities to handle flat areas. In addition, these algorithms require a threshold for the critical contributing drainage area (CDA) when determining channel heads. Due to the lack of reliable information, this threshold is usually set to a constant value. Using a constant threshold results, however, in a uniform drainage density over the entire study area.

A major challenge in deriving a river network at continental scale, therefore, is to establish a methodology that reflects the natural variability in drainage density. In order to comply with this need, a landscape stratification was prepared that reflects the landscape aptitude to develop different drainage densities. The landscape strata, therefore, needed to be based on a combination of environmental factors with a strong influence on channel development.

The resulting landscape types are assumed to be homogeneous with respect to drainage density and to exhibit a characteristic relationship between local slope and contributing drainage area. As a consequence, the threshold for the minimum contributing drainage area necessary to start a drainage channel can be varied in space, thus producing different drainage densities for different landscape types.

In order to develop a simplified parametric model for the continental landscape stratification, we formalise the relationships between drainage density and environmental parameters through a set of scores. The established relationships (and resulting scores) are based on published results from field studies and model simulations, predicting drainage density from a-priori knowledge of the main hillslope processes. Climate, relief type, vegetation cover, soil transmissivity, and rock erodibility were retained as the most important factors determining drainage density (Vogt et al. 2003a; Vogt et al. 2003b). They have been parameterised by the following variables:

- Mean annual precipitation (1975 – 1999) served as climate indicator (Moglen et al. 1998),
- The influence of the terrain morphology was parameterised by relative relief, defined as the maximum altitude difference in a moving window of 3 by 3 grid cells (Oguchi 1997; Roth et al. 1996),
- The percentage of surface covered by vegetation has a strong effect on critical shear stress and thus on channel initiation (Tucker et al. 1997;

Foster et al. 1995). Consequently, land cover data were reclassified into 14 classes and monthly cover percentages were assigned to each class according to Kirkby (1999),

- As a proxy indicator of soil transmissivity, soil texture has been chosen as the main soil factor affecting drainage density (e.g., Dietrich et al. 1992; Tucker and Bras 1998). Dominant topsoil texture for each soil mapping unit (SMU) was derived from the Soil Geographical Database for Europe, SGDBE (ESBSC 1998),
- For bedrock erodibility the dominant lithology for each SMU was extracted from the SGDBE. Based on these data a rock erodibility factor was assigned according to the scale proposed by Gisotti (1983).

Different methodologies to combine these variables for deriving a drainage density related index have been tested (Vogt et al. 2003a). Finally, a simple scoring technique has been retained as the most appropriate for a continental application. Annual precipitation, relative relief, bedrock erodibility and soil transmissivity have been classified into 5 classes each, each class receiving a different score. Vegetation cover was retained as a secondary variable, which modifies the scores for annual precipitation. The choice to apply a simple scoring system is justified on the basis that we are not deriving a physical value of drainage density but rather defining areas with specific environmental conditions. The sum of all scores for each grid-cell determines an index, which we call Landscape Drainage Density Index (LDDI). This index was classified into ten classes as shown in Figure 5.

The number of landscape classes to some degree depends on subjective judgement, which indicates that the methodology could be improved by implementing a continuously varying threshold, depending on the LDDI and not related to specific landscape classes. This is, however, difficult to achieve, since the derivation of the relationship between local slope and contributing drainage area requires a sample of pairs which need to be related to a geographical entity.

The derived European landscape stratification appears to be strongly controlled by relief and geology. A visual comparison between the derived map and external data shows that the regions with low LDDI values are mainly located in the Pannonian basin, the North European plains and the Fennoscandian shield, while middle values correspond to Hercinian ranges and the highest values are to be found in the Pyrenean and Alpine regions. Improvements could be expected from giving more weight to the climate factor, especially in the Mediterranean, where a distinction between perennial and intermittent flow paths is important.

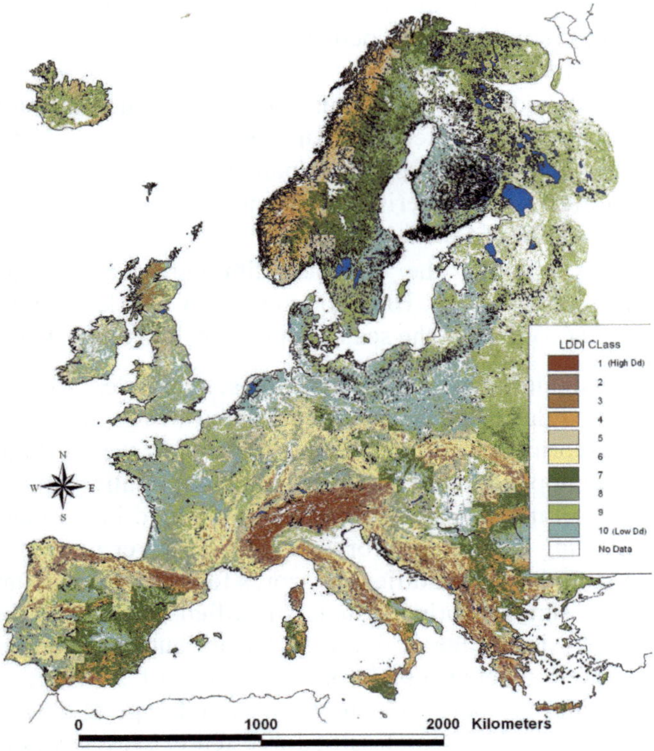

Fig. 5. Landscape stratification into ten landscape drainage density index (LDDI) classes

Areas for which channel initiation could be considered as extremely improbable were not included in the stratification procedure. In particular, all areas with a local slope of less than 2% were excluded. Urban areas, marshes, lakes, lagoons, and glaciers were considered as impervious surfaces, where no new channels start.

4.2 Threshold Definition

For each LDDI class a critical contributing drainage area was determined by interpreting the diagram of local slope versus contributing drainage area. Many efforts have been made to infer an adequate threshold by using the log-log relationship between local slope and contributing drainage area, computed from DEMs. Different inflection points can be observed in the log-log plot derived from high-resolution DEMs and many studies suggest that these enable the distinction between various geomorphic and hydrologic regimes (e.g., Tarboton et al. 1991; Montgomery and Foufoula-Georgiou 1993; Willgoose 1994; Montgomery and Dietrich 1994; Ijjász-Vásquez and Bras 1995; Tucker and Bras 1998; Ibbit et al. 1999; McNamara et al. 1999; Montgomery 2001; Whipple and Tucker 2002; Hancock 2005). Thresholds derived from the log-log analysis of coarser resolution DEMs (250 to 1000 meter grid cell size) and their usefulness for deriving river networks at medium to small scales have, for example, been explored by Wolock and Price (1994), Walker and Willgoose (1999) and Ibbit et al. (1999).

For the threshold definition local slope and contributing drainage area were computed using the D∞ algorithm (Tarboton 1997), which allows for flow dispersion. The log-log plot of local slope versus contributing drainage area was then generated from a random sample of grid-cells for each of the landscape classes. The number of samples varied between 40,000 and 1,000,000 samples, depending on the area extent of the individual classes. Subsequently, samples have been aggregated by binning 600 to 800 samples and calculating the average and standard deviation for each bin.

In Figure 6 the local slope – contributing drainage area (CDA) relationship is shown for four out of the ten landscape classes at continental scale. The remaining classes fall within the shown range. In general, three sections with different scaling responses can be distinguished in each of the graphs shown. A first section with a trend to an increasing slope can be observed. This part of the graphs ends with a gradient change at a contributing drainage area of about 0.15 km^2 for all classes. The second section of the graph is characterised by constantly decreasing slopes. With increasing class number the shape of the graph comes closer to the theoretical form with a steeper curve at the beginning of section two, which then flattens off before the slope of the graph slightly increases again to approach a theoretical straight line with a negative slope around 0.5. This latter (straight) part of the graph characterises section 3, which is the so-called fluvial scaling line that represents areas of pre-dominant fluvial transport. The point where the slope of the graph starts to increase again is defined as the breakpoint between sections two and three. This point is varying in its position along the x-axis with an increasing contributing drainage area from class one to class ten. It allows for the definition of different critical contributing drainage areas for our landscape classes. The shift of this point is coherent with our hypothesis that the landscape classes represent areas of different drainage density at medium to small scale.

These last inflection points have been selected by visual inspection of the different graphs and they are considered as the critical contributing drainage area for each landscape class at the given spatial resolution. The resulting thresholds per landscape class are shown by the arrows in Figure 6 and given for all classes in Table 1.

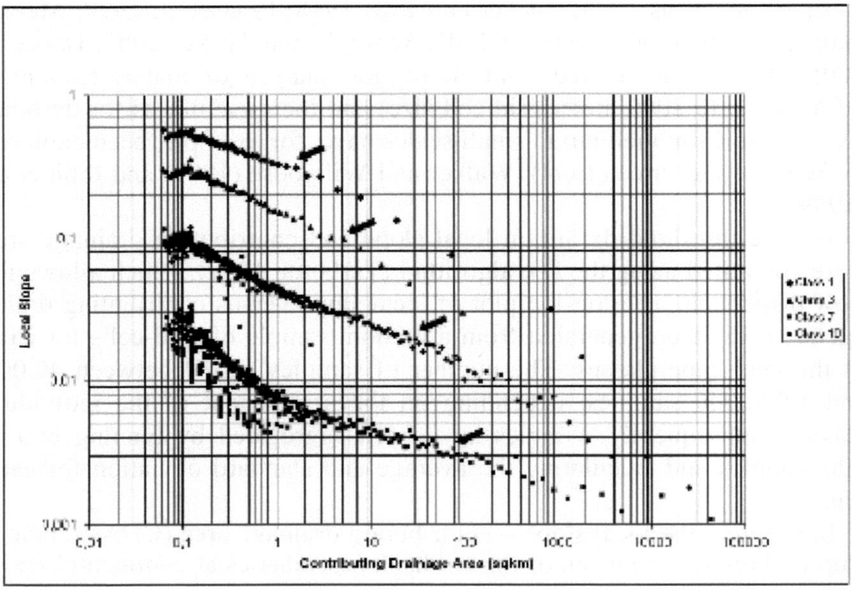

Fig. 6. Local slope versus contributing drainage area relationship for LDDI classes 1, 3, 7 and 10, plotted on log-log scale. Each point represents the average of several hundred original samples. Arrows indicate final CDA thresholds (see Table 1)

Table 1. Contributing Drainage Area (CDA) Threshold per LDDI Class

LDDI Class	CDA Threshold (km^2)
1	1
2	3
3	6
4	15
5	20
6	30
7	35
8	50
9	60
10	80

4.3 River Network Extraction

The primary DEM-hydrological quantities (local slope, flow direction, and contributing drainage area) were derived by a suite of algorithms based on the concepts of mathematical morphology (Soille 2003). These algorithms allow for a fast and efficient removal of spurious pits by carving the DEM. Carving is a new methodology by which each pit is suppressed by creating a descending path from it to the nearest grid-cell having a lower elevation value. This is achieved by lowering the elevation values along the detected path (Figure 7). An improvement to methods for routing flow over flat regions is also introduced. The algorithms are described in detail in Soille et al. (2003).

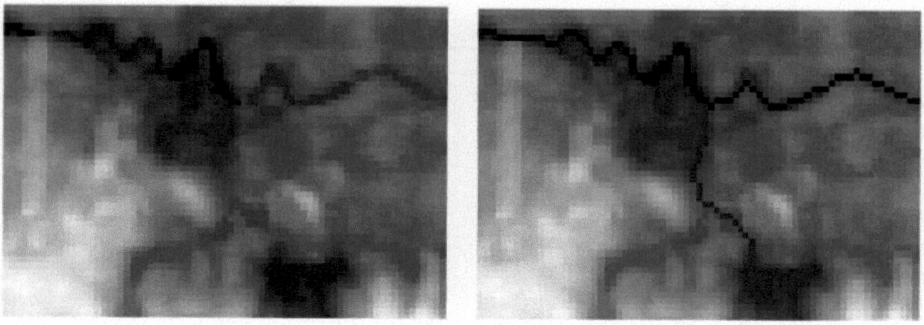

Fig. 7. DEM processing by carving. Left: Original DEM. Right: Result after carving. Grid-cell size is 250 meters. Window size is 17.5 x 12.5 km

The carved DEM was then input for an adaptive drainage enforcement process that aims at creating a more precise flow direction path, especially in flat terrain (Soille et al. 2003). This adaptive stream burning algorithm uses as input selected segments of a reference river network and, by an iterative process, defines the places where the reference network deviates substantially from the automatically detected river networks. While stream burning can itself create some artefacts (e.g. Saunders 1999) this procedure ensures that rivers are burned only where it is really necessary, preserving the topography of the carved DEM. It reduces co-registration problems that produce double streams and alleviates discrepancies in scale or generalisation level between the DEM and external streamlines that may lead to the removal or creation of features such as meanders. In addition, priority queue data structures allow for an efficient implementation of the algorithm, which in turn enables the processing of files such as the complete pan-European DEM. Details about the algorithm can be found in Soille (2002) and Soille et al. (2003).

In order to arrive at a realistic drainage density, the contributing drainage area (CDA) threshold for each landscape class was applied in the course of this process.

Giving as further input information on the position of lakes, coastal lagoons, estuaries, intertidal flats and glaciers, the algorithms produce a flow direction that is consistent with these water classes. For the classes connected to the sea (coastal lagoons, estuaries, intertidal flats), the flow direction is stopped at their border (i.e., before reaching the actual coastline), while in all the other cases the flow path is always connected with the coastline (i.e., continuing through the lakes). This allows for the derivation of a fully connected and hierarchically structured river network, which is coherent with lakes and the other mentioned water types. A view on a subset of the resulting drainage network is shown in Figure 8.

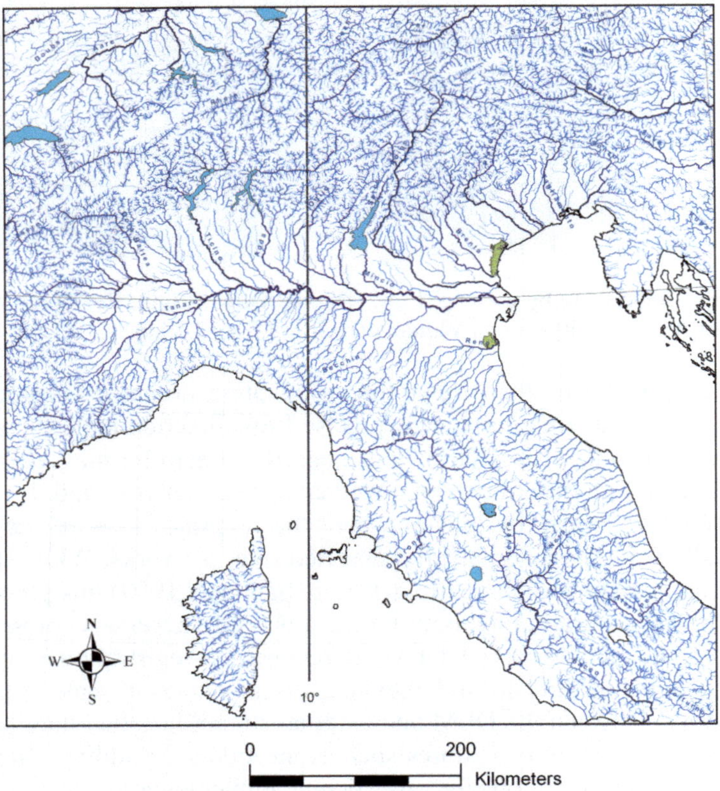

Fig. 8. Example of the CCM River Network for northern and central Italy. (Note the variation in drainage density according to the landscape classes)

4.4 Drainage Basin Delineation

Basins and sub-basins are delimited by automatically identifying all outlet points (confluences, lake inflows, lake outflows, sea outlets) in the drainage network and building the catchment for each outlet point from the flow direction grid. Rivers are further ordered according to the Strahler system (Horton 1945; Strahler 1952; Strahler 1957), which introduces a hierarchy into the river and catchment system. In version 1.0 of the CCM River and Catchment database, the Strahler hierarchy arrives at an order of eight, for example for the Danube and the Rhône while, for example, the Po, Rhine, Ebro, Douro and Garonne rivers belong to the 7th Strahler order.

A vector database was finally generated from the river network and catchment grids. Vectors were deliberately not smoothed, in order to make the user aware of the grid nature of the data. A further correction was introduced in the case of sub-basins intersecting lakes. These were recalculated so that the outlet of each basin lies along the lake perimeter. Tests in alpine areas showed that the overestimation of sub-catchment areas without this correction could be as large as 10%.

The resulting continental map of river networks together with the major (highest order) drainage basins is shown in Figure 9.

5. Data Validation

Validation has been performed qualitatively and quantitatively against independent European and national datasets. The position of river reaches was, for example, validated by overlaying them to reference datasets and evaluating the degree of correspondence through buffers of varying size. Catchment perimeters were evaluated in a similar way by overlaying them with digital data from Finland and Spain. The derived length of major rivers in Europe was compared to officially published figures. Catchments could further be validated through a large sample of data on catchment size in EEAs Eurowaternet database (Nixon et al. 1998, Boschet et al. 2000). The different methodologies and results are described in detail in Colombo et al. (2001), Vogt et al. (2003b,c), and Colombo et al. (2005). Summarising figures are given in Table 2.

Table 2 shows that length of the DEM derived large rivers deviates by about 10% from the officially reported values. In general, the length of the flow paths is underestimated. This is partly due to the fact that estuaries and deltas are not considered in the CCM flow path. The main reason for the underestimation is, however, due to underestimation of meanders, which results from the 250m grid-cell size of the DEM.

Fig. 9. CCM River network (2nd Strahler order and higher) with the major European drainage basins

Table 2. Validation results

Validation Type	Sample Size	Percentage Deviation	
		Mean	Standard Dev.
Length of large Rivers	48	10.0	8.0
Size of large River Basins	48	3.4	4.6
Size of Eurowaternet station catchments	1600	3.3	3.9
Catchment boundaries Finland[1]	21	10.0	6.1
Catchment boundaries Spain[2]	26	5.0	3.0

[1] *total perimeter 9906 km.* [2] *total perimeter 2346 km.*

The validation against published catchment size is a standard analysis for continental and global datasets (Verdine and Jenson 1996; Oki and Sud 1998; Graham et al. 1999; Renssen and Knoop 2000; Döll and Lehner 2002). The comparison between derived catchment area and reference area for large river basins showed a good correspondence with a mean discrepancy of 3.4%, without prevalence of under/over estimation.

A large sample of catchments of widely variable size from all parts of Europe could be drawn from the Eurowaternet database of the EEA. Table 2 reports the results for the area comparison for a sample 1600 catchments on continental Europe. With a mean error of 3.3% and a standard deviation of 3.9% the results indicate a high quality of the DEM derived catchments.

In order to estimate the potential errors due to area over and underestimation along the catchment perimeter, an overlay analysis has been carried out. A simple comparison of catchment size may not reveal these errors since underestimation and overestimation errors along the catchment perimeter may cancel out. This kind of evaluation could be performed for Finland, a country with low relief energy and a complex drainage system with many lakes, and Spain, a country characterised by more accentuated relief. Catchments were automatically compared in order to produce a table of underestimation and overestimation areas. The total error was computed by summing all the discrepant areas divided for the catchment reference area. A buffer strip of 500m along the catchment perimeter was excluded from this analysis. This buffer serves to eliminate the errors due to co-registration problems and due to different scales of reference and simulated datasets. The discrepancy along the catchment boundaries (measured as the deviation in area along the catchment perimeter) was found to be of a magnitude of 10.0% (± 6.1%) for Finland and of 5.0% (±3.0%) for Spain.

6. Additional Features

6.1 Pfafstetter Coding

In order to add specific topological analysis functionalities to the network, Pfafstetter hydrological feature codes have been assigned to all river reaches and catchments. The system has been developed in 1989 by Otto Pfafstetter, a Brazilian engineer. The strength of the Pfafstetter system is that it assigns catchment (or river reach) IDs based on topology and size of area drained and that it is strictly hierarchical. Its numbering system is systematic and self-replicating, making it possible to automatically assign IDs down to the level of the smallest sub-basin. For a given location it is possible to automatically identify all upstream catchments, all upstream river reaches, or all downstream reaches, which would be affected, for example, by a pollutant introduced to a system at the given location. These analyses can be done solely on tabular data that include the Pfafstetter code. The basic Pfafstetter principle is illustrated in Figure 10.

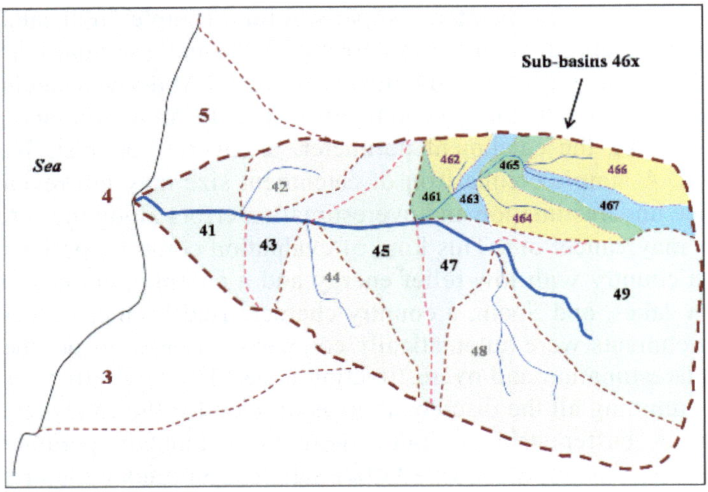

Fig. 10. Principle of the Pfafstetter Hydrological Feature Coding System. Subdivisions for a theoretical basin with code 4

For any given basin, a sub-division is achieved by identifying the four largest sub-basins (in terms of area) along the main river stem. These are assigned even numbers following the main stem in upstream direction (i.e. 2, 4, 6, 8). The remaining inter-basins are assigned odd numbers (i.e. 3, 5, 7, 9). Inter-basin 1 is the area drained by the main stem between mouth of basin two and the mouth of the main stem. Inter-basin nine always consists of the headwater areas of the main stem. This system is replicated for all smaller entities until all sub-basin and river reaches are assigned a code.

It should be noted that the river reaches along the main stem are identified by the inter-basin codes. Ultimately, the inter-basin codes define the full river network. They also identify the areas that drain in a diffused manner into that network. Detailed explanations of the Pfafstetter system can be found in Verdin (1997), Verdin and Verdin (1999), and Furnans and Olivera (2001). Britton (2002) provides an overview of different hydrological feature coding systems, including an extensive description of the Pfafstetter system.

The implementation of the Pfafstetter system provides at the same time a check for the consistency of the network and gives the total upstream flow path and total upstream area for each individual river reach. Both are interesting features of the system and are exemplified for the Danube drainage basin in Figure 11. Also the main stem for each basin in the hierarchy is identified on the basis of the longest flow path.

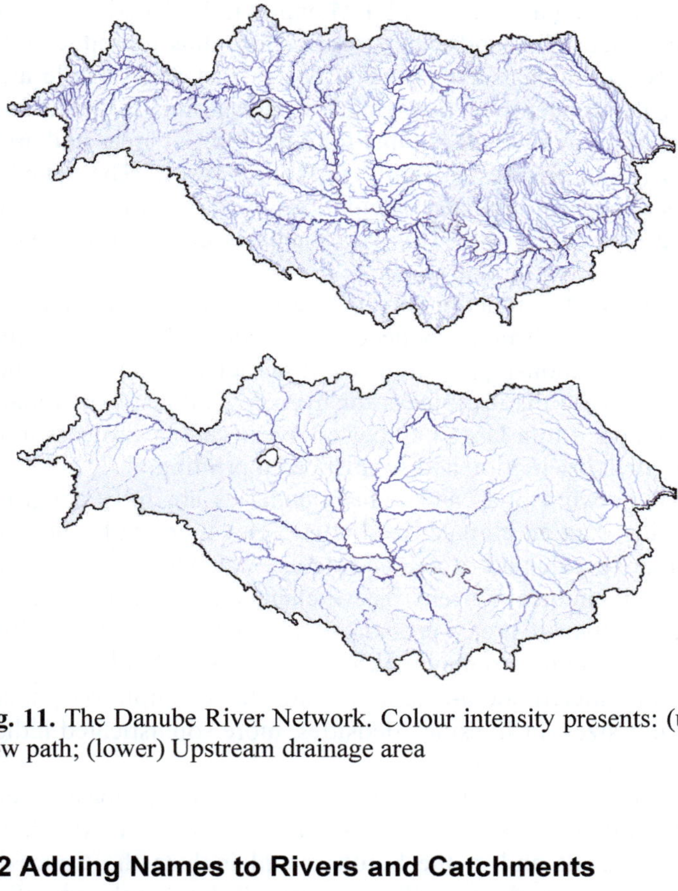

Fig. 11. The Danube River Network. Colour intensity presents: (upper) Upstream flow path; (lower) Upstream drainage area

6.2 Adding Names to Rivers and Catchments

In order to facilitate the location of individual rivers and catchments, the database has been filled with the names of some 600 rivers and more than 2000 drainage basins. Where possible, the English name has been retained together with local names. It should be noted that in Europe many rivers change their name along their course while crossing linguistic boundaries.

7. Conclusions and Outlook

In this paper we presented a methodology for the derivation of a high quality pan-European dataset of river networks and associated catchments, useful in support to hydrological analyses at medium to small scales. The approach is unique in the sense that it introduces a set of new and advanced algorithms for processing an extended DEM and that the location of channel heads is controlled by a landscape stratification, thus reproducing the natural variability in drainage density. The use of fast and reliable algorithms based on the concepts of mathematical morphology together with the development of innovative algorithms for handling flat areas and artificial depressions (pits), ensures a highly satisfying result. The speed of these algorithms allows for repeated calculations even for large areas such as the entire European continent. This is a major asset for implementing corrections after further validation steps.

In order to improve the functionality of the system, Pfafstetter hydrological feature codes have been introduced. These codes allow for fast queries on topological relationships within the flow network and catchment system and thus increase the usefulness for hydrological analysis. Finally, names have been added to a large number of rivers and catchments in order to allow for quick orientation and sophisticated printing.

Since the flow network and catchment boundaries are fully congruent with the underlying data on climate, land cover, elevation, soils and geology, it is possible to calculate a set of primary characteristics for each catchment in the hierarchical system. These characteristics can serve as a basis for the calculation of proxy pressure indicators, useful for environmental risk analysis and thus policy making. Together with administrative boundaries and distributed socio-economic data, for example on population densities, farm sizes, or livestock densities, more sophisticated indicators and evaluations are possible.

A second version of the CCM River and Catchment database is currently under preparation. This new version will cover an area extended to the East up to the Urals and to the South it will include the Caucasian states and Turkey. It is based on a 100 meter grid, following the specifications for European reference grids (Annoni 2005), which in the long-term will facilitate the combination with a large number of data layers covering both environmental and socio-economic themes. In developing this new version, we take advantage of newly available data such as a 100m grid-cell DEM based on SRTM data and the full TM mosaic for the European Union provided by the Image2000 and CLC2000 project [4]. The DEM based on SRTM is providing us with better resolution and positioning accuracy in all areas where variations in terrain elevation exceed the variations of elevation due to the presence of forests, cities, and other large man made structures. Since SRTM elevation data are based on microwave technology, they represent a digital surface model (as opposed to a digital terrain model), which includes surface variations due to forests and large build-up structures. In these latter cases as well as in areas with low relief energy, pan-European Landsat imagery made available through the Image2000 project will help us in deriving the precise location of large rivers as well as lakes, lagoons, and a high resolution coastline (Bagli and Soille 2003).

It is further expected that the new DEM will improve the determination of the CDA thresholds and together with CLC2000 data will allow for an improved landscape stratification and LDDI derivation.

In the long-term the database will be coupled with sophisticated hydrological and land use change models, in order to provide the basis for evaluating the impact of different policy scenarios. More information on the CCM activity and the CCM River and Catchment database can be found at [5].

References

Annoni A, Luzet C, Gubler E, IHDE J (eds) (2003) Map Projections for Europe. European Commission, Joint Research Centre, Ispra (Va), EUR 20120 EN. 131p.

Annoni A. (ed.) (2005) European Reference Grids. Proposal for a European Reference System. Proceedings of the European Reference Grid Workshop, 27-29 October 2003, Ispra (VA), Italy. European Commission, Joint Research Centre, Ispra (Va), EUR 21494 EN, 189p.

Bagli S, Soille P (2003) Morphological automatic extraction of coastline from pan-European Landsat TM images. In: Proceedings of the Fifth Int. Symp. on GIS and Computer Cartography for Coastal Zone Management, GISIG and ICCOPS, Genova, 58-59

Boschet AF, de Paepe V, Lack TJ (2000) Inland Waters. Annual topic update. European Environment Agency, Topic Report 1/2000, Luxembourg, Office for Official Publications of the European Communities, 30 p.

Britton P (2002) Review of Existing River Coding Systems for River Basin Management and Reporting. WFD GIS Working Group, October 2002, 29p. http://193.178.1.168/ River_Coding_Review.htm, 29 p.

Colombo R, Vogt JV, Bertolo F (2001) Deriving drainage networks and catchment boundaries at the European scale. A new approach combining Digital Elevation Data and environmental characteristics. EUR 19805 EN, Luxembourg, Office for Official Publications of the European Communities, 58 p.

Colombo R, Vogt JV, Soille P, Paracchini ML, de Jager A (2005) On the derivation of river networks and catchments at European scale from medium resolution digital elevation data. CATENA (submitted)

Dietrich WE, Wilson CJ, Montgomery DR, McKean J, Bauer R (1992) Erosion thresholds and land surface morphology. Geology 20:675-679

Döll P, Lehner B (2002) Validation of a new global 30-min drainage direction map. J Hydrol 258:214-231

EEA, European Environment Agency (1997) CORINE Land Cover Technical Guide. European Topic Centre Land Cover (ETC/LC), http://image2000.jrc.it/reports/technical_guide.pdf.

ESBSC European Soil Bureau Scientific Committee (1998) Georeferenced soil data-base for Europe, Manual of Procedures. EUR 16393 EN, European Commission – JRC, Ispra, 184 p.

Foster GR, Flanagan DC, Nearing MA, Lane LJ, Risee LM, Finkner SC (1995) Hillslope erosion component.- In: Flanagan DC, Nearing MA (eds) WEPP: USDA-Water Erosion Prediction Project: 11.1-11.12; NSERL Report No. 10, USDA ARS, Lafayette, IN

Furnans F and Olivera F (2001) Watershed topology: The Pfafstetter system. Proc. 21st ESRI User Conference, ESRI, San Diego, CA.

Gisotti G (1983) Geologia e Pedologia nell'assetto del territorio. Bologna (Ed. Edagricole)

Graham ST, Famiglietti JS, Maidment DR (1999) Five-minute, 1/2 and 1 degree data sets of continental watersheds and river networks for use in regional and global hydrologic and climate system modelling studies. Water Resour Res 35:583-587

Hancock GR (2005) The use of digital elevation models in the identification and characterization of catchments over different grid scales. Hydrol. Proc. DOI: 10.1002/hyp.5632

Horton RE (1945) Erosional development of streams and their drainage basins; hydrophysical approach to quantitative morphology. Bull Geol Soc Am 56:275-370

Ibbit RP, Willgoose GR, Duncan M (1999) Channel network simulation models compared with data from Ashley River (New Zealand). Water Resour Res 35:3875-3890

Ijjasz-Vasquez EJ, Bras RL (1995) Scaling regimes of local slope versus contributing area in digital elevation models. Geomorphology 12:299-311

Kirkby MJ (1999) Definition and practical demonstration of a pre-operational system for desertification monitoring in the Mediterranean Basin based on remote sensing methods. 1st Annual Report, MODEM Project, JRC-Space Applications Institute, Ispra (Va), Italy.

McNamara JP, Kane DL, Hinzman LD (1999) An analysis of an arctic channel network using a digital elevation model. Geomorphology 29:339-353

Moglen GE, Eltahir EAB, Bras RL (1998) On the sensitivity of drainage density to climate change. Water Resour Res 34:855-862

Montgomery DR, Foufoula--Georgiou E (1993) Channel network source representation using Digital Elevation Models. Water Resour Res 29:3925-3934

Montgomery DR, Dietrich WE (1994) A physically-based model for the topographic control on shallow landsliding. Water Resour Res 30:1153-1171

Montgomery DR (2001) Slope distributions, threshold hillslopes, and steady-state topography. American J Science 301:432–454

Nippel T and Klingl T (1998) Swiss Land Use in the European context. Integration of Swiss Land Use statistics with CORINE Land Cover. Swiss Federal Statistical Office & Swiss Agency for the Environment, Neuchâtel, ISBN 3-303-02045-0, 41 p.

Nixon S, Grath J, Bogestrand J (1998) Eurowaternet. The European Environment Agency's monitoring and information network for inland water resources. European Environment Agency, Technical Report No. 7, Copenhagen (EEA) 47 p.

Oguchi T (1997) Drainage density and relative relief in humid steep mountains with frequent slope failure. Earth Surf Processes and Landforms 22: 107-120

Oki T, Sud YC (1998) Design of Total Runoff Integrating Pathways (TRIP) - a global river channel network, Earth Interact 2 (2-001):53 pp.

Renssen H, Knoop J (2000) A global river routing network for use in hydrological modeling. J Hydrol 230:230-243

Roth G, La Barbera P, Greco M (1996) On the description of the basin effective drainage structure. J Hydrol 187:119-135

Saundres W (1999) Preparation of DEMs for use in environmental modelling. http://www.esri.com/library/userconf/proc99/proceed/papers/pap802/p802.html

Soille P (2002) Advances in the analysis of topographic features on discrete images. Lecture Notes in Computer Science 2301:175-186

Soille P, Vogt JV, Colombo R (2003) Carving and adaptive drainage enforcement of grid digital elevation models. Water Resour Res 39:1366

Soille P (2003) Morphological image analysis: Principles and applications. 2nd Edition, Springer, New York [Corrected 2nd printing, 2004].

Strahler AN (1952) Hypsometric (area-altitude) analysis of erosional topography. Bull Geol Soc Am 63:1117-1142

Strahler AN (1957) Quantitative analysis of watershed geomorphology. Trans Am Geophys Union 38: 913-920

Tarboton DG (1997) A new method for the determination of flow directions and upslope areas in grid digital elevation models. Water Resour Res 33:309-319

Tarboton DG, Bras RL, Rodriquez-Iturbe I (1991) On the extraction of channel networks from digital elevation data. Hydrol Proc 5: 81-100

Tucker GE, Gasparini NM, Lancaster ST, Bras RL (1997) An integrated hillslope and channel evolution model as an investigation and prediction tool. Technical Report prepared for the U.S. Army Corps of Engineers Construction Engineering Research Laboratories

Tucker GE, Bras RL (1998) Hillslope processes, drainage density, and landscape morphology. Water Resour Res 34: 2751-2764

Verdin KL (1997) A System for topologically coding global drainage basins and stream networks. Proc. ESRI Users Conference, ESRI, San Diego, CA. http://lpdaac.usgs.gov/ gtopo30/hydro/p311.asp

Verdin KL, Jenson SK (1996) Development of continental scale DEMs and extraction of hydrographic features. Third International Conference/Workshop on Integrating GIS and Environmental Modeling. Santa Fe, USA, http://edcdaac.usgs.gov/gtopo30/papers/ santafe3.html

Verdin KL, Verdin JP (1999) Topological system for delineation and codification of the Earth's river basins. J Hydrol 218: 1-12

Vogt JV, Colombo R, Paracchini ML, Soille P, de Jager A, Folving S (2003a) A European landscape stratification reflecting drainage density In: Helming K, Wiggering H (eds) Sustainable Development of Multifunctional Landscapes. Springer, Berlin, Heidelberg, New York, pp 95-110

Vogt JV, Colombo R, Bertolo F (2003b) Deriving drainage networks and catchment boundaries. A new approach combining digital elevation data and environmental characteristics. Geomorphology 53:281-298

Vogt JV, Colombo R, Paracchini ML, de Jager A, Soille P (2003c): CCM River and Catchment Database, Version 1.0. EUR 20756 EN, European Commission – Joint Research Centre, Ispra (Varese), Italy, 30p. (http://agrienv.jrc.it/activities/pdfs/CCM-Report-EUR20756EN.pdf)

Walker JP, Willgoose GR (1999) On the effect of digital elevation model accuracy on hydrology and geomorphology. Water Resour Res 35:2259-2268

Whipple KX, Tucker GE (2002) Implications of sediment-flux dependent river incision models for landscape evolution. J Geoph Res 107(B2):310-320

Willgoose GR (1994) A statistic for testing the elevation characteristics of landscape simulation models. J Geoph Res 99(B7):13987-13996

Wolock DM, Price CV (1994) Effects of digital elevation model map scale and data resolution on a topography based watershed model. Water Resour Res 30:1665-1680

Referenced Web Sites

[1] http://srtm.usgs.gov/
[2] http://www.geo-informatie.nl/projects/pelcom/
[3] http://agrifish.jrc.it
[4] http://image2000.jrc.it
[5] http://agrienv.jrc.it/activities/catchments

Decision Supporting Hydrological Model for River Basin Flood Control

János Adolf Szabó

1. Introduction

Flooding is the major disaster affecting many countries in the world year after year. It is an inevitable natural phenomenon occurring from time to time in all rivers and natural drainage systems. Floods not only damage lives, natural resources and the environment, but they also cause losses to economy and health. Damage from flooding has been increasing each year resulting in loss of lives, property and production as well as affecting activities in the flooded areas. Large scale and long duration floods can be considered as the cause for economic loss to the concerned countries.

Flooding cannot be completely avoided, but damages from severe events can be reduced if effective flood prevention steps are implemented. The responsibility for flood prevention is always at policy level. In order to reduce damaging influences of floods the decision-maker should weigh many different, difficult circumstances and factors as well as the economic situation of the country. In order to analyze the numerous alternative strategies for effective flood prevention, one of the most useful tools is an up-to-date Spatial Decision Support System (SDSS) for river basin flood control. An SDSS typically consists of:

- *Databases*, which may include among other things geographical information systems (GIS), hydro-meteorological and river geometry database, spatially distributed watershed-attributes and parameters, etc,
- *Modelling tools*, to analyze unique scenarios of different meteorological, hydrological conditions, dike breach and/or using emergency flood reservoirs,

- *Artificial intelligence techniques* for accessing information in a timely manner, and decision analysis techniques for assessing trade-offs among alternatives,
- *Documentation* on the decision making process.

All of this information and model-running are organized according to the needs of the users within an up-to-date, user-friendly, interactive graphical software application.

An SDSS does not replace the decision-maker. It rather places the decision maker at the centre of the decision-making process so that information and timely assistance can be effectively utilized. As such, an SDSS should allow people to combine personal judgment with the results of analysis.

In other words, decision support tools are applications or programs that offer the resources necessary to make a decision. The tool *does not make* the decision. Instead, a decision support tool provides the necessary information in a timely manner so that a *person* can make a qualified judgment.

One of the most important parts of an SDSS is the modelling tool used to analyze the different scenarios. The less successful statistical and/or empirical model-experiments of earlier decades have highlighted the importance of numerical methods based on physics and distributed catchment-attributes, which describe the forming and transformation of flood waves even on catchments of tens of thousands of square km. Answers to questions like what are the effects of human actions in the catchment area (e. g. forestation or deforestation) or the changing of climate on the flood level and the speed of its occurrence, or what is the role of catchment soil and morphology in forming floods, have become increasingly important. Nowadays the answers to these kinds of questions can be easily provided. The progress of mathematical methods, the advanced state of computer technology as well as the development of remote sensing and meteorological radar technology have accelerated the research capable of answering these questions using physically based, distributed watershed models.

This study presents the physically based distributed rainfall-runoff model DIWA (DIistributed WAtershed) as a part of an SDSS. The model DIWA developed by the author at the Research Group for Water Management of the Hungarian Academy of Sciences in Budapest University of Technology and Economics, Department of Sanitary and Environmental Engineering, is designed to model the following: rainfall-runoff processes, flood events analysis, regional land-use change scenario analysis, real-time hydrological forecasting and control. The application of this model has been carried out on River Tisza, which is the largest sub-basin of the River Danube. Some interesting results for the Upper-Tisza basin are also presented. Those simulation results can demonstrate the usefulness of the DIWA model for supporting decisions on river basin flood control.

2. The DIWA Model

The DIWA model is based on the distributed rainfall-runoff model philosophy (e.g. Dolman et al. 2001). According to this approach the catchment is divided into basic elements, cells (Figure 1) where the basin characteristics, parameters, physical properties, and the boundary conditions are applied in the centre of the cell, and the cell is supposed to be homogenous between the block boundaries.

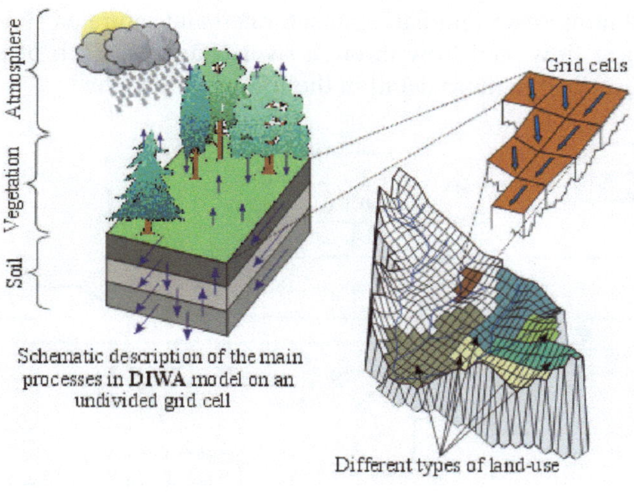

Fig. 1. Schematic description of a distributed hydrological model

The neighbouring cells are connected to each other according to runoff hierarchy (see section 2.1). Applying the hydrological mass balance and the adequate dynamic equations to these cells, the result is a 3D distributed description of the runoff. Within the simulation each element (cell) is represented by three fundamental layers:

- The *atmosphere* is a boundary condition of the system described by precipitation, air temperature, and potential evapotranspiration data based on calibrated radar data and/or interpolated meteorological fields (see section 3.2.2),

- The *vegetation* layer is the major link between the atmosphere and soil, defined by vegetation type and its parameters (estimated by bibliography), LAI, and water potential of cells,
- The *soil* basically has three categories of layers: The first is the so-called '*upper dynamic layer*', which is a special layer at the top of the surface containing the inequality of surface (small hummocks and holes), and mixture of litter and fallen leaves. This layer is equivalent to the organic horizon of the soil, which gives surface runoff and percolation as well. The second and the third categories are the *unsaturated* and the *saturated soil* layers (see more details in section 2.5).

The fundamental processes simulated by the model (Figure 2) include interception of precipitation, snow-accumulation and snow-melting, infiltration, water intake by vegetation and evapotranspiration, vertical and horizontal distribution of soil moisture, unsaturated and saturated flow of water in soil, surface flow, and flow through river channels. Each of these submodels is described in more detail in the following chapters.

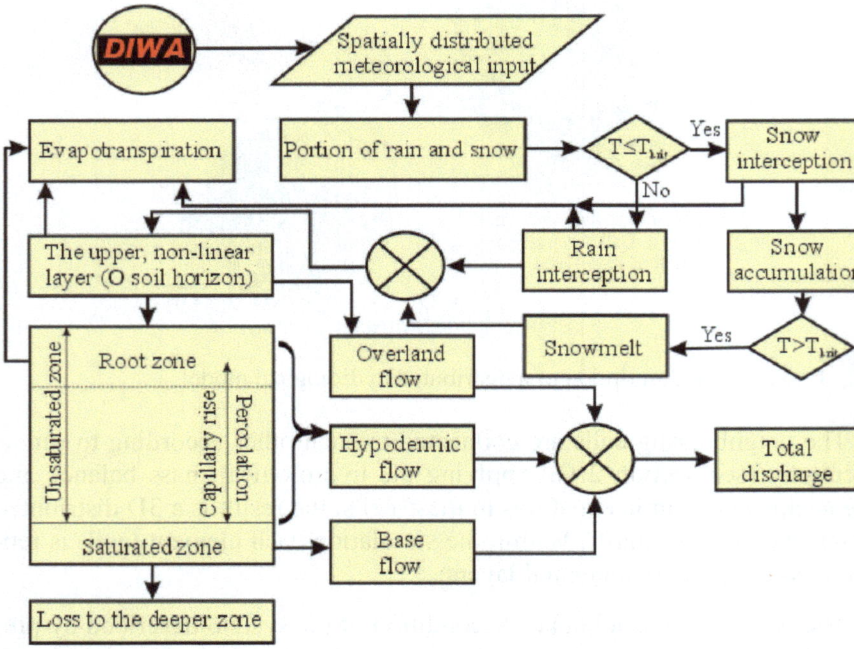

Fig. 2. Simplified flowchart of DIWA model

According to the basic concept of DIWA, the model does not support the cell-based, 3D, distributed mathematical description of the runoff processes on such sub-basins of a watershed, where the runoff coefficient is 'too small', especially in case of lowlands or flat sloped rolling ground, because the uncertainty of the 3D calculations are greater then than in the case of using 1D runoff-runoff model on a section of the lower-course type. For that very reason it is not the aim of DIWA to model the groundwater movement of permanent ground water in the above mentioned part of a catchment.

2.1 Cell Link Network Definition Based on Digital Elevation Model (DEM)

Digital elevation data are used to create both a cell link network and a cell-slope map. As a result of this 'central point' approach, the precipitation on a block will flow in the direction of the adjacent cell with the lowest altitude (Figure 3). The DIWA model calculates the gradient along the main diagonals too, and these values are also considered during the runoff calculation. The figure demonstrates the rules of runoff: Precipitation accumulates mainly on the so-called '0th order' elements (indicated with red colour in Figure 3b) until it reaches the stream channel. From that point it is transported in the channel (indicated with blue colour in Figure 3b). An element is considered a '0th-order' element, if there is only one other element from which it receives stream input. The routes connecting the '0th-order' elements are special, not stream-like, as they do not form stream channels according to the given coordinates. It is possible of course that a new stream channel would appear along the 0th-order elements if a finer resolution would be chosen, or some stream channels would 'disappear' when using larger grid cells.

The rank of rivers in the runoff hierarchy represented by graphs can be denoted by an ordinal number. This number increases by one if the river is the recipient of two or more rivers with the same ordinal number, otherwise the resultant river 'inherits' the highest ordinal number. So the DIWA model considers the graph edges with ordinal numbers higher than zero to be rivers and – as it is explained later in the mathematical description – runoff is calculated by different principles in these cases.

The runoff hierarchy, created by using the system described above, results in a tree structured graph. This graph 'automatically' designates the sub-basin area belonging to any grid element, thus it is possible to find the border of the area easily. On the other hand the graph simplifies the horizontal runoff to one dimensional sub-processes along the edges.

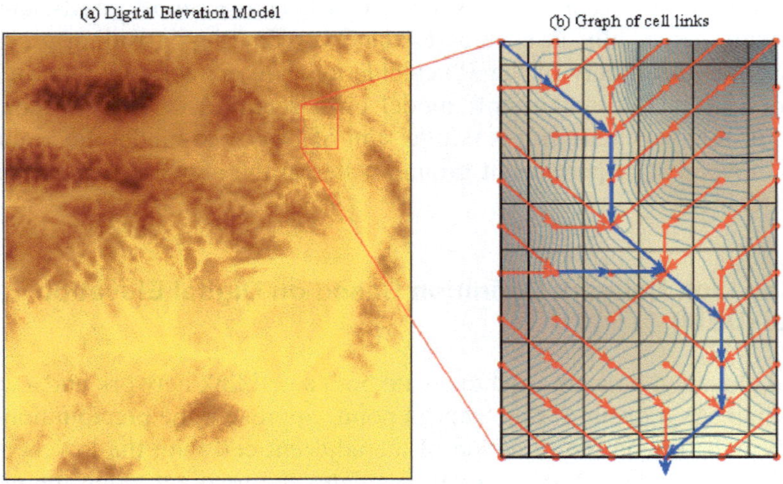

Fig. 3. A. DEM of study area. B. Example of cells and the river routing

The 1 km DEM for the DIWA model was downloaded from the homepage of U.S. Geological Survey (USGS), Earth Resources Observation and Science (EROS). This DEM is base on Digital Terrain Elevation Data. The following is a summary of the map projection parameters used for this projection:

- Projection Type: Lambert Azimuthal Equal Area (optimized for Europe),
- Units of Measure: meters,
- Pixel Size: 1000 meters,
- Radius of sphere: 6370997 m,
- Longitude of origin: 20 00 00 E,
- Latitude of origin: 55 00 00 N,
- False easting: 0.0,
- False northing: 0.0,
- XY corner coordinates (centre of pixel) in projection units (meters):
  - Lower left: (-3000000, -4999000)
  - Upper left: (-3000000, 8000000)
  - Upper right: (9999000, 8000000)
  - Lower right: (9999000, -4999000).

2.2 Rain or Snow, and Snowmelt

A critical temperature ($T_{crit,S}$ [°C]) determines whether a precipitation event will be treated as rain ($T_{mean} \geq T_{crit,S}$) or as new snow ($T_{mean} < T_{crit,S}$), where T_{mean} [°C] is the daily average temperature. When the precipitation event is determined to be snow, it accumulates on the surface until it melts. In DIWA $T_{crit,S}=0$.

The rate of snowmelt is estimated using the so-called 'degree-day factor' method (Martinec 1960; Martinec and Rango 1986), which usually takes the following simple linear form:

$$M = a \cdot \left(T_{mean} - T_{crit,M}\right) \tag{1}$$

where M [mm] is the daily rate of snowmelt depth and 'a' [mm•°C^{-1}•day^{-1}] is the degree-day factor and $T_{crit,M}$ [°C] is some critical temperature of melting. Currently in DIWA it is supposed that the critical temperature is a regional model-parameter depending on the elevation, and the degree-day factor depends on the momentary water content of snow (section 3.3).

2.3 Interception and Through-fall Estimation

Watershed surfaces are covered with vegetation, and this vegetation cover has considerable effect on the hydrological cycle. Only a part of precipitation reaches the surface directly (through-fall), some of it becomes temporarily stored in the vegetation (interception). One part of the retained water will start to evaporate after the rainfall event, while the other part will 'drip' to the surface. It is particularly important to estimate the through-fall which arrives directly to the land surface, because it significantly determines the volume of runoff in a catchment.

Supposing that the vegetation canopy is a linear reservoir, the process of interception is mathematically describable using the following equation system:

$$\left.\begin{aligned} &\frac{dS_I}{dt} = QP - QP_{th} \\ &QP_{th} = QP - \sigma \cdot \left(1 - \frac{S_I}{S_{I,\max}}\right) \cdot QP \\ &S_I(0) = S_{I,0} \end{aligned}\right\} \tag{2}$$

where S_I [mm] is the intercepted water volume on the vegetation, QP [mm/Δt] is the flow rate (intensity) of the precipitation, QP_{th} [mm/Δt] is the flow rate (intensity) of the through-fall, $S_{I,max}$ [mm] is the maximum storage capacity of the vegetation surface, σ is the retention capacity of the vegetation under dry condition [-] (around 0.05), and $S_{I,0}$ [mm] is the amount of water stored on the vegetation surface at the beginning of the calculations.

The $S_{I,max}$ value indirectly depends on the vegetation type, and can be estimated by the empirical equation: $S_{I,max}=h_{max}(1+LAI)$, where h_{max} [mm] is the maximum storage capacity of the vegetation surface (vegetation-type dependent model-parameter), LAI [m^2/ m^2] average Leaf Area Index of the cell.

Solving the (2) ordinary differential equation concerning the Δt time interval, the results can be expressed as follows:

$$S_{I,\Delta t} = S_{I,\max} - (S_{I,\max} - S_{I,0}) \cdot \exp\left(-\frac{\sigma \cdot QP_{\Delta t}}{S_{I,\max}} \cdot \Delta t \right) \tag{3}$$

and here $S_{I,\Delta t}$ is the amount of water stored on the vegetation surface at the end of the time interval.

The stored water amount is assumed to be decreasing because of evaporation. The intensity of evaporation is supposed to be constant within the Δt time interval. The calculation of evaporation is detailed in the section 2.4.

2.4 Evaporation and Evapotranspiration

The procedure is used for calculating evaporation of intercepted water and evapotranspiration from the vegetation, according to the method of Varga-Haszonits (Varga-Haszonits 1969).

2.4.1 Evaporation of Intercepted Water

Evaporation of intercepted water occurs at the maximum rate of evaporation from an open water surface (E_W [mm]), which is expressed with the daily mean relative humidity (RN_{mean} [-]) and daily mean temperature (T_{mean} [°C]; when T_k<1 than T_k=1+ T_k/273) as follows:

$$E_W = \left[\frac{1 - RN_{mean}}{2 - RN_{mean}} T_{mean} \right] \cdot \frac{\Delta t}{86400} \tag{4}$$

The water amount stored on the forest canopy is assumed to be decreasing in each precipitation-free day, subsequently for $S_{I,\Delta t}$ (equation 3):

$$S_{I,\Delta t} = \begin{cases} S_{I,\Delta t} - E_W & (S_{I,\Delta t} \geq E_W) \\ 0 & (S_{I,\Delta t} < E_W) \end{cases} \tag{5}$$

2.4.2 Evapotranspiration (Water Intake by Vegetation)

Not all of the water held in the soil is equally available to the plants. Plants are able to extract water easily from soils that are near field capacity (when all the water that can has drained from the soil by gravity) water contents. As the soil dries, the plants must work progressively harder to extract water, until finally, the soil is so dry that the plant can no longer expend enough energy to extract any more water. All soils retain water that plants are unable to extract (clays hold the most unavailable water while sands hold the least). This is the concept of wilting point. In spite of the fact that wilting point is affected strongly by soil structure, the *wilting point is not the same for all plants*. It varies for different types of plants and for different types of soils. Though clays hold the most water at both field capacity and wilting point, silt loams hold the most plant available water.

DIWA uses a two-step approach to calculate daily evapotranspiration. The first step involves calculation of the potential evapotranspiration rate according to the Varga-Haszonits approach, using daily value of evaporation from an open water surface (equation 4) and the biological properties of the evaporating surface:

$$PE_{\Delta t} = \kappa(t) \cdot E_W \tag{6}$$

where $PE_{\Delta t}$ [mm] is the average potential evapotranspiration within the Δt time interval, $\kappa(t)$ [-] is a dimensionless biological factor depending on time as the vegetation-density (LAI) changing in the time. In the model this $\kappa(t)$ factor is estimated as:

$$\kappa(t) = \kappa \cdot LAI_{\Delta t} \tag{7}$$

and here κ is a dimensionless vegetation-type dependent model-parameter ($0 \leq \kappa \leq 1$), and $LAI_{\Delta t}$ [m^2/ m^2] is the average leaf area index of the cell within the Δt time interval.

In the second step the actual evapotranspiration rate is determined taking the restrictive effect of the water capacity of the soil:

$$E_a = f(VDRS_t) \cdot PE_{\Delta t} \tag{8}$$

where E_a [mm] is the actual evapotranspiration, $f(VDRS_t)$ is a dimensionless reduction-factor, which is function of actual value of the vegetation-type depended relative saturation rate $VDRS_t$ [-], defined as follows:

$$VDRS_t = \begin{cases} \dfrac{\Theta_t - \Theta_{R,WP}}{\Theta_S - \Theta_{R,WP}} & (\Theta \geq \Theta_{R,WP}) \\ 0 & (\Theta < \Theta_{R,WP}) \end{cases}, \text{ where } \Theta_{R,WP} = \max\{\Theta_R, \Theta_{WP}\} \tag{9}$$

and here Θ_t [m^3/m^3] is the volumetric water content, Θ_R [m^3/m^3] is the residual water content, Θ_{WP} [m^3/m^3] is the water content at the vegetation-type dependent wilting point, and Θ_S [m^3/m^3] is the saturated water content.

In several hydrological models the $f(\cdot)$ reduction-factor function is constant or at most linear. However this should be non-linear because of the water retention in a porous medium is a strongly non-linear function of saturation. When soil becomes drier, the soil's hydraulic conductivity decreases, which may reduce E_a to a lower actual evapotranspiration rate. In DIWA the $f(\cdot)$ function is defined according to the structure of the soil pore system:

$$f(VDRS_t) := \begin{cases} 1 - \dfrac{pF\left(VDRS_t \cdot [\Theta_S - \Theta_{R,WP}] + \Theta_{R,WP}\right)}{pF(\Theta_{R,WP})} & (\Theta \geq \Theta_{R,WP}) \\ 0 & (\Theta < \Theta_{R,WP}) \end{cases} \tag{10}$$

where $pF(\cdot)$ symbolizes the soil water retention curve of the given soil-type (detailed in the next, section 2.5).

Figure 4 demonstrates two different non-linear reduction-factor function of potential evapotranspiration.

2.5 Modelling of the Subsurface Run-off Processes

The concept behind the DIWA model is that the ground water flow is not a major factor in forming floods because of the difference in time scale, therefore the movement of ground water can be considered in a simplified way in the model. It is more important to have a most accurate description about the processes in the unsaturated zone, because these processes determine the ratio between the runoff and the total water amount reaching the watershed, which is a fundamental factor determining the size of the flood.

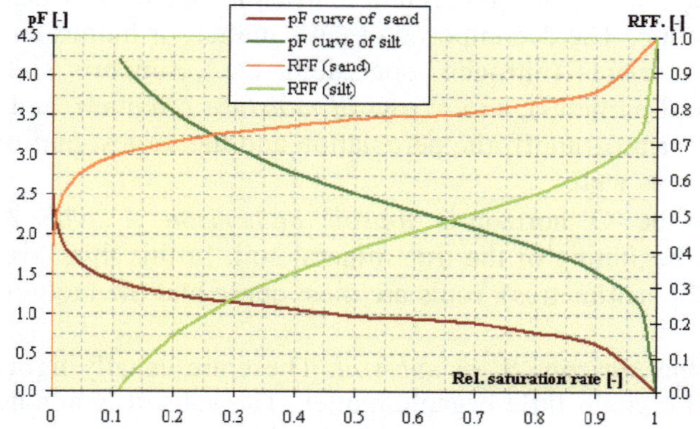

Fig. 4. Two examples for pF curves and reduction-factor functions of potential evapotranspiration as a function of relative saturation rate

2.5.1 Typical Soil Profiles

Soils develop over long periods of time, perhaps ten thousand years, as a response to the soil forming factors. In general, soils develop into layers. These layers (called horizons of a typical soil profile) are usually seen along road cuts and other areas where a cross section of the soil is exposed. In a hypothetical situation, there are four soils horizons in a soil profile above bedrock (Oberlander and Muller 1987) (Figure 5).

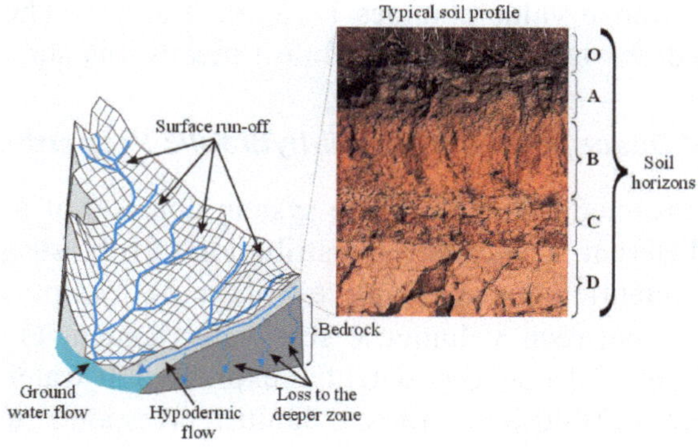

Fig. 5. Typical soil profile

The thickness of each layer varies with location and conditions. In a 'disturbed' condition not all horizons will be present.

- At the top of the horizon is the *horizon O*. The 'O' horizon is primarily composed of organic matter. Fresh litter is found at the surface, while at lower depths all signs of vegetation structure have been destroyed by decomposition. The decomposed organic matter, or humus, enriches the soil with nutrients (nitrogen, potassium, etc.), supports soil structure (acts to bind particles), *enhances soil moisture retention*, and this layer gives the surface runoff, the percolation and the largest part of the total evaporation as well,
- Just below the 'O' horizon is *'topsoil' or horizon 'A'*. The 'A' horizon marks the beginning of the true mineral soil. In this thin layer (usually less than 20 centimetres), roots are most dense and the biological activity is high,
- Below is the *horizon 'B' or sub-soil*. These are usually light coloured, dense, and contain little organic matter. The subsoil is a zone of accumulation since most of the materials leaching from the topsoil accumulate here. This layer can vary in thickness from a few centimetres to a meter or more,
- *C horizon* - This horizon lies under the sub-soil and is called the *'parent material'*. This is the original material from which the soil has developed. This layer has deposits of sand, gravel, pebbles, boulders and rock in various mixtures. 'C' horizons are typical of poorly drained areas,
- *Bedrock (D horizon)* - Under the layer of parent material is the layer of bedrock. It is fairly typical in higher mountainous regions that the bedrock is often visible on the surface. The different types of bedrocks are typical of very poorly drained areas, indeed it is often causing impermeability. In the DIWA model the user can determine (even cell by cell) that the bedrock is permeable or not, defined by a map of the permeability factor, whose value changes between 0 and 1. The 0 value means that the bedrock is impermeable while 1 means it is not.

2.5.2 Mathematical Description of the Soil-hydraulic Properties

Theoretically it is assumed that a soil pore system consists of a series of capillary tubes of different sizes, and the distribution of tube sizes follows from the soil water retention curve, pF curve (Figure 6). The pF curve defines the relationship between volumetric soil water content Θ [m^3/ m^3] and negative logarithm of the suction potential expressed in water column-cm. For example pF=3 (1000 [cm] water column) corresponds to 1 [atm] suction potential. Soil matrix potential is defined in such way that it is zero when all the pores are filled with water; it is negative in the unsaturated soil above the groundwater table and positive (hydrostatic pressure) below the groundwater level.

The DIWA model uses the generalized pF curve collection constructed by Van Genuchten (1999) based on actual measurements, but these can be replaced with actual measurement curves.

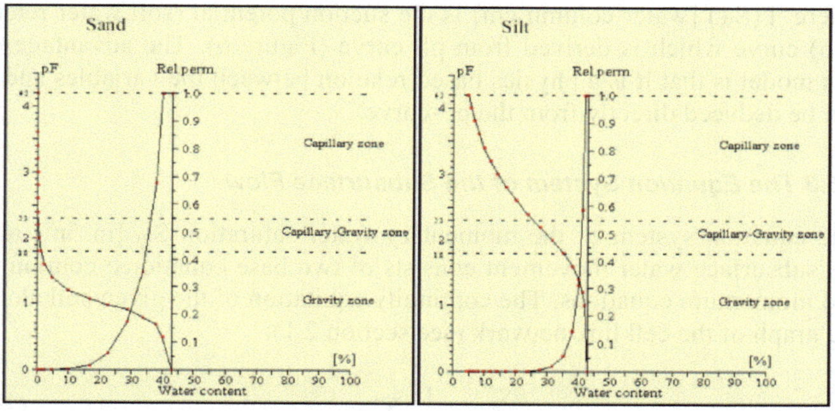

Fig. 6. pF-curves for two different types of soil (saturation is expressed in volume per cent on the horizontal axis)

Permeability is an essential parameter of unsaturated flow. Different permeability belongs to different porosity fractions (Burdine 1953; Phillips 1991). This is explained by the so-called 'capillary tube bundle' approach, according to which the porous medium is described as a bundle of tubes of different diameters, where the distribution of tube diameters is identical to the pore-fraction distribution of the soil. According to the theory, unsaturated flow means that the aqueous phase is not continuous in every tube. If there is a potential difference between the two ends of any tubes of finite length, then flow motion can be observed mainly in the tubes with greater diameter, while the tighter tubes will join in later.

Unsaturated hydraulic conductivity is considerably smaller than the saturated value 'k' [m/s], which is defined as the conductivity value when all the pores are filled with water. When the soil gets unsaturated, larger pores will be emptied first and therefore $k(S_W)$ [m/s] unsaturated hydraulic conductivity decreases very rapidly when the momentary saturation S_W [m^3/m^3] decreases.

In order to find a physically established function between saturation and relative conductivity ($k_r(S_W)$ [-]) the following (11) relation was deduced from the laws of Darcy, Jurin and Poiseuille:

$$k_r(S_W) = \frac{k(S_W)}{k} = \frac{\frac{1}{S_W} \cdot \int_0^{S_W} \frac{dS_W}{\Psi^2(S_W)}}{\int_0^1 \frac{dS_W}{\Psi^2(S_W)}} \tag{11}$$

where $\Psi(S_W)$ [water column cm] is the suction potential (soil water retention) curve which is derived from pF curve (Figure 6). The advantage of this model is that it is a physics based relation between the variables and it can be deduced directly from the pF-curve.

2.5.3 The Equation System of the Subsurface Flow

The equation system of the momentary water saturation S_W [m^3/m^3] and the subsurface water movement consists of two base equations: continuity and momentum equations. The continuity equitation of the given cell along the graph of the cell link network (see section 2.1):

$$\frac{dS_W}{dt} = \frac{1}{V_0 \Phi} \sum q_i \tag{12}$$

where Σq_i [m^3/s] symbolizes the algebraic sum of the net material stream of the cell taking all input and output components into consideration namely precipitation, evapotranspiration, infiltration, horizontal and vertical soil moisture redistribution driven by gravity and capillary rise, V_0 [m^3] is the volume of this cell, Φ [-] is the porosity of the soil.

The momentum equation of the model is the so-called Darcy relation based on Darcy's law which expresses the linear relation between the $d\Psi/dh$ tension-gradient and the $q_{d\Psi}$ [m^3/s] water discharge moving in the direction of a unit vector with α angle (with the horizontal, referential plane) resulting from the $d\Psi$ tension difference:

$$q_{d\Psi} = -k \cdot A_{d\Psi} \left(\frac{d\Psi}{dh} + \sin \alpha \right) \tag{13}$$

and here $A_{d\Psi}$ [m^2] is the area of the surface element perpendicular to the flow direction resulting from the $d\Psi$ tension difference.

According to (13) there are three different cases regarding the direction of the flow:

- $d\Psi/dh + \sin(\alpha) > 0 \Rightarrow$ gravity and the capillary potential act in the same direction, therefore there is a flow into a direction opposite to the given unit vector.
- $d\Psi/dh + \sin(\alpha) = 0 \Rightarrow$ state of capillary equilibrium, no movement;
- $d\Psi/dh + \sin(\alpha) < 0 \Rightarrow$ the capillary force is greater than gravity, so there is a flow into the same direction as the given unit vector.

According to the equations 11, 12 and 13 and discussions above in this section, the complete Darcy flow model can be defined as follows:

$$\left.\begin{aligned} &\frac{dS_W}{dt} = \frac{1}{V_0\Phi}\sum q_i \\ &q_{d\Psi} = -k(S_W)\cdot A_{d\Psi}\left(\frac{d\Psi}{dh} + \sin\alpha\right) \\ &\Psi = \Psi(S_W) \\ &k(S_W) = k\cdot k_r(S_W) \end{aligned}\right\} \tag{14}$$

Boundary conditions:

- The upper boundary conditions are determined by the hydro-meteorological processes: The flow rate of through-fall (see equation 1) and melted snow (see equation 3),
- The lower boundary condition is determined by the permeability of the bedrock. The lower boundary conditions in DIWA are optional and can be defined by the user:
  - Closed lower boundary condition, which corresponds to the case when the bedrock is presumably impermeable below the horizons of soil,
  - The user can define an open, or partially open lower boundary condition with a map of permeability factor, whose values change between 0 and 1. If this value is greater than zero, it means that the bedrock is permeable. A fictive grid element is added below the bottom cell if the lowest cell is considered permeable. The saturation of these fictive elements is identical to the last (real) element multiplied by the user defined permeability factor.

2.5.4 Modelling Infiltration Process

The DIWA model describes the infiltration process in a physically well-established way which is described by Darcy relations using equation 14. The so-called pF curve, describing the water hold-up capacity of the soil, plays a major role during this calculation. In DIWA we assumed that the top of the soil ('O' horizon) can be characterized with a pF curve and storage capacity which occurs until the through-fall and the snowmelt begin infiltrating or flowing further on the surface. The thickness of this layer is a parameter of the model.

In DIWA the user can define arbitrarily the number and the thickness of the soil-layers above the bedrock (even cell by cell) with different soil-properties: pF curve, porosity and saturated hydraulic conductivity. According to the way we designed the basic concept of DIWA the main run-off concentration below the surface is the same as it is on the surface defined by the surface gradient, derived from DEM.

2.6 Surface Run-off Calculation

The DIWA model divides runoff elements in two different groups – stream and surface runoff, and the calculations of surface runoff are based on these two categories along the graph of the cell link network (section 2.1). Runoff is reduced by infiltration on the elements with surface runoff. The water left after the infiltration travels on the so-called '0th order' elements in the graph of the cell link network until it reaches a stream channel. The water hereafter flows in the channel without further loss.

2.6.1 The Top Layer and its Motion Equations:

The DIWA model simulates the surface runoff as flow motion taking place in the 'O' horizon of the soil with user-defined properties: cell-thickness, pF curve, porosity and saturated hydraulic conductivity. The motion equations of the topmost zone are the same as in the non-linear Darcy model (equation 14).

2.6.2 Runoff in Channels (Flood Routing)

In the DIWA model runoff in stream channels is described by a hydrologic flood routing method based on the 'linear reservoir' approach (Kalinin-Miljukov approach): the river is described as sequential series of reservoirs. The outflow of a reservoir does not depend on the bottom boundary conditions; it is a linear function of the actual state of the reservoir. The basic equations for the *(i,j)* element of the river are:

$$\left.\begin{aligned} \frac{dV_t^{i,j}}{dt} &= \sum q_l^{i,j} \\ q_{out}^{i,j} &= K^{i,j} \cdot V_t^{i,j} \end{aligned}\right\} \qquad (15)$$

where Σq_l [m^3/s] symbolizes the algebraic sum of the net material stream of the cell taking all input and the single output discharge into consideration, V_t [m^3] is the water storage, $q^{i,j}_{out}$ [m^3/s] is the outflow discharge from the element and $K^{i,j}$ [s^{-1}] is the storage coefficient ($1/K^{i,j}$ [s] is the residence time).

2.7 Numeric Solution

The calculations are executed recursively by time steps in the model. The *initial conditions* for each $t+\Delta t$ moment are the storing characteristics in the t moment interpreted in each grid cell on the:

- vegetation canopy,
- upper nonlinear dynamic layer,
- each layer of the soil,
- stream channel,
- snow.

The distributions of the meteorological characteristics (see section 3.2) serving, as *boundary conditions*, for each time step should be provided according to the grid density of the DIWA model for each Δt time step:

- Sum of precipitation within the given time interval,
- Daily average air temperature,
- Daily average potential evapotranspiration.

These characteristics are considered to be constant within the given time step. The numerical algorithm can be divided into three separable phases:

1. Calculation of the state variables:

- interception,
- evaporation of the water, stored on the vegetation canopy,
- snow accumulation,
- snowmelt.

2. Solving the rainfall-runoff system for:

- actual evapotranspiration,
- interaction between the upper nonlinear dynamic layer and the soil layers,
- percolation,

- horizontal components of the runoff in the direction of the next stream channel:
  - surface runoff through the upper nonlinear dynamic layer,
  - lateral flow (unsaturated flow),
  - groundwater flow (saturated flow),

3. Calculation of the runoff in stream channels.

2.8 Some Numerical Aspects

The processes in step 1 can be handled separately from all other processes. Using the provided relations all the calculation can be executed for all (i,j) grid elements and for all time elements before the other two phases, because the base equations are algebraic.

The channel-soil relation can only be unidirectional, because of the unknown geometrical characteristics of the stream channel. According to this, the calculation of the runoff in stream channels should be separated from the rainfall-runoff calculation.

With regard to the linearity assumption for the channel routing (see in section 3.6), the calculations of the runoff in stream channels are independent from the lower boundary conditions. Therefore, each calculated outlet discharge of a certain piece of the streams will be the input for the next, and so on.

An implicit predictor-corrector iteration process is organized for estimating the value of '*k*' conductivity in the time element by solving the Darcy system in DIWA. The non-linearity in the solution is the result of this '*k*' conductivity. The linear equation systems resulting in each iteration step are directly integrated.

3. Application of the DIWA Model in the Upper Part of the Tisza River Basin

3.1 Background and Some Characteristics of the Tisza Basin

3.1.1 Location

The Tisza River basin (966 km, 154,039 sq km) is located almost in the geographical centre of Europe (Figure 7), which is with respect to its length and catchment area, the largest tributary of the Danube.

Fig. 7. Location and general view of the Tisza River Basin

Five countries fall within the basin: Ukraine, Romania, Slovakia, Hungary and Serbia-Montenegro. From its total length, about 160 km lies in the Ukraine and Romania and about 800 km in the Great Hungarian Plain (650 km in Hungary, 150 km in Serbia-Montenegro). Romania has approximately 47 percent of the catchment area, Hungary 29 percent, Slovakia 10 percent; Ukraine 8 percent and Serbia-Montenegro 6 percent.

The Tisza rises in the Zakarpathian Mountains in North-western Ukraine and is formed from the confluence of the Belaya Tisza and the Chiornaya Tisza, and it enters Hungary at the North-East border of the country (*Upper-Tisza basin:* UTB).

The other main branch of the headwaters rises in the mountains of east Slovakia. The Uzh and Latoritsa tributaries pass from Ukraine into Slovakia where they join the Bodrog before it enters Hungary. The Szamos (Somes), the Körös (Cris) and the Maros (Mures) all rise in the Romanian Carpathians and drain west into Hungary before they join the Tisza *(Transylvanian basin)*. The river then enters Serbia where it joins the Danube.

3.1.2 Some Geographic Features of the Basin

The Tisza is a river of definitely lowland character - the proportion of mountain areas on its catchment area is about 1%, while for those below 200m it is 46% - and the highest point of the catchment area (2506m) is below the permanent snow-line.

On the spring region of the Tisza (in UTB) in its valleys of 20-50 m/km decline we cannot speak about a definite river bed. After larger rainfalls floods run through the valley bottom on its whole width, while in a dry weather period the water gurgles only on the stairs made by the rocky bottom and hides between huge rock-glaciers. Below the union of the Black and the White Tisza we can speak about a wild mountain stream abounding in water, which already has a shallow bed. From the inflow of the Visó the original water output doubles, and at the same time the decline of the bed decreases to 2 m/km. Here the bed is a few decimetres deep, with gravelly bottom, the river meanders among islands and gravel shoals, continuously changing its route. Below the inflow of river Borsa the decline of the bed further decreases, then the river reaching the Great plain runs in a deep bed. Its decline decreases quickly, in the height of the Szamos mouth it shows a value around 8.8 cm/km and further down it decreases from 3.7 cm/km to 2.5 cm/km. On the Tisza lowland the river bed changes to more or less 200 m wide, however in certain sections it is only half of that or even less, while elsewhere it becomes 300m or 420m wide. It should be mentioned here, that flood protection embankments on the lowland course are 1400-1800 m wide in average, but in many places they are less than the half of the average. Winding is a characteristic feature of the lowland course. Flat curves of wider arch are characteristic to this course.

3.1.3 Climatic Conditions

Among the natural forces shaping the water regime of the Tisza, the temperature and precipitation relations and outflow characteristics are basic. The Tisza catchment area has a temperate continental climate. In general mean annual temperature is 10-11 °C on the lowland parts of the Tisza catchment area, while 8-9 °C within the Transylvanian basin, and 6-9 °C on middle elevation mountains. Mean annual temperature values themselves show 3-4 degrees of fluctuation, but differences are significantly greater for certain months, and these can exceed 10 degrees. Of course the largest differences are from the temperature extremities. Thus the greatest value of the fluctuation of the temperature is 70 degrees (max. 41.3 °C; min. -32.2 °C), which proves the continental character of the climate.

The Tisza catchment area is poor in precipitation. About 60% of the upstream area gets more than 1000 mm precipitation annually. The valley of the Black Tisza (part of UTB) is in the shadow of the Hoverla, and the annual precipitation does not exceed 1200-1300 mm at the spring region of the White Tisza. Compared to their high elevation the valleys of Visó and Iza are definitely dry. Coming out to the Great Hungarian Plain, the river gets a rapidly decreasing amount of precipitation, and this stays below 600 mm per year everywhere on the Tisza lowland. Distribution of the precipitation within a year is worthy of attention. Since on the whole catchment area temperate continental climate is a dominating factor, the early summer precipitation maximum is characteristic. About 25% of the annual precipitation falls in May and June. This image is modified by the air currents of the Mediterranean Sea, in consequence of which the summer precipitation maximum is being completed with a secondary autumn maximum around October-November. Annual distribution of precipitation therefore is quite uneven; it can be characterized by two peaks and a dry season between them. Distribution of precipitation of the mountain courses is greatly influenced by the relief; here it is more balanced to a certain extent. Significant differences show also in distribution of precipitation of certain years, which is not only the characteristic feature of the last decades in this region.

The other important element of the water regime is evaporation. Relative humidity on the Tisza catchment area is 80-85% in the winter months and 60-65% in summer. On the Great plane in the summer months - primarily around noon - humidity decreases even on 40%. From a hydrological point of view evaporating capacity of the atmosphere is very important, which shows the potential maximum of evaporation. The real value of evaporation stays below the possible values on the lowland areas which are poorer in precipitation, but in part within the Transylvanian basin too, the reason for which is, that evaporation exhausts the water reservoir of the soil. Assuming the relationship of the potential evaporation capacity and precipitation the received image well represents the arid character of the Tisza catchment area. According to these annual evaporation on the Great Hungarian plane exceeds annual precipitation even in mean annual amounts, namely on significant part of the territories beyond the Tisza, in a measure exceeding 40%. Total water deficit of the vegetation period (April to September) on the Great Hungarian plain reaches 160-180mm.

3.2.4 Hydrological Overview

The runoff over the Tisza basin varies within very wide ranges. In general on the territory of the catchment area spring months are the most abundant in water. Floods are quite violent, their drainage is significantly influenced by floods of tributaries. During long lasting rainy periods floods can be congested. It is a peculiarity that ice-floes are of much less significance on the Tisza, than on the Danube, because in the deep bed piling ice cannot raise the water level significantly. The amount of outflow is influenced by impermeability of the catchment area too. From this point of view it can be stated, that semi-permeable surfaces dominating in the catchment area of the Upper Tisza, the Szamos and the Bodrog are able to moderate violence of floods, and they serve as some reserves in the low water periods. There is an impermeable surface on the Great Hungarian plain, and the same thing is characteristic of the Körös region. On these territories this is the main reason for inland waters. After draining the water stored in the snow cover discharge decreases rapidly, i.e. reserves of the river exhaust soon. Low water periods appear primarily in summer and in the autumn, mainly between August and October, which state, besides shortage of precipitation, is further worsened by high values of evaporation.

3.1.4 Some Characteristics of the Upper-Tisza Basin

The DIWA model was calibrated against time series of discharges at 11 measurement stations along the frontier of Hungary. This study puts into the focus the results only for the Upper-Tisza basin (UTB) which consists of the area from the spring region of the river and its tributaries in the Carpathian Mountains up to the discharge station at Tiszabecs/Hungary (Figure 8).

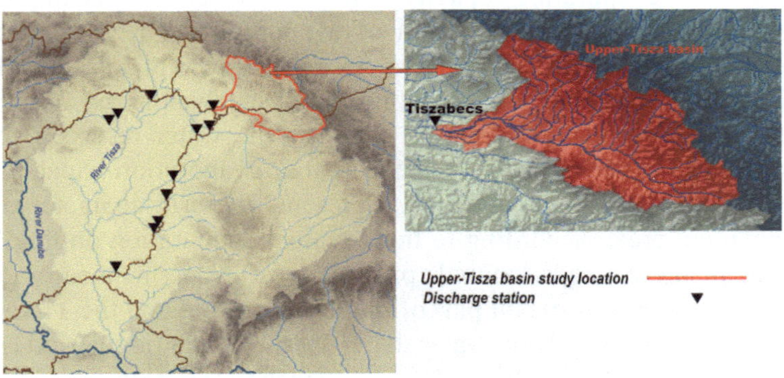

Fig. 8. Tisza River basin. Discharge stations belonging to the up-stream areas and the Upper-Tisza basin

The UTB has the surface area of 9707 [km^2], and the lowest/mean/highest discharges of the flow (recorded during the last 40 years) are 70/350/4200 [m^3/s].

Some important direct derivatives of DEM which are interesting for the hydrological modelling point of view were estimated (Figure 9). Figures 9a and 9b. present the frequency and the distribution functions of the elevation and the slope of the UTB, where the direction of the slope was defined by the local drainage direction (see in Section 1.1 in this study). From this analysis it can be seen that about 30% of the basin area is higher than 1000 m and it is rapidly increasing from this altitude up to 2256 m high, which is the highest point of the UTB (Figures 9a). This fact is in complete agreement with the results of the statistical analysis for the slope. About 30% of the basin area has steeper gradient than 0.1 m/m and it further increases from this value up to 0.33 m/m (Figures 9b).

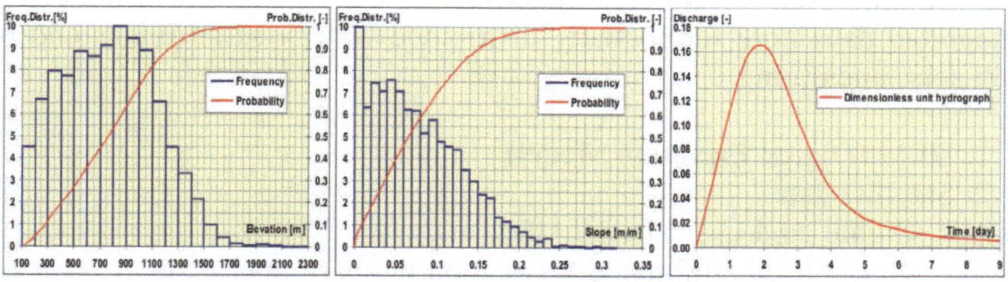

Fig. 9. A and B. Some important direct derivatives of DEM. C. The dimensionless unit-hydrograph of Upper-Tisza basin

Another important hydrological attribute of a watershed, the dimensionless unit-hydrograph (as a unit-response of the system for the unit step of precipitation) function was also estimated for UTB using numerical simulation by DIWA. For this analysis it was assumed ‘typical’, average summer conditions for the soil moisture, for the vegetation, for the degree of channel filling and for the meteorological parameters as well, and then we ran DIWA on UTB with these conditions and 100 mm unit step of precipitation over the whole basin. The Fig. 9/c. demonstrates the normalized unit-response in case of these conditions. As it was suggested from the above outlined direct derivatives of DEM, the unit-hydrograph shows a ‘fast responded’ system with about 42 hours response-time.

Moreover, about 70% of the UTB gets more than 1000 mm precipitation annually. The average rainfall varies from 700 mm in the lowlands and the lowest situated valleys and more than 2000 mm in the high mountain area.

Due to these particular topographic and climate conditions flash floods are common in spring and summer with river stages increasing 1.5 to 2.5

meters within four or five hours, further intensified by the low infiltration capacity of the soils in the Carpathian Mountains. Rainfall, snowmelt produced and mixed floods all are frequent. These floods cause enormous inundation in the vast lowland areas. During the last 30 years the river has been effected by more than 100 flood events. Moreover, the rate of occurrence and degree of floods has continuously increased and reached its maximum with two flood events of 1998 and 2001. That's just all the more reason why we focused our attention on this sub-basin of River-Tisza.

3.2 Data Preparation and Pre-processing

Data is likely to be the most important part of any SDSS for large-scale river basin flood control, and for water resources management and planning. Before the actual modelling, the data should be collected, analyzed and pre-processed attentively. Underpinning the DIWA model there is an integrated data store for spatially distributed watershed-attributes and parameters, integrating:

1. DEM (Digital Elevation Model).
2. Satellite data:
   - digital land-use data (DLUD),
   - Normalized Difference Vegetation Index (NDVI).
3. Spatially distributed meteorological data:
   - temperature,
   - precipitation,
   - potential evapotranspiration.
4. Soil data (layer by layer):
   - thickness,
   - type,
   - hydraulic properties: pF curves, porosity, saturated conductivity.

3.2.1 Generating the Characteristics of Vegetation Density Using Remote Sensing Data

As discussed above, the vegetation influences the water budget of the watershed in two ways: by interception and by evapotranspiration. Both of these factors depend on the type and actual condition of the vegetation cover. The vegetation type and its regional distribution are determined by the DLUD, while the vegetation density – which is constant neither in time, nor in space – can be obtained by using the NDVI distributions. (Both the DLUD and the NDVI satellite data were collected and downloaded from the home-page of USGS, EROS with the same geometric characteristics as the DEM (see in section 2.1).

One of the two necessary components for determining the vegetation density is the distribution of land use. The DLUD is a grid-map of land use categories. The grid elements in a DLUD file contain those codes which are typical of the land use for the area. The different categories are presented with different numbers, and the same numbers represent hydrological properties.

The second information we need for each grid element to determine the condition of the vegetation is the NDVI values. The NDVI is obtained by measuring, the long-wave radiation of the earth with remote sensing. Its value is the $\mathrm{NDVI} = \frac{\mathrm{NIR} - \mathrm{VIS}}{\mathrm{NIR} + \mathrm{VIS}}$ normalized value, which was generated by using the values of the (2nd channel: 0.73-1.1 μm) radiation domain close to infrared (NIR) and the (1st channel: 0.55-0.68 μm) visible (VIS) spectrum ($-1 \leq \mathrm{NDVI} \leq 1$) (Kidwell 1997).

The NDVI is very sensitive to the condition of the plant-chlorophyll; hence it is suitable to describe the vegetation density. A positive value indicates green vegetation. The higher this value is, the greener and denser is the surface vegetation. Negative values indicate the presence of barren areas, water bodies or ice, snow and cloud respectively. Although arithmetically $-1 \leq \mathrm{NDVI} \leq 1$, in practice the NDVI usually falls between -0.1, and 0.7, where $-0.1 \leq \mathrm{NDVI} \leq 0.05$ indicates the presence of water, ice or snow.

The fact mentioned above indicates that the NDVI behaves like a variable corresponding to the vegetation period. It is sufficient to refresh the distribution of the vegetation parameters once a month, considering that the 'primary consumers' (trees and undergrowth) in the watershed are capable of developing their maximal foliage over a very short time.

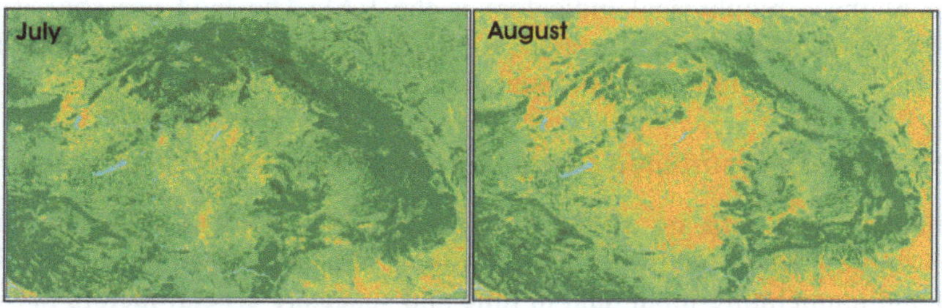

Fig. 10. Monthly NDVI composite over the Carpathian Basin in July and August

The next step is estimating the LAI distribution for each month as a function of the NDVI values. The LAI is a good characteristic of the vegetation variable with climate from the hydrological point of view, because the precipitation retention potential of the vegetation basically depends on the total leaf area. The typical LAI values for different vegetation types can be found in standard tables. For deciduous forest $3 \leq LAI \leq 7$, while for the pine needle forest $10 \leq LAI \leq 47$ values can be assumed. The direct method used for measuring the LAI is to determine the area of the vertical projection of the tree. The total weight of the leaves is compared to the leaf-weight that covers an area unit, and then the LAI can be calculated. This method is obviously not suitable for measurements on hydrological scale, so indirect measuring methods are used. The NDVI is probably the best parameter for this.

Based on theoretical consideration, Sellers (1987) demonstrated that the NDVI must be in a non-linear relation with LAI. Later on Ishii et al. (2001) find a strong correlation between LAI and NDVI as:

$$NDVI = 0.29 * \ln(LAI/0.577)$$

Presumably there is a (theoretical) maximal LAI value for every vegetation culture that from a statistical point of view can not be exceeded by the vegetation in question. Furthermore, we can presume that such a 'mature' stock has the maximum NDVI value, that is NDVI=0.7. From the consideration detailed above it results that to determine the LAI distributions the land use maps and the NDVI maps must be conferred, and furthermore, the maximal LAI value (LAI_{max}) must be provided for every land use category. If all these factors are known, the LAI distribution can be generated for every NDVI distribution.

Finally, it can be ascertained that the NDVI is a very useful indicator for describing the biophysical condition of the vegetation over a large area. One of these biophysical indicators is the LAI. The land use should be taken into account during the evaluation of the NDVI distributions in order to reduce the possibility of potential errors.

3.2.2 Spatially Distributed Hydro-meteorological Data Pre-processing

Meteorological data are a key input for any hydrological model because they mainly determine the volume of runoff in a catchment. Accurate rainfall and temperature input data in time and space are crucial factors in runoff modelling. In particular it is important to know the amount and spatial distribution of precipitation as well as to have reliable estimation of temperature-fields, because this information strongly influences all hydrological elements and processes.

Climatic data are usually measured at point locations, whereas it is necessary to know these values everywhere; therefore we have to use an interpolation method for mapping. In distributed hydrological models, spatial interpolation is typically applied to a grid with estimates made for all cells. The most promising stochastic tool may be the variogram analysis and Kriging to estimate values of the precipitation at an unsampled location in the field (Cressie 1985).

A new algorithm for interpolation of high resolution precipitation fields has been developed. An approved geostatistical method was also applied in the case of temperature (Szabó 2005).

The main advantage of using these methods is that they allow the estimation of uncertainty of interpolated fields and consequently the uncertainty in flood simulation and forecasting can also be estimated (Szabó-Kalaš 2004).

It has been proved statistically that by applying these methods the uncertainty is smaller compared with other popular methods (Szabó 2005).

3.2.3 Digital Soil Map Pre-processing

Similarly to the climatic data, the soil properties data are also given as measured values at point locations on the Tisza basin. Therefore it was necessary to use a spatial interpolation method to characterize patterns in soil-hydraulic properties and soil-type texture over the basin as well, provided that the data were abundant and spatially dependent.

Several researchers have studied the spatial variability of soil properties for mapping, for instance Burgess et al. (1980), or Schloeder et al. (2001). From those studies, one of the most powerful approaches to estimate values of the soil data at an unsampled location in the field may be the variogram analysis and ordinary Kriging. Following several tests, the spherical semi-variogram model was proposed.

The above outlined interpolation technique is a built-in function of the programme-system of the DIWA model, and it was used for interpolating the soil properties data on Tisza basin, based on 1010 soil samples sites.

3.3 Calibration and Validation

3.3.1 An Overview of the Calibrated Parameters and the Method for Calibration

The distributed DIWA model has 5 important parameters, and they should be calibrated by considering their spatial variations.

The critical temperature of snowmelt model $T_{crit,M}$ [°C] in DIWA model is a spatial distributed parameter. For instance, it is roughly supposed that this value varies linearly with the altitude of the grids, having a larger value in higher mountainous areas and a smaller value in lower parts of the basin:

$$T_{crit,M}^{i,j} = T_{crit,M}^{\min} + \left(Z^{i,j} - Z^{\max}\right) \cdot \frac{T_{crit,M}^{\max} - T_{crit,M}^{\min}}{Z^{\max} - Z^{\min}}, \qquad (16)$$

where $T_{crit,M}^{i,j}$ and $Z^{i,j}$ are the critical temperature and the elevation at any grid, $Z^{\min}$ and $Z^{\max}$ are the lowest and highest elevation values in all grids, $T_{crit,M}^{\min}$ and $T_{crit,M}^{\max}$ parameters represent the smallest and largest critical temperature in all grids. $T_{crit,M}^{\min}$ and $T_{crit,M}^{\max}$ model-parameters are determined by calibration procedure.

Long term experiments of Martinec verified, that the value of the degree-day factor is between 3.5 and 6.5 [mm•°C^{-1}•day^{-1}]. The degree-day factor (a) is not a constant. It changes according to the changing snow properties and depth during the snowmelt season (Martinec 1960; Martinec-Rango 1968). According to Martinec's results, the DIWA model assumes that the 'a' factor depends on snow-depth, which means that this factor indirectly correlates with the water content of snow. Under an assumption of linear dependency, we can write:

$$a = \begin{cases} a^{\max} - \dfrac{a^{\max} - a^{\min}}{WC_{tv}} \cdot WC & \left(WC \le WC_{tv}\right) \\ a^{\min} & \left(WC > WC_{tv}\right) \end{cases} \qquad (17)$$

where WC [mm] is the water content of snow, a^{min} is the smallest possible value of degree-day factor (recommended value is 3.5 mm•°C^{-1}•day^{-1}), a^{max} is the greatest possible value of degree-day factor (recommended value is 6.5 mm•°C^{-1}•day^{-1}), and WC_{tv} is the upper threshold value of the water content of snow [mm], which is a regional model-parameter determined by calibration procedure.

The maximum storage capacity of the vegetation surface h_{max} [mm] is depending on both, the type and the seasonal density of vegetation. Some basic values of h_{max} are available in the research literature for all different types of vegetation. In order to realize a fine-tuning on this parameter-distribution (keeping the basic relations between the different types of vegetation), in DIWA has been introduced a regional scale factor (C_h [-]). This factor is determined through calibration.

As discussed above, the so called '0' horizon of the soil gives the surface runoff and the percolation as well. The k_o [m/s] (equation 14) saturated hydraulic conductivity of this layer is a spatially distributed parameter under an assumption of linear dependency with the slope of the neighbouring grids defined by the cell link network, having a larger and smaller value in steeper and gently sloping grids in the basin:

$$k_0^{i,j} = k_0^{\min} + \left(SL^{i,j} - SL^{\max}\right) \cdot \frac{k_0^{\max} - k_0^{\min}}{SL^{\max} - SL^{\min}} \qquad (18)$$

where $k_0^{i,j}$ and $SL^{i,j}$ are the saturated hydraulic conductivity and the slope at any grid, $SL^{\min}$ and $SL^{\max}$ are the smallest slope and steepest slope values in all grids, $k_0^{\min}$ and $k_0^{\max}$ parameters represent the smallest and largest saturated hydraulic conductivity in all grids. $k_0^{\min}$ and $k_0^{\max}$ model-parameters are determined by way of calibration.

The $K^{i,j}$ [s^{-1}] storage coefficient (equation 15) is, similarly to the case of k_o, supposed to be proportional to the slope of the neighbouring grids. A grid with larger slope usually has shorter residence time (residence time = $1/K$) and opposite:

$$K^{i,j} = K^{\min} + \left(SL^{i,j} - SL^{\max}\right) \cdot \frac{K^{\max} - K^{\min}}{SL^{\max} - SL^{\min}} \qquad (19)$$

where $K^{\min}$ and $K^{\max}$ are the smallest and largest coefficient values. $K^{\min}$ and $K^{\max}$ model-parameters are determined through calibration procedure.

Since there is no way to identify by measurements the model parameters discussed above, it is necessary to determine them via calibration, using 'hand fitting' procedures, or applying optimization algorithms. The DIWA provides a built-in automatic calibration routine based on an efficient hybrid optimization procedure of adaptive partition-based search (APS) and downhill simplex algorithms (DSA) for calibrating parameters of DIWA model. The elaborated hybrid algorithm basically consists of two phases. Firstly, the APS algorithm offers an adaptive global search (branch-and-bound) on the whole closed set of the feasible parameters (Pintér and Szabó 1985a, 1985b; Pintér et al. 1986) Secondly, the incorporated DSA

(Nelder and Mead 1965) offers a local-search scheme starting from the 'best' parameter vectors, which were detected in the previous phase by the APS. The combined (APS-DSA) algorithm does not require derivative information: the solver operations are based exclusively on the computation of goal function values at each algorithmically selected search point.

In DIWA, the objective of the automatic calibration is to maximize the so called 'Modified Nash-Sutcliffe Coefficient' as a function of a certain combination of the parameters under calibration.

3.3.2 Results of the Calibration and Validation

The DIWA model was calibrated against time series of discharges. The period of 01.05.1991–30.04.1995 and the period of 01/05/1995–30/04/2001 have been chosen for calibrations and for the validation. Calibrated results are listed in Table 1.

Table 1. Parameters and calibrated results for UTB

Parameter		Values		
Symbol and unit	Short name	Bounded types		Constant types
		Lower	Upper	
C_h [-]	Regional scale factor			1.3
$T_{crit,M}$ [°C]	Critical temperature	0.5	1.8	
WC_{tv} [m]	Thresh. value of the wat.cont.of snow			1.12
k_o [m/day]	Sat. hydr. Cond. of '0' soil-horizon	0.01	64.5	
K [day^{-1}]	Storage coefficient of the channel flow (in residence time [s])	45.8 (≈1886)	86.6 (≈998)	

Figures 11 to 13 present a comparison between flows observed and those simulated with DIWA model, using the parameter values in Table 1. In order to evaluate the results, four statistical coefficients were calculated after calibration.

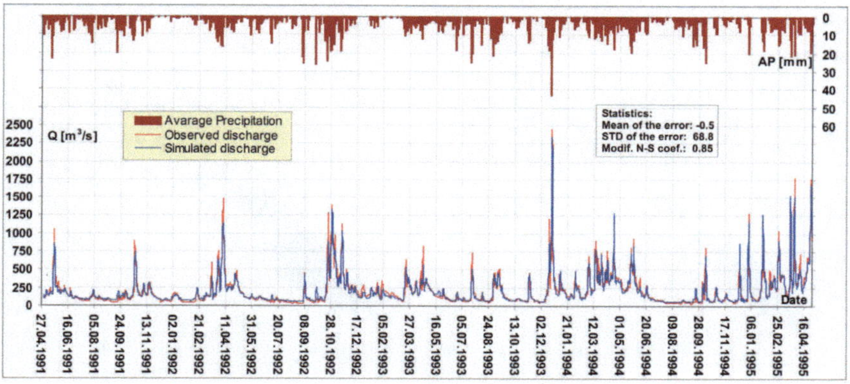

Fig. 11. Graphical comparison between observed and simulated daily hydrograph at Tiszabecs gauge in period of calibration: 01.05.1991 – 30.04.1995

Following calibration, the estimates had to be validated in a different time period using the same parameter set. The most significant events in this period in terms of maximum flood peaks for Tiszabecs gauge occurred on 06.11.1998 and 07.03.2001. Figure 12 shows reconstruction of the discharge history as run by DIWA model.

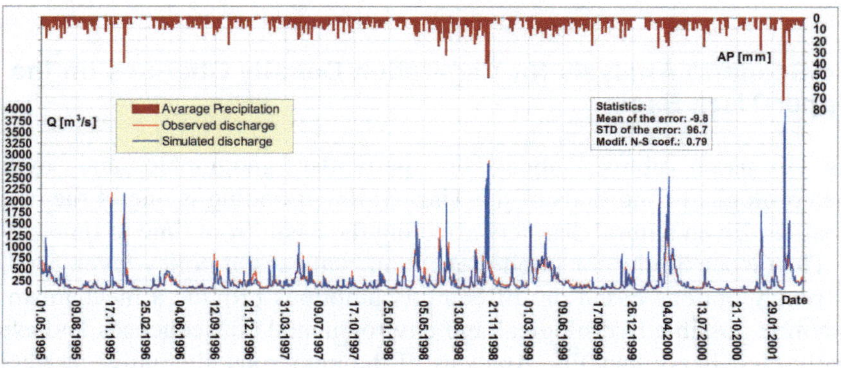

Fig. 12. Graphical comparison between observed and simulated daily hydrograph at Tiszabecs gauge in period of validation 01.05.1995–30.04.2001

Figure 13 shows the scatter plot of simulated versus observed discharge for both periods: calibration and validation, which graphically compares the distribution of the discharge to the 'perfect' model. The straight line represents what our data would look like if the simulation were perfect. The discharge data is represented by the squares plotted along this line.

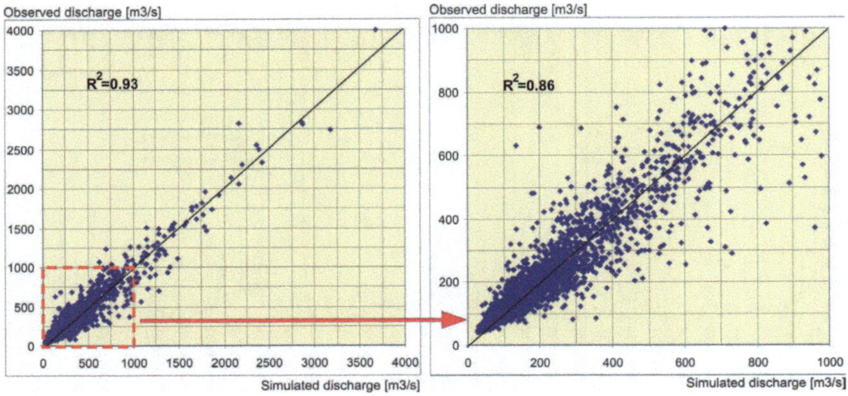

Fig. 13. Scatter plot of simulated versus observed discharge at Tiszabecs gauge for periods, calibration and validation 01.05.1991–30.04.2001

As it can be traced by means of the graphical comparisons and by the calculated statistical parameters as well: application of the DIWA model to the UTB demonstrates that the model successfully simulates the flow regime at the gauge of Tiszabecs, even on the most significant maximum flood peaks events: 06.11.1998, and 07.03.2001. The model is efficient in term of calibration and validation as well.

3.4 Scenario Analysis for Vegetation Density Changes on the Upper-Tisza Basin

As discussed above, under the particular geographic and hydrometeorological conditions of the Tisza-valley, flooding is one of the major weather driven natural disasters affecting the basin from time to time.

The responsibility for flood prevention is always at policy level. Studies of policy options based on different assumptions provide a mechanism to examine possible hydrological and environmental consequences and especially their likely benefits. Analysis of the numerous alternative strategies for effective flood prevention can be realized through scenario analysis using SDSS (Spatial Decision Support System). Following some successful studies using DIWA, the upsurge of interest in floodplain management and planning caused by the floods has increased the demand for user friendly, interactive methods for analysing influences of the use of emergency flood reservoirs and/or land use changes, and managers can use these to evaluate the economic and environmental consequences of alternative management schemes.

The objective of this chapter is to highlight efforts made in incorporating SDSS technology into the policy decision making process in the wake of the early spring flood event of 2001 in UTB, based on scenario studies.

The early spring flood of 2001 was the record flood in UTB history. The maximum flood peak for Tiszabecs gauge occurred on 07.03.2001. The extraordinary flood wave developed from two causes: extremely rising gradient of temperature induced rapid snowmelt and heavy rain.

Figure 14 shows the history of the flood and the average precipitation (melted snow plus rain), and it also demonstrates that DIWA model simulated the flood wave successfully in this critical period.

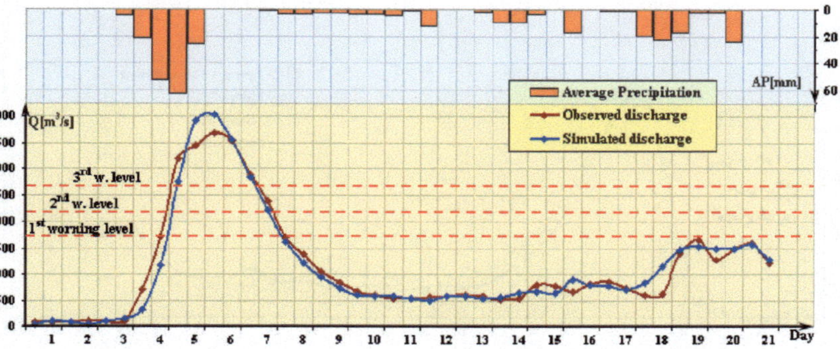

Fig. 14. Graphical comparison between observed and simulated daily hydrograph including the different levels of warning at Tiszabecs gauge in period 01.03.2001-21.03.2001

In that time, following the flood event it was generally held that the real reason for increasing volume of floods during the recent decades was the decreasing crop cover caused by the intensive deforestation actions on the mountainous area of UTB.

In order to test this opinion of the public and a large number of environmentalists as well, two different vegetation density change scenarios were developed and the effects on simulated discharges and water balance were investigated:

- *Scenario I.* Under the assumption that the flood event occurred in summertime in July when the vegetation density is highest. Therefore, the distribution of LAI with reference to July, typical daily temperatures for this month, and the actual rain increased by melted snow was taken into account for this scenario,
- *Scenario II.* In addition to the conditions of the previous scenario, an unrealistic, extremely high density of the vegetation has been supposed. Particularly, instead of the original land cover pattern pertaining to July

we assumed for all grids above 200 metre altitude on UTB the same land cover properties: fully-developed deciduous broadleaf forest. With this assumption we increased the original average LAI value from 5.12 to 7 (in NDVI: from 0.51 to 0.68).

Figure 15 shows the results of the three different simulations:

- original conditions (Simulated-0),
- scenario I (Simulated-I),
- scenario II (Simulated-II).

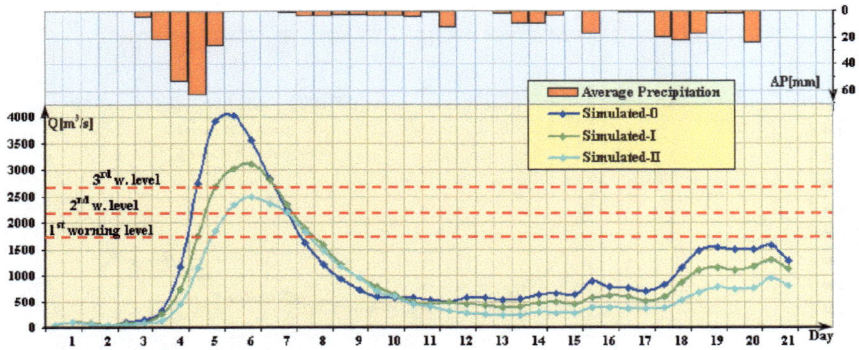

Fig. 15. Simulated hydrographs of the scenario analysis on vegetation density changes at Tiszabecs gauge in period 01.03.2001-21.03.2001

The results demonstrated that because of the extremely high intensity of the rain, there are no 'normal', natural conditions to keep this flood wave below the 3rd warning level. In other words, there is no natural protection against such an extraordinary extreme hydro-meteorological condition.

The responsible policy authorities should explain somehow to the public and to the 'green activists', that the inherent principles behind the forming of catastrophic floods are highly complicated, rather than declaring the existence of only one cause. Although conscious, well-thought-out land use management is fundamentally important, it can not be the only option in the complex strategies required to be effective against flood impact damages.

4. Conclusions

Flooding cannot be completely avoided, but damages from severe events can be reduced if effective flood prevention steps are implemented. The responsibility for flood prevention is always at policy level. In order to reduce damaging influences of floods, or to provide answers to questions like what are the effects of human actions taken on the catchment area (e.g. deforestation, forestation, development of urbanization, etc.) or the changing of the climate on the flood level, duration and frequency, the decision-maker should weigh many different, difficult circumstances and factors. With the aim of analyzing the numerous alternative strategies for effective flood prevention, one of the most useful tools can be a Spatial Decision Support System (SDSS) designed for integrated river basin management and flood control.

The objective of this paper is to highlight efforts made in development of the physically based distributed rainfall-runoff model DIWA (DIistributed WAtershed) as one of the most important parts of an SDSS. The model DIWA is designed to model the following: rainfall-runoff processes, flood events analysis, regional land-use change scenario analysis, climate-change scenario analysis, real-time hydrological forecasting and control. All the fundamental hydrological processes are simulated by the model including interception, snow-accumulation and melting, infiltration, water intake by vegetation and evapotranspiration, vertical and horizontal distribution of soil moisture, unsaturated and saturated flow of water in soil, surface flow, and flow through river channels. Each of these sub-models is presented in more detail.

In addition to the algorithms of the DIWA model, the developed software system provides some useful built-in functions, such as:

- an automatic calibration routine based on an efficient hybrid optimization procedure for adaptive partition-based search, and downhill simplex algorithms for calibrating model parameters,
- modified universal kriging based interpolation method for mapping temperature data,
- special interpolation method for mapping precipitation data based on combination of kriging and an algorithm for automatic recognition of local anisotropy.

In order to demonstrate the efficiency of the DIWA model, a model calibration and validation was presented using the upper part of the Tisza River Basin. The model was calibrated against time series of discharges. The periods of 01.05.1991–30.04.1995 and 01.05.1995–30.04.2001 were chosen for calibrations and for the validation. All the 5 parameters of the distributed DIWA model were calibrated by considering their spatial variations with the altitude/slope of the grids, using the built-in automatic calibration routine. The application of the DIWA model to the River Tisza demonstrates that the model successfully simulates the rainfall-runoff processes.

Two scenario analyses for vegetation density changes on the early spring flood of 2001 of Upper-Tisza basin are also presented. The results demonstrated that because of the extremely high intensity of the rain, there are no 'normal', natural conditions to keep this flood wave below the 3rd order flood warning level. In other words, there is no natural protection against such an extraordinary extreme hydro-meteorological condition.

The evidence from these two scenario-analyses shows that the inherent principles underlying the formation of catastrophic floods are highly complicated and it is not correct to declare the existence of only one cause for development of extreme floods. Although the conscious, well-thought-out land use management is fundamentally important, it can not be the only option in the complex strategies needed to be effective against flood impact damages.

Acknowledgments

This research was funded by the Water Centre of Hungary (Hungarian abbreviation VKK) under a cooperative agreement with the Budapest University of Technology and Economics, Department of Sanitary and Environmental Engineering. For that, the author would like to thank Mr. György Jakus, head of the directorate of water.

Thanks to our editors, Robert Peckham and Győző Jordán for their patience and support, and for their fantastic working ability. It was a pleasure working with you on this volume. I would also like to thank to my friend and colleague Katalin Bódis for the GIS support.

Huge thanks to my wife Zsófia, who, for many months, got to hear, 'But honey, I'm still working on the model and on some studies'. I am grateful to her for being incredibly understanding, supportive, and most of all, patient.

References

Amani A, Lebel T (1997) Lagrangian kriging for the estimation of Sahelian rainfall at small time steps. Journal of Hydrology 192:125-157

Box GEP, Jenkins GM (1970) Time series analysis, forecasting and control. Holden-Day, San Francisco

Burdine NT (1953) Relative permeability calculation from pore size distribution data. Trans. AIME 198:71-78

Burgess TM, Webster R (1980) Optimal interpolation and isarithmic mapping of soil properties. I. The semi-variogram and punctual kriging. J. Soil. Sci. 31:315–331

Cressie NAC (1985) Fitting variogram models by weighted least squares. Journal of the International Association for Mathematical Geology 17:563-586

Dolman AJ, Hall AJ, Kavvas ML, Oki T, Pomeroy JW (eds) (2001) Soil, vegetation, atmosphere transfer schemes and large-sale hydrological models. IAHS Publications No. 270, Wallingford, UK

Van Genuchten MTh, Leij F J, Wu L (eds) (1999) Characterization and measurement of the hydraulic properties of unsaturated porous media. Parts 1 and 2. University of California, Riverside, CA

Kidwell KB (1997) NOAA Global Vegetation Index. User's Guide 1997. U.S. Department of Commerce NOAA National Environmental Satellite data and Information Service

Martinec J (1960) The degree-day factor for snowmelt runoff forecasting. IAHS Publications 51:468-477

Martinec J, Rango A (1986) Parameter values for snowmelt runoff modelling. Journal of Hydrology 84:197-219

Nelder JA, Mead R (1965) A simplex method for function minimization. Computer Journal 7:308-313

Oberlander TM, Muller RA (1987) Essentials of physical geography today. 2nd ed. Random House, New York

Pintér J, Szabó JA (1985) Global optimization algorithms: theory and some applications. In: Prékopa A, Szelezsán J, Straziczky B (eds) Proceedings of the 12th IFIP Conference on Modelling and Optimization, Budapest, 1985. Springer Verlag, Heidelberg

Pintér J, Szabó JA (1985b) Multiextremal (global) optimization algorithms for engineering applications. In: Adey RA (ed.) Proceedings of the Fourth International Conference on Engineering Software (ENGSOFT'85). Springer Verlag, Kensington, pp. 7-17

Pintér J, Szabó JA, Somlyódy L (1986) Multiextremal optimization for calibrating water resources models. Environmental Software 1:98-105

Phillips FRS (1991) Flow and reactions in permeable rocks. Cambridge University Press, Cambridge

Sanders R, Tabuchi S (2000) Decision support system for flood risk analysis for River Thames. Photogrammetric Engineering and Remote Sensing 66

Schloedera CA, Zimmermanb NE, Jacobsc MJ (2001) Comparison of methods for interpolating soil properties using limited data. Soil Science Society of America 65:470-479

Sellers PJ (1987) Canopy reflectance, photosynthesis, and transpiration II. The role of biophysics in the linearity of their interdependence. Remote Sensing of Environment 21:143-183

Szabó JA (1985) A class of global optimization procedures and its application for determining model parameters. M.Sc. Thesis, Eötvös University, Budapest (in Hungarian)

Szabó JA (2005) Estimation of high resolution meteorological fields based on geostatistical approaches. European Geosciences Union (EGU) General Assembly, 24-29 April 2005, Vienna, Austria

Szabo JA, Kalaš M (2005) Influence of uncertainty of interpolated meteorological inputs for distributed hydrological models. European Geosciences Union (EGU) General Assembly, 24-29 April 2005, Vienna, Austria

Takashi I, Makoto N, Hisashi S (2001) Large-Scale mapping of leaf area index using remote sensing data. Vegetation-atmosphere transfer schemes and large-sale hydrological models. IAHS Publications No. 270, Wallingford, UK, pp. 287-290

Todini E (2001) Influence of parameter estimation uncertainty in Kriging. Part 1 Theoretical development. Hydrol. Earth Systems Sci. 5:215-232

Varga-Haszonits Z (1969) Determination of the water content and of the evaporation of bare soil. Időjárás 73:328-334

Wheather HS, Abelink R, Perry R. (1986) Parameter identification in models of unsaturated solute transport. Proceedings of the International Conference on Water Quality Modelling in the Inland Natural Environment, Bournemouth

Yao H, Hashino M (2001) A completely-formed distributed rainfall-runoff model for the catchment scale. IAHS Publications No. 270, Wallingford, UK

Potential Flood Hazard and Risk Mapping at Pan-European Scale

Ad De Roo, Jose Barredo, Carlo Lavalle, Katalin Bodis and Rado Bonk

1. Introduction

In the last decade Europe has experienced a number of unusually long-lasting rainfall events that produced severe floods, e.g. in the Netherlands, Belgium, France and Germany in 1993 and 1995, the Czech Republic, Poland and Germany in 1997, in North Italy (in 1994 and 2000), in the UK (e.g. in 1998 and 2000), Tisza (in 2000 and 2001), in the Elbe and Danube in 2002, and in 2005 in Romania and the northern Alpine region. There seems to be a trend of increasing flood hazard. This may support projections of future climate indicating that further increases in severe floods in North and Northwest Europe are likely (e.g. IPCC 2001). The Joint Research Centre of the European Commission (JRC) provides EU policy support on flood issues, especially focused on cross-border river basins. In addition to the work on the development of an early warning system at the pan-European scale (EFAS, the European Flood Alert System), JRC carries out flood mitigation and forecasting case studies in the Elbe and Danube catchments, and flash floods, climate change effects and flood risk mapping at European scale [1].

There is an increasing need for accurate information on the potential hazard and risk of flooding. The general public needs to be informed if their property is prone to flooding, and consequently which actions to take. The insurance industry needs to know which premium to charge their customers that want to insure themselves for flood risk. Governmental bodies need to know the flood hazard to decide to give out new building permits or take adequate measures to reduce the damage and prepare evacuation schedules for possible flood disasters.

In the USA, the Federal Emergency Management Agency (FEMA) is establishing national flood hazard mapping with the Federal Insurance and Mitigation Administration's Hazard Mapping Division through their National Flood Insurance Program [2]. At present, only scattered information on flood hazard and flood risk is available within Europe. Similar to FEMA, the Environment Agency of England and Wales is producing flood maps available on-line to the general public [3]. For the Rhine, an atlas with flood hazard maps has been produced by a team of hydrologists, economists, and GIS experts at a scale of 1:100,000 and published in 2001 (ICPR 2001), available online as pdf files from [4] and [5]. As for the Meuse river, the Dutch State Water Authorities published maps of flood extent in 1995 (RWS 1995).

The European Commission is currently preparing a Directive on Floods [6] in which European Member States are called upon to provide information about the areas at risk of flooding. The objective of the Floods Directive will be to create obligations for Member States to manage risks of floods to people, property and environment by concerted, coordinated action at river basin level and in coastal zones in order to reduce the risks of floods to people, property and environment. It would be gradually developed with a focus on particular regional conditions in order to ensure that local and regional conditions are taken into account in:

- the analysis of present and future flood risk through flood mapping,
- information on flood risk and its effects which should be made available to citizens, involved parties and relevant authorities,
- the elaboration and implementation of flood risk management plans.

In addition to the Flood Directive, envisaged future regional policies of the Commission more and more take natural risks into account. The INTERREG program of DG Regional Policy has covered and funded many projects related to floods. The large INTERREG III initiative ESPON (European Spatial Planning Observation Network [7]) aims to diagnose the principal territorial trends at EU scale such as land use changes, population dynamics, etc, and floods are a major factor in there.

In support to this process, the Commission's Joint Research Centre is developing flood hazard and flood risk maps at the European scale, to provide a harmonized overview of flood prone areas.

2. Flood Hazard Mapping Using DEM

Hydrological processes such as flooding and soil erosion are determined by topographic parameters, and therefore digital elevation data play a crucial role in modelling. An overview of hydrological applications of DEMs can be found in Gurnell and Montgomery (1999) and De Roo (1998). Wilson (2004) and Wilson and Atkinson (2003) showed the effect of data and data uncertainty on predictions of flood inundation and thus flood hazard, using the LISFLOOD-FP model (Bates and De Roo, 2000; De Roo et al. 2000). The accuracy of the DEM is of utmost importance for the accuracy of the final flood hazard map. Inaccuracies in the final predicted flood extent arise from unknown surface roughness values, or linear features on floodplains such as roads, ditches etc. that are not well incorporated in the DEM.

With the aim to provide an overview of potential areas susceptible to flood risk, the JRC attempts to produce a pan-European flood hazard and risk map. Flood risk is defined as a product of hazard, exposure and vulnerability (Kron 2002; Barredo et al. 2005):

- hazard: the threatening natural event including its probability/magnitude of occurrence,
- exposure: the values/humans that are present at the location involved,
- vulnerability: the lack of resistance to damaging/destructive forces.

Digital Elevation Models are the essential element to produce a flood hazard map. The vertical accuracy is of crucial importance to achieve accurate results, as demonstrated by Wilson (2004) and numerous others.

At JRC, a large GIS database has been established containing the base products to calculate hazard, exposure and vulnerability. The major difficulty lies in the need for spatially and thematically homogenous data for the 27 countries that cover the study area. So far, the database contains several layers with information on Gross Domestic Product (GDP), population density, land use, etc. The sources of information are EUROSTAT [8], CORINE 1990 [9], CORINE 2000 [10] and GISCO (GISCO 2001). It should be made clear that the study presented here focuses on a regional and (trans)national scale using relatively low DEM resolutions, which is a key limitation when working at these scales.

3. Methodology

To estimate flood hazard at a European scale, the following stepwise approach is followed:

- Floodhazard at 1km grid scale: overview potential flood hazard using a 1km Digital Elevation Model (source: GTOPO30 (Global 30 Arc Second Elevation Data) from USGS [11]), a global digital elevation model with a grid spacing of 30 arc seconds. GTOPO30 was derived from several raster and vector sources of topographic information. and the CIS (European Catchment Information System) river network at European scale (Hiederer and De Roo 2003), using waterlevel assumptions based on upstream catchment area (completed),

- Floodhazard at 250m grid scale: overview potential flood hazard using a 250m Digital Elevation Model (resampled DEM mosaic based on commercial national products compiled for the CCM project) and the CCM [12] (Catchment Characterisation and Modelling project) river network at European scale (Vogt 2003), using waterlevel assumptions based on upstream catchment area (ongoing),

- Floodhazard at 250m grid scale: overview potential flood hazard using a 250m Digital Elevation Model and the CCM river network (the resulting river network of the above mentioned Catchment Characterisation and Modelling project) at European scale (Vogt 2003), using waterlevel estimations derived with the flood simulation model LISFLOOD (planned for 2006/2007),

- Floodhazard at 100m grid scale: overview potential flood hazard using a 100m Digital Elevation Model derived from the SRTM (Shuttle Radar Topography Mission) [13] data, and the CCM2 river network (a methodologically improved result of Catchment Characterisation and Modelling project using processed SRTM data, Image2000 imageries and classified land use maps, Vogt, 2006) at European scale using waterlevel estimations derived with the flood simulation model LISFLOOD (planned for 2007/2008); alternatively, the LISFLOOD-FP model (Bates and De Roo 2000) will be used to estimate flood hazard.

This study shows preliminary results of the work done on the 1km and 250m scale flood hazard mapping, using the first two datasets. It is expected that each step will produce a more accurate product, by increasing the grid resolution and the quality of the underlying DEM, and using water levels derived from a hydrological simulation model instead of an assumption of the water level.

3.1 Step 1: Defining the Elevation Difference of Each Pixel with the River

The flood hazard used in this study is obtained using a 1km and 250m grid digital elevation model and the 1km (CIS – Hiederer and De Roo 2003) and 250m (CCM; Vogt 2003) European flow networks developed at JRC. An algorithm has been developed to find the elevation difference between a specific grid-cell and its closest neighbouring grid-cell containing a river, while respecting the catchment tree-structure: in this way a grid cell can never be linked to a river in another (sub)-catchment.

Using the CRaster software [14] for dynamic modelling in a GIS environment the rivers, taken from the CIS and/or CCM river networks, are defined as 'pits' – outlets of individual small local catchments. For each of those pits, the local catchment draining into this pixel is defined using the 'catchment' function (Figure 1).

Next, the elevation of the river pixel (the pit) is given to this entire sub-basin: the river pixel is assigned the elevation from the DEM. The other pixels are assigned a zero elevation. The function 'areatotal' is used to assign the sum of all elevations (= the elevation of the river pixel) to the entire local sub-basin.

Finally, the difference between the original DEM and the DEM with the local river altitude is estimated (Figure 2).

3.2 Step 2: Defining the Critical Water Levels

As mentioned above, the current EU flood hazard map is an intermediate product towards up-graded versions based on the use of higher resolution DEMs (250m and 100m grid size) and estimation of water levels from the LISFLOOD model (De Roo et al. 2000; De Roo et al. 2003; Van der Knijff et al. 2005). Therefore, considering that at the current scale of analysis (1km grid) there are no reliable estimates of extreme waterlevels and information on flood protection measures covering the whole Europe, two simple and direct approaches have been implemented.

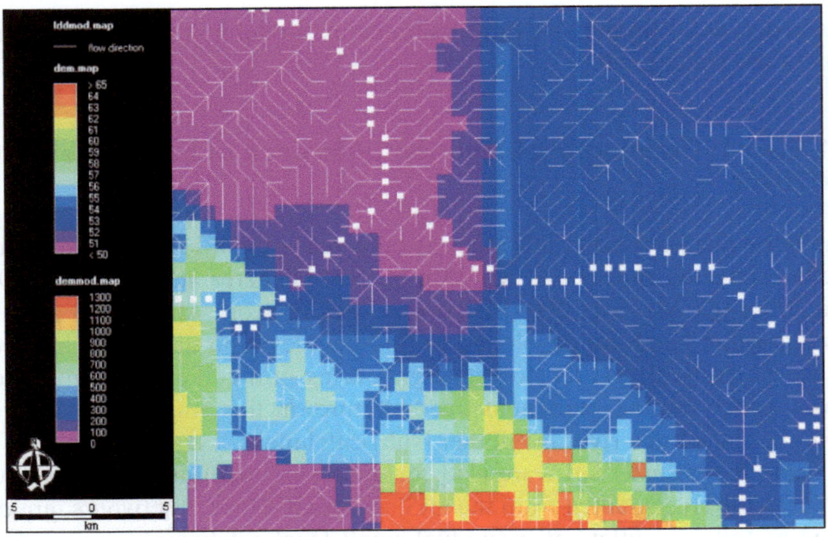

Fig. 1. Digital elevation map with the local drainage direction overlaid, including the pits (white boxes) assigned to each river pixel

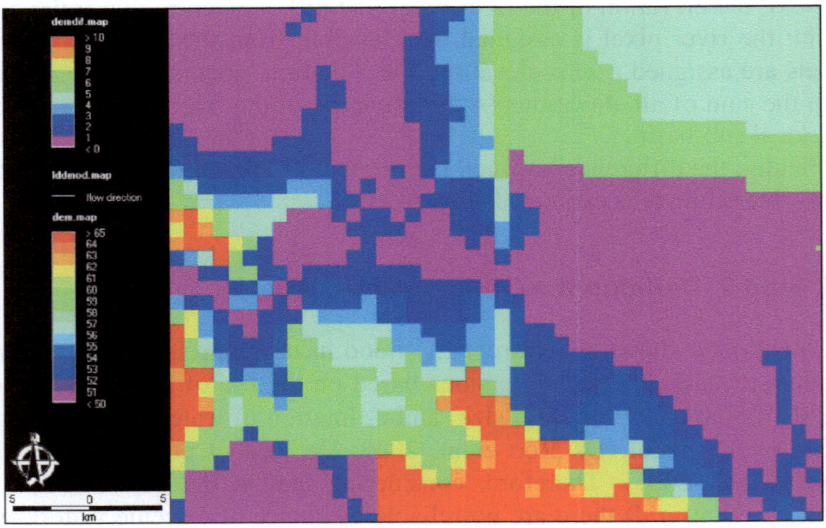

Fig. 2. Map showing the elevation difference of each pixel with its nearest river pixel

The source of the applied 1 km resolution DEM was the well documented GTOPO30 (Global 30 Arc Second Elevation Data) [15] from the U.S. Geological Survey. GTOPO30 is a global digital elevation model with a grid spacing of 30 arc seconds. GTOPO30 was derived from several raster and vector sources of topographic information [16]. The original projection and coordinate system of the downloaded dataset was geographic based on WGS84 datum and spheroid. In order to obtain a grid layer fitting into the analysis environment of our other data layers the European subset of GTOPO30 was re-projected into a metre-based Lambert Azimuthal Equal-Area projection and re-sampled into the 1000 m that is approaching to the original 30 Arc Second on the concerning latitudes. This DEM provides a useful basis for modelling different surface processes but besides the moderately coarse 1 km resolution, even visible systematic errors and irregular noises (Figure 3a) disturb the results of modelling of potential inundation.

The sources of the restrictedly available 250 m resolution DEM were different providers (e.g., Nigel Press Associates UK, Geosys F, national mapping agencies). The acquired Digital Elevation Models (DEMs) with grid-cell dimensions of 100 or 250 metres were merged in order to cover the entire European continent. For areas where DEMs at this resolution were not available (e.g. Iceland, Russia), data from the HYDRO1K global digital elevation dataset [9] with 1000 m grid-cell resolution were used. The DEMs of different locations and in variant resolution and reference systems were sub-sampled, projected and merged. The data quality was only visually checked and in spite of cliff effects being sometimes observed (Figure 3b) along the boundaries between adjacent DEMs no filtering or edge matching techniques have been done since the algorithm used in the original purpose (river extraction) of this compiled DEM did not required (Vogt 2003) that preparation. At the same time the residual systematic errors can be misleading in the calculation of the potentially flooded areas.

The future flood mapping activities are based on the finest (100 m) available DEM. The source of this global dataset is the SRTM-3 (Shuttle Radar Topography Mission) product [17]. The SRTM-3 data are sampled at three arc-seconds, that is corresponding to approximately 100 m. The original SRTM product needs a serious pre-processing (filling voids, complement the polar altitudes, projection, re-sampling, delineating coastlines, etc.). The Jet Propulsion Laboratory has released the full SRTM accuracy report [18].

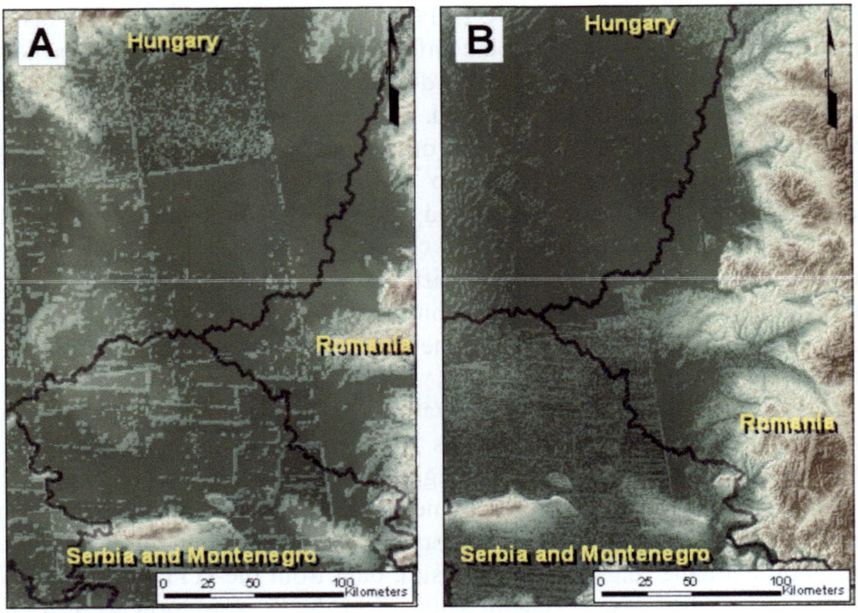

Fig. 3. A. Systematic and irregular error in the 1 km DEM. B. Visible map sheets in the 250 m DEM

Approach A:
As a first approach, extreme waterlevels have been estimated using a simple function based on upstream catchment size:

Critical waterlevel.map=min(log10(upstream area.map)*3,10)

The waterlevel in the rivers above which flooding is assumed, is estimated as the 10log of the upstream catchment area of a particular pixel (in km^2), multiplied by 3. This waterlevel is assumed not to be larger than 10 m. This formula is based on trial and error experiments analyzing several cross section data available for larger European rivers.

Approach B:
As an alternative, an even simpler approach has been chosen solely classifying the altitude differences. The following simple lookup table (Table 1) is used for the critical water levels and associated flood hazards:

Table 1 Flood Hazard Classes

Elevation Difference with River (m)	Flood Hazard Class
River itself	6
< 0.0 m	5
0.0 – 3.0 m	4
3.0 – 6.0 m	3
> 6.0 m	2
Coastal areas (elevation < 3m above sealevel)	7
Lakes	1

Maps derived from both methods have been validated against actual maps of floods which occurred in Europe in the last years. Examples of this validation are presented hereinafter.

4. Potential Flood Hazard Maps of Europe

Using the methodology described above, a pan-European flood hazard map for Europe has been produced at 1km grid scale (Figure 4). A zoomed-in part of the map, showing parts of The Netherlands, Belgium and Germany and the Rhine-Meuse delta, is shown in Figure 5.

Alternatively, using the 250m DEM, a pilot flood hazard map has been produced for the Rhine-Meuse. Figure 6 shows the same area as in Figure 5, but now based on the 250m DEM.

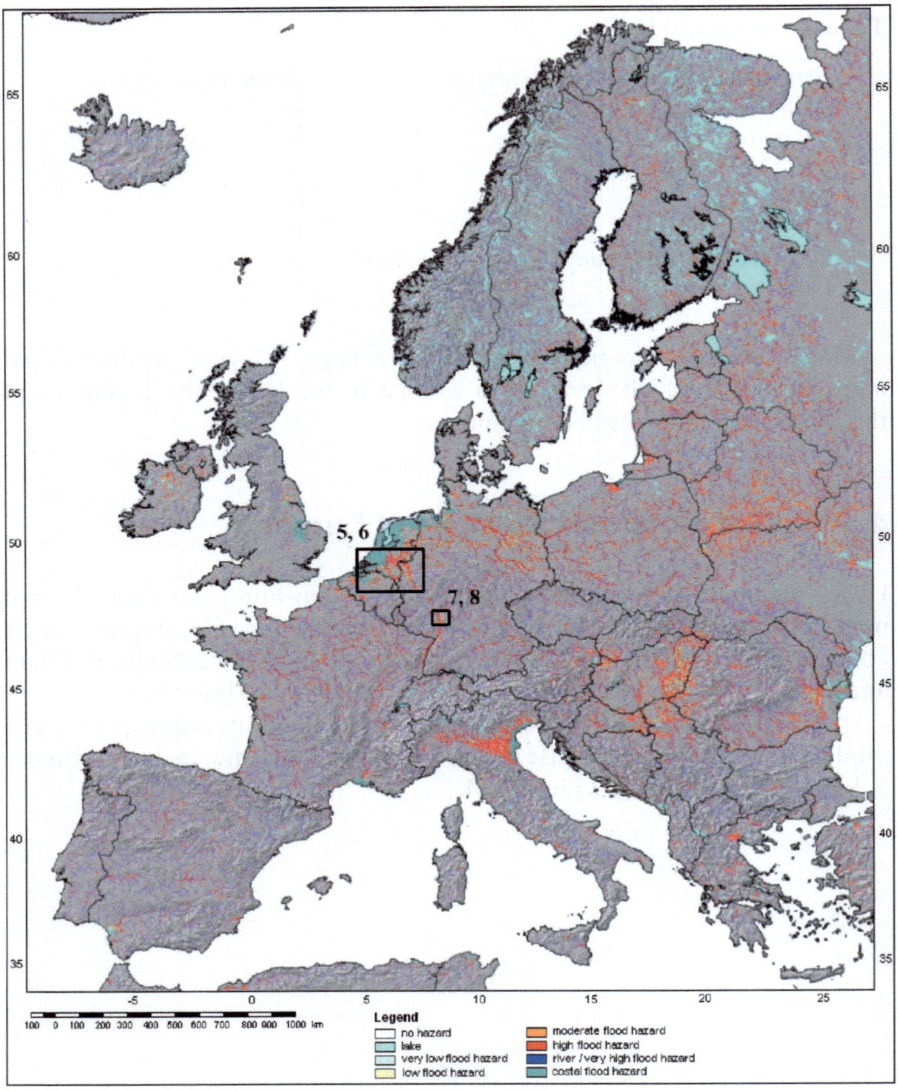

Fig. 4. Potential Flood hazard map of Europe at 1km grid scale. Potential flood risk is estimated using 1 km Digital Elevation Model and the European Flow Network developed in CIS. Information on flood defence structures (dykes, etc.) is not included. The enclosed boxes show the areas discussed below

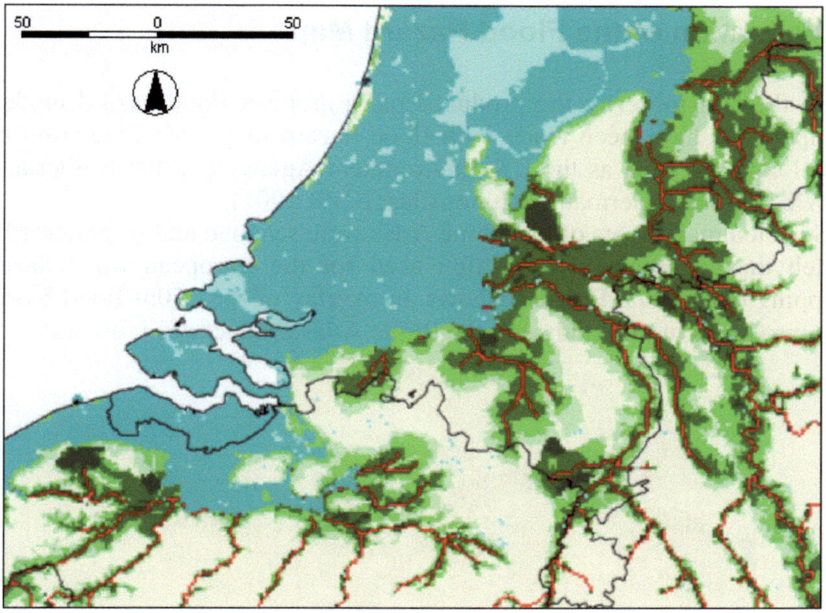

Fig 5. Potential flood hazard map of the Rhine/Meuse delta at 1km grid scale

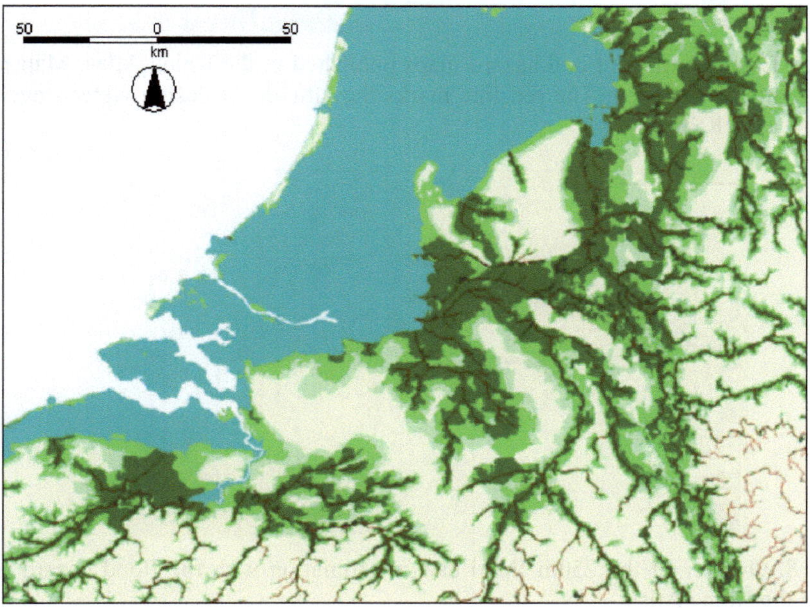

Fig. 6. Potential flood hazard map of the Rhine/Meuse delta at 250m grid scale

5. Validation of the Flood Hazard Map

With the aim to verify the produced pan-European flood hazard models, comparisons have been made with flood hazard maps available from national sources, such as the UK Environment Agency [3], the Netherlands (RWS 1995) and Germany (Rhine Atlas; ICPR 2001).

Selected map sheets of the Rhine Atlas were scanned and re-projected to match the geographical projection used for the European flood hazard mapping (Figure 7). Figure 8 shows an overlay of the 250m flood hazard map produced in this study and the Rhine Atlas product of Figure 6.

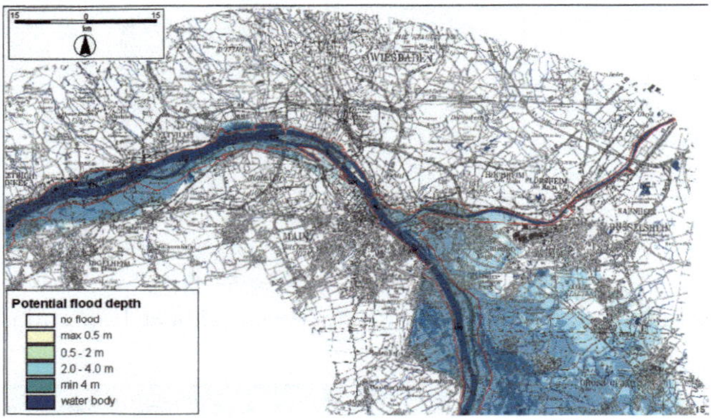

Fig.7. Example of the flood hazard maps published in the Rhine Atlas. Mapsheet 15, Mainz-Wiesbaden. The red line marks the altitude of design water stage for 100 years (IKSR 2001)

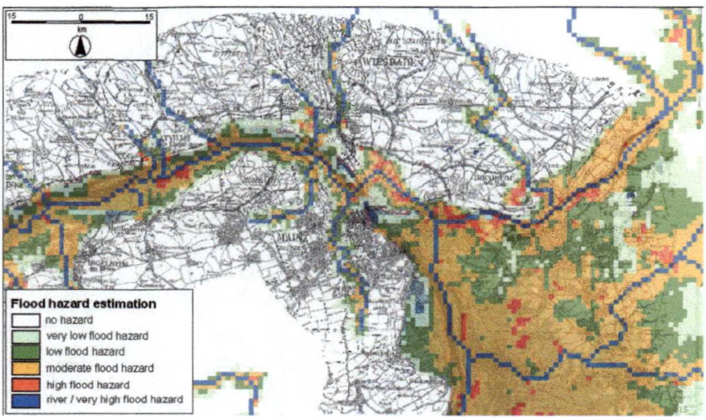

Fig. 8. Overlay of the 250m flood hazard estimation with the flood hazard map from the Rhine Atlas

From the scanned maps like the one shown in Figure 7, the flood hazard extent has been extracted by digitizing flood depths classes, and it was compared to the results obtained in this study. Tables 2 and 3 show a summary of this comparison, achieved from the analysis of three map sheets, including the one shown in Figure 7, and two others from the same atlas.

Table 2. Comparison of 250m flood hazard mapping results (based on the 250 m DEM) with three mapsheets in the Rhine Atlas

	Missed	Moderate flood hazard in both	Correct Flood Hazard
MAP15	0.043	0.845	0.597
MAP09	0.010	0.732	0.445
MAP06	0.124	0.627	0.460

'Missed' = areas indicated as flood hazard areas in the Rhine Atlas, but not detected by our methodology; 'moderate flood hazard in both' = both the Rhine Atlas and the flood hazard map of the authors indicate that an area is susceptible to low, moderate or a high flood hazard.; 'correct flood hazard' = areas with low flood hazard in the Rhine Atlas and in our map + areas with moderate and high flood hazard in the Rhine Atlas and in our map.

Table 2 shows that for mapsheet No. 15, 4.3% of the flood hazard area indicated in the Rhine Atlas was not 'picked-up' using the method of 250 m flood hazard mapping in this study. For the two other mapsheets (No.09 Strassbourg and No. 06 Breisach), missed flood hazard areas were 1% and 12%. Around 73% of the study area corresponds and is classified in the Rhine Atlas and in the 250m flood hazard map with moderate or high flood hazard. However, if a more detailed comparison is done with the individual flood hazard classes, on average around 50% corresponds to the flood hazard areas indicated in the Rhine Atlas, as shown in Table 2 in the column 'correct flood hazard'.

Table 3. Comparison of 250m flood hazard mapping results with mapsheets in the Rhine Atlas (detailed)

	Missed	River	Low flood hazard	High flood hazard
MAP15	0.043	0.062	0.092	0.225
			0.073	0.505
MAP09	0.010	0.006	0.124	0.534
			0.004	0.321
MAP06	0.124	0.022	0.243	0.368
			0.026	0.214

Table 3 shows these results in more detail. In mapsheet No. 15 for example, 50.5% of the area is correctly predicted as a high flood hazard area, and 9.2% of the area is indicated as low flood hazard area. 7.3% (0.073) of the flood hazard area in the Rhine Atlas is classified low flood hazard area in our model, whereas it is indicated a high flood hazard area in the Rhine Atlas. Vice versa, 22.5% of the area is indicated as high flood hazard area in our methodology, whereas it is indicated as a low flood hazard area in the Rhine Atlas. In conclusion, the methodology described here seems to overestimate the high flood hazard areas. This can be corrected by applying calibration. It will be investigated if the SRTM based 100 m DEM yields better results. However, it is known that the SRTM dataset has many types of uncertainties (Rodriguez at al. 2005) that need to be corrected or will influence the accuracy of the resulting flood hazard map. Taking into account the European continental scale dimension of the analysis however, the initial results achieved here are fairly promising.

6. From Regional Flood Hazard to Regional Flood Risk

Using the flood hazard map derived above using the 1km DEM, a flood risk map of Europe has been produced for the EU Member States plus Romania and Bulgaria.

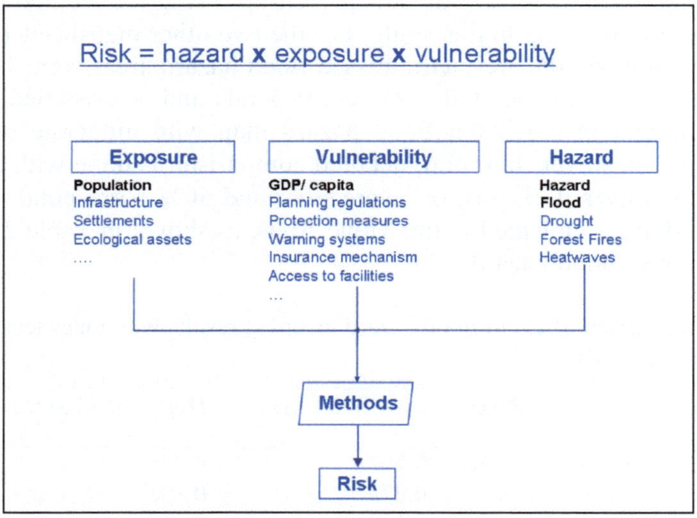

Fig. 9. Calculating flood risk from flood hazard

Figure 9 shows the approach followed for the definition of the flood risk index. Risk is calculated by multiplying flood hazard with exposure and vulnerability. The exposure factor (Figure 10) is based on population density and the potential flood damage related to the local land use (Van der Sande 2001). For each CORINE land use class a function estimating the flood damage relative to the water depth is used to assess the monetary costs.

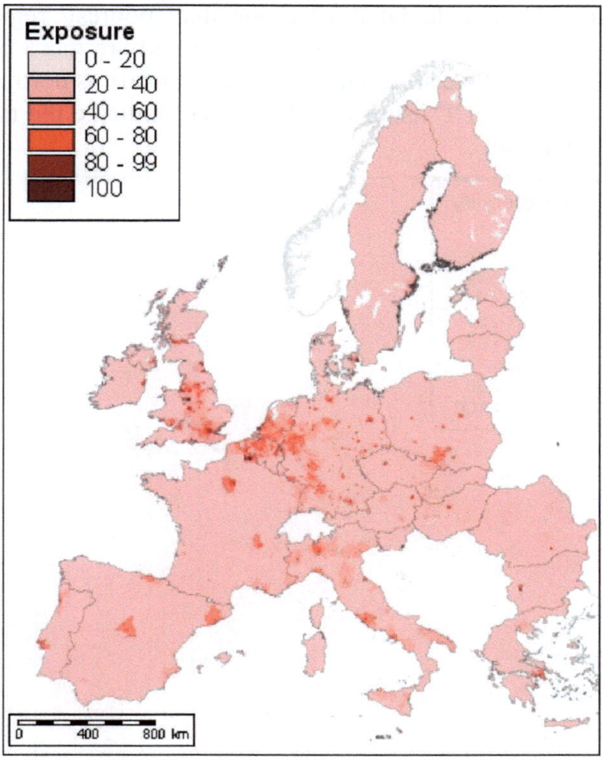

Fig. 10. Flood exposure factor (based on population density)

Several assumptions have been made while creating the flood exposure map:

- Damage is related here only to water depth. However there are a number of other factors that contribute to increased damage, e.g. flow velocity, duration, sediment amount, etc.,
- For the price level the level of 1995 for The Netherlands was used to calculate the maximum financial damage per land use class (costs for reconstruction),

- Cars are not taken into account in the damage assessment, only fixed buildings and infrastructure,
- The average damage per land use class in CORINE varies considerably. Moreover, important differences do exist between countries with a lower and higher GDP.

GDP per capita is so far the only proxy data for vulnerability assessment available for the whole EU plus Bulgaria and Romania. The vulnerability, shown in Figure 11, is calculated assuming that 'poorest' regions – with the lowest GDP per capita - are worst prepared and are more 'vulnerable' to the effects of natural hazards. The insurance industry is normally using a linear function of the GDP itself, and thus it will indicate the richest areas as the most vulnerable.

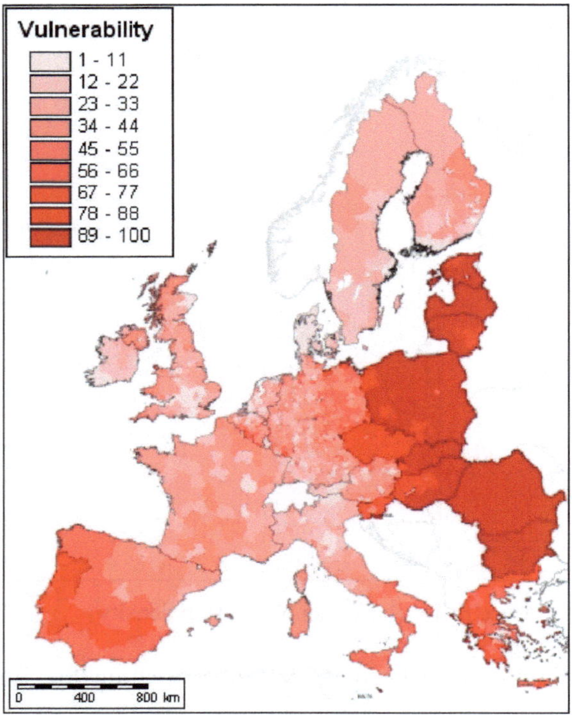

Fig. 11. Flood vulnerability (based on an inversely scaled GDP)

To avoid 'artificial' weighting assignments among the input factors, a non-compensatory approach is used for the production of the standardised index map for flood risk. Initially, flood hazard, flood exposure, and flood vulnerability have been standardised to a range of 1 to 100 – 1 being the lower risk contribution and 100 the maximum. For GDP per capita and population density a linear standardisation with saturation is used. A saturation of 5% is used in both layers in order to avoid the effect of extreme values (low and high). The flood hazard has been standardised with a linear function considering its more homogenous histogram.

The integration of the standardised factors is done by simple multiplication of the corresponding input layers. In addition a risk-no risk mask layer has been included in order to define risk areas in only those pixels in which hazard > 0 is reported.

7. A Flood Risk Map of Europe

Multiplication of the flood hazard, flood exposure and flood vulnerability maps resulted in the flood risk map of Europe (Figure 12). The results are aggregated for NUTS3 level.

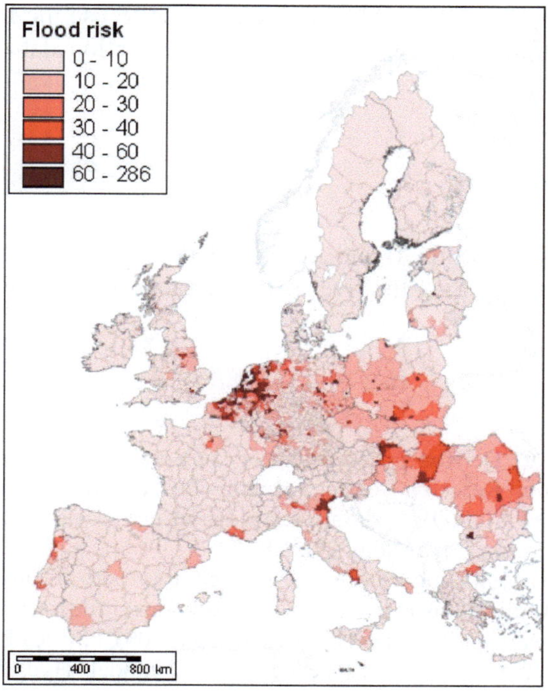

Fig. 12. Potential flood risk map of Europe: higher values indicate a higher flood risk, as caused by either high flood hazard, a high flood exposure, or a high vulnerability to floods

8. Conclusions

This paper summarizes the first results obtained for pan-European flood risk mapping. Initial results showed that 75% of the area indicated as high or moderate flood hazard area using the methodology described here matched with the Rhine Atlas flood hazard area. Taken into account the European continental scale dimension of the analysis however, the initial results achieved here are fairly promising.

Using the flood hazard map, the flood exposure map (derived using population density map, and flood vulnerability map (derived using GDP data) resulted in a flood risk map of Europe, aggregated at NUTS3 level.

The methodology will be calibrated using existing flood hazard maps, such as the ones of the Rhine atlas. Furthermore, the use of higher quality DEMs is envisaged to produce more accurate results. More accurate water-level assumptions will be used in future studies obtained from the pan-European flood simulation work done with the European Flood Alert System (EFAS) and LISFLOOD. Also, it is envisaged to use the LISFLOOD-FP flood inundation model with SRTM based elevation data to estimate flood hazard.

Acknowledgements

R. Hiederer of JRC (CIS data) and J. Vogt of JRC (CCM data) are greatly acknowledged for providing the river network data used for this study.

References

Barredo J, Lavalle C, De Roo A (2005) European flood risk mapping. European Commission Joint Research Centre. Special Publication S.P.I.05.151.EN

Bates P, De Roo APJ (2000) A simple raster-based model for flood inundation simulation. Journal of Hydrology 236:54-77

De Roo APJ (1998) Modelling runoff and sediment transport in catchments using GIS. In: GIS Applications in hydrology. Hydrological Processes 12:905-922

De Roo APJ, Wesseling CG, Van Deursen WPA (2000) Physically-based river basin modelling within a GIS: The LISFLOOD model. Hydrological Processes 14:1981-1992

De Roo A, Bates P, Van Der Knijff J, Schmuck G (2000) A simple floodplain inundation model to assist in floodplain management. In: Maione, Majone Lehto and Monti (eds) New trends in water and environmental engineering for safety and life. Balkema Publishers, Rotterdam

De Roo A, Gouweleeuw B, Thielen J, Bates P, Hollingsworth A (2003) Development of a European flood forecasting system. International Journal of River Basin Management 1:49-59

GISCO (2001) The GISCO Database manual. Eurostat, GISCO Project, Rue Alcide Gasperi, Batiment Bech D3/704 L-2920 Luxembourg, edition November, 2001. http://eusoils.jrc.it/gisco_dbm/dbm

Gurnell A, Montgomery D (1999) Hydrological applications of GIS. John Wiley and Sons, New York

Hiederer R, De Roo A (2003) A European flow network and catchment data set. Report of the European Commission, Joint Research Centre, EUR 20703 EN.

ICPR (2001) Atlas der Uberschwemmungsgefährdung und möglichen Schäden bei Extremhochwasser am Rhein. International Commission for the Protection of the Rhine River. ISBN 3-935324-43-X (in German)

IPPC (2001) Third Assessment Report – Climate Change. Intergovernmental Panel on Climate Change (IPCC). http://www.ipcc.ch/

Kron W (2002) Flood risk = hazard x exposure x vulnerability. In Wu et al. (eds) Flood Defence, Science Press, New York.

Rodriguez E, Morris CS, Belz JE, Chapin EC, Martin JM, Daffer W, Hensley S (2005) An assessment of the SRTM topographic products. Technical Report JPL D-31639, Jet Propulsion Laboratory, Pasadena, California

RWS (1995) De Maas Hoogwater januari/februari 1995. Ministerie van Verkeer en Waterstaat Directoraat-Generaal Rijkswaterstaat Directie Limburg, Maastricht (in Dutch)

Van der Knijff J, De Roo A, Franchello G, Thielen J (2005) LISFLOOD user manual

Van der Sande C (2001) River flood damage assessment using Ikonos imagery. DG-JRC, Ispra, S.P.I. 01.147

Vogt JV et al. (2003) CCM River and Catchment Database for Europe, Version 1.0. EC-JRC, EUR 20756 EN

Vogt JV et al. (2006) Development of a pan-European river and catchment database. In: R. Peckham and G. Jordan (eds), Digital elevation modelling. Development and applications in a policy support environment. Joint Research Centre, European Commission, Ispra. (in press)

Wilson MD (2004) Evaluating the effect of data and data uncertainty on predictions of flood inundation. PhD thesis, University of Southampton

Wilson MD, Atkinson PM (2003) Prediction uncertainty in elevation and its effect on flood inundation modeling. Proceedings of the 7th International Conference on GeoComputation, University of Southampton, United Kingdom, 8-10 September 2003. http://www.geocomputation.org/2003/

Referenced Web Sites

[1] http://natural-hazards.jrc.it/

[2] http://www.fema.gov/nfip/

[3] http://www.environment-agency.gov.uk/subjects/flood/?lang=_e
[4] http://www.iksr.org/
[5] http://www.iksr.de/rheinatlas/atlas/flood_1.pdf
[6] http://www.europa.eu.int/comm/environment/water/flood_risk/consult.htm
[7] http://www.espon.lu
[8] http://europa.eu.int/comm/eurostat/
[9] http://dataservice.eea.eu.int/dataservice/metadetails.asp?id=820
[10] http://dataservice.eea.eu.int/dataservice/metadetails.asp?id=822
[11] http://edcdaac.usgs.gov/gtopo30/hydro/
[12] http://agrienv.jrc.it/activities/catchments/
[13] http://www2.jpl.nasa.gov/srtm/mission.htm
[14] http://www.pcraster.nl/
[15] http://edc.usgs.gov/products/elevation/gtopo30/README.html
[16] http://edc.usgs.gov/products/elevation/gtopo30/README.html#h17
[17] http://www2.jpl.nasa.gov/srtm/
[18] http://www2.jpl.nasa.gov/srtm/SRTM_D31639.pdf

High-Resolution DEM for Design of Flood Emergency Reservoirs

Katalin Bódis

1. Introduction

The EU Water Framework Directive – integrated river basin management for Europe – puts great emphasis upon conservation considerations, the maintenance and development of water ecosystems, terrestrial ecosystems dependent on water and wetlands, focusing on the increasing demand from citizens and environmental organisations for sustainable water management [1]. The Hungarian New Vásárhelyi Plan that called to life by the Hungarian Government in order to reduce and overcome the danger posed by the flooding of the Tisza River gives a feasible opportunity for flood protection and nature conservation to assert EU directives simultaneously in the given region [2]. According thereto, in the frame of the New Vásárhelyi Plan several emergency reservoirs would be built along the Hungarian Tisza section to improve flood safety in the region [3]. In relation to the location and the possible capacity of those reservoirs the emerging problem is whether the calculations in the plan are accurate and what effective tool could demonstrate corrected results. In the preliminary phases of the decision making procedure the sophisticated, time consuming and expensive approaches (such as high resolution surface models derived from stereo-imagery or laser scanning) are usually not available, but a proper basis for the essential calculations is still required. Results based on the optional surrogates, but possibly not appropriate technology could be misleading. In our case an obvious mistake of capacity calculation in published plans of flood emergency reservoirs was discovered and a simple, but useful GIS technique was presented as a possible tool.

The continuous flood control activities go back to the Middle Ages in Hungary. The first embankments were built for protection of fortresses and settlements, but the mediaeval water resources development was focusing on mainly land drainage and fishery in the 12th - 15th centuries. The next two centuries (the Turkish occupation and the following period) had very negative impacts on the water ecosystems: since the settlements depopulated and disappeared most of the existing waterworks were destroyed, the anthropogenic deforestation was very significant and large area became waterlogged (Gruber et al. 1996). After slow and usually local improvements the large river regulations that had effect on the present river system started at the middle of the 19th century. Figure 1 shows the area under permanent and regular (3-4 times/year) inundation along the Tisza River before the regulations.

The regulation efforts have changed the entire hydrological system of the Great Hungarian Plain. The length of the Tisza used to be 1419km. After the changes the new length of the river was 966km (68% of the original), with 589km of cut-off oxbows and 136km of new riverbed. (As a comparison: the total area of the protected lands in the Tisza Valley is larger than in the Netherlands and almost 3 times larger than in the Po Valley.) The length of the flood-control works comprises 2940km (out of 4220km of all Hungarian embankments) which forms one of the largest flood protection systems in Europe; larger than that of the Netherlands' 1500 km, the Po River's 1400km or the Loire Valley's 480km (Süli-Zakar 2001). The embankments resulted about 3 metres accretion of flood level in the regularized Tisza Valley, but the discharge capacity became more than twice as much as the natural channel had (Ando 2002). The increasing quantity (deforestation) of deposing sediment also raises the flood level. The embankments did in fact accelerate the river run-off, but since the regularization works began in the upper part of the valley the flood summit had tragic consequences in the lower regions during the flood of 1879 (Mezosi 2004), when the city of Szeged had been practically devastated (only 265 out of 5723 pre-flood buildings remained).

The hydrology of Hungary can be characterised by its basin-like situation (tectonically preformed main river network, hydrological dominance in surface processes, typical water balance). Calculating a multi-annual average of precipitation, transpiration, incoming and outgoing volume of water more than 95% (3602m^3/s) of the outgoing discharge (3782m^3/s) comes from beyond the country borders and less, than 5% (180m^3/s) has local origin (Mezosi 2004).

High floods may occur on the Tisza River in any season of the year, but the nature of the floods and their determinant circumstances are very variant and unpredictable (initial status for retention of water in the river basin, peaks coinciding at the same time, hydraulic conditions, backwater, distribution and intensity of precipitation, etc.). After a 17-year-long dry period four extremely high flood waves have occurred successively between 1998 and 2001 (Figure 2). The main river and a few tributaries have exceeded the previous highest high water (HHW) levels year by year (Vagas 2001). The flood in April, 2006 rewrote the records along the Middle- and Lower-Tisza and the Koros Rivers. The previous highest high water levels were exceeded with 32-62 cm [9].

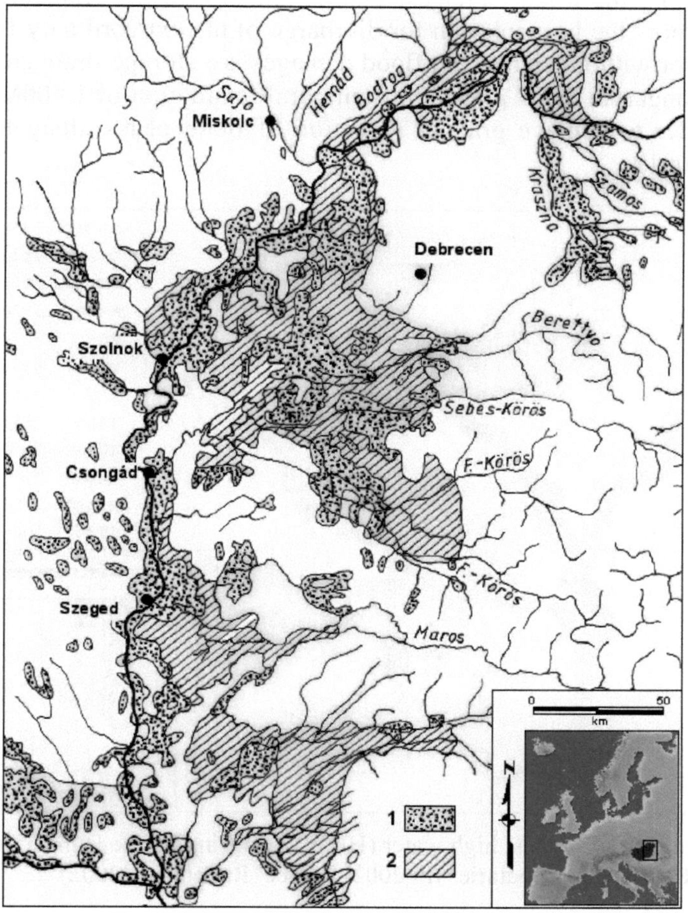

Fig. 1. Hydrographical overview along the Tisza River before the river regulations (mid 19th century), 1: area under permanent inundation, 2: area under regular flood. Source: Kaan (1939), Ando (2002)

According to the plans for reinforcement on the existing dam-system [2003/2002 (I.11.) Government Act], the capability of the flood-control works satisfies the legal requirements [3]. However a feasible solution is needed for embanked discharge and water storage in the case of floods exceeding the accepted peak stage of design flood. As mentioned above the reason for floods in Hungary is basically the water which comes from abroad and the most effective solution would be a well-designed flood protective system on the upper reaches of the rivers as well. For mitigation of this very exposed situation there are strong efforts also in Hungary aimed at decreasing the flood damages. This is one aim of the New Vásárhelyi Plan [4] too: "the best solution for discharge of the extraordinary floods of Tisza River without significant flood damages is a storage-drainage system on the Hungarian flood plains" for temporal retardation of 1,500Mm^3 water in order to achieve gradual reduction of flood peaks along the river (Vagas 2001).

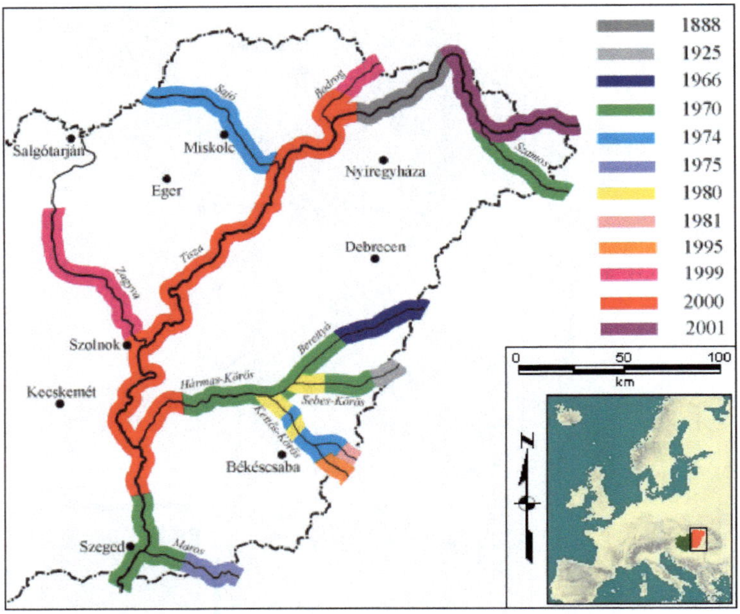

Fig. 2. The year of highest high water (HHW) occurring on the Hungarian section of Tisza River and its tributaries till 2002. Source: Rakonczai (2002)

GIS tools and Digital Elevation Models (DEM) can be applied in several points of the flood reservoir design process, such as the following:

- Multiphase evaluation for selection of the most feasible areas,
- Description of the planned reservoirs, their borders, drafting the necessary earthworks and new dams, mapping the location of natural elevated river banks,

- Description of the current land use and ownership of the selected properties,
- Calculation of the capacity of planned emergency reservoirs and feasibility-analysis.

Our research was therefore focusing mainly on technical questions of capacity calculation of reservoirs on the flat area. The study was carried out in the case of the so-called Szeged Reservoir (Figure 3). The selection of the study area was induced by a remarkable calculation in the summarised report of hydrological plans. The Szeged reservoir would have the third largest extent, but the first the biggest reservoir capacity (Figure 3), but according to the plans the estimated costs of this large reservoir were significantly lower, than in the case of the other reservoirs but with definitely lower capacity (Table 1). This assumed contradiction in the expected capacities and the estimated costs has attracted our attention to the calculation methods behind the numbers.

Table 1. The estimated costs of emergency reservoirs based on the published Conception plan. The costs of the reservoir with the largest capacity (Szeged reservoir) were significantly lower than some similar or smaller reservoirs

ID	Emergency reservoirs	Area (km^2)	Capacity (Mm^3)	Estimated costs (2001) (M HUF)
1.	Felső-túr reservoir	5,7	9,5	900
2.	Szamosköz reservoir	54,5	102	3308
3.	Inérhát reservoir	42	100	1420
4.	Hortobágy középső reservoir	61,8	135,5	5423
5.	Nagykunság reservoir	44	100	2995
6.	Hany-Jászság reservoir	60	113	3666
7.	Szeged reservoir	60,6	187	1700
8.	Szamos-Kraszna-köz reservoir	58,4	100	2099
9.	Dél-Borsod reservoir	131	170	6990
10.	Tiszacsege reservoir	52	105	5923
11.	Cserőköz reservoir	19,7	40	3196
12.	Tiszanána reservoir	37,5	72	5469
13.	Hany-Tiszasüly reservoir	57	132	6492
14.	Kőrőszug reservoir	68	160	3252
	Total:	752,2	1526	52833

Source: The new Vásárhelyi Plan (Conception Plan) [10]

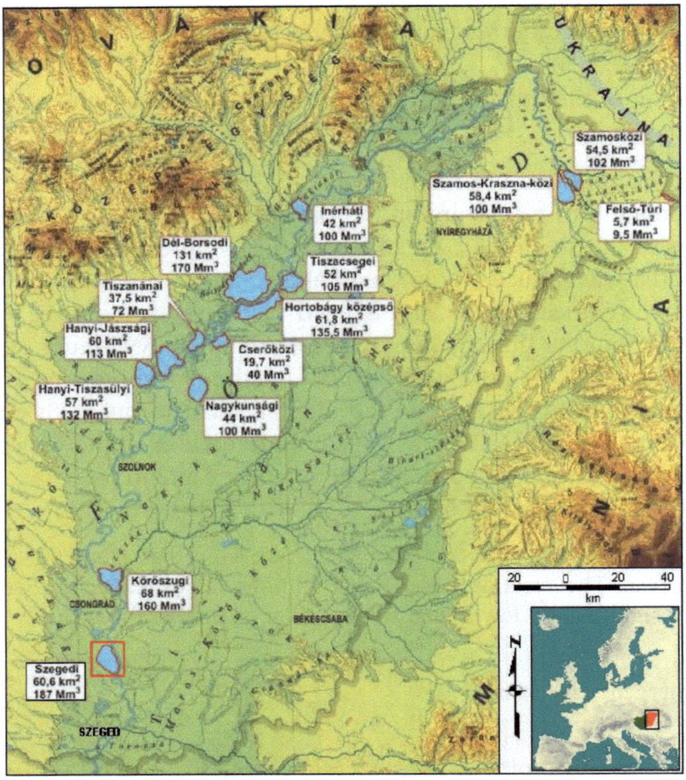

Fig. 3. The selected location of reservoirs and their most important parameters (water surface [km^2], planned reservoir capacity [Mm^3]). Source: The new Vásárhelyi Plan (Conception Plan) [11]

2. Materials and Methods

Based on the map of the planned reservoir (Figure 4), new dams are intended only on the northern edge of the area. The reservoir would be bordered from east by the existing dams of the Tisza River and some heightening earthworks are planned only at the south. Comparing the map (Figure 4) to a more detailed geomorphic map we can see that the western border of the reservoir is marked by an old, naturally elevated river bank along the 80m contour line. According to the information in the map (Figure 4) the planned 61km^2 reservoir should be inundated by more than three-metre-deep water to reach the intended 187Mm^3 capacity. In this case the west border would be similar to a stretched-surface swimmingpool and the possible islands, which are higher than 80m within the reservoir have not been taken into consideration at the calculation of the volume.

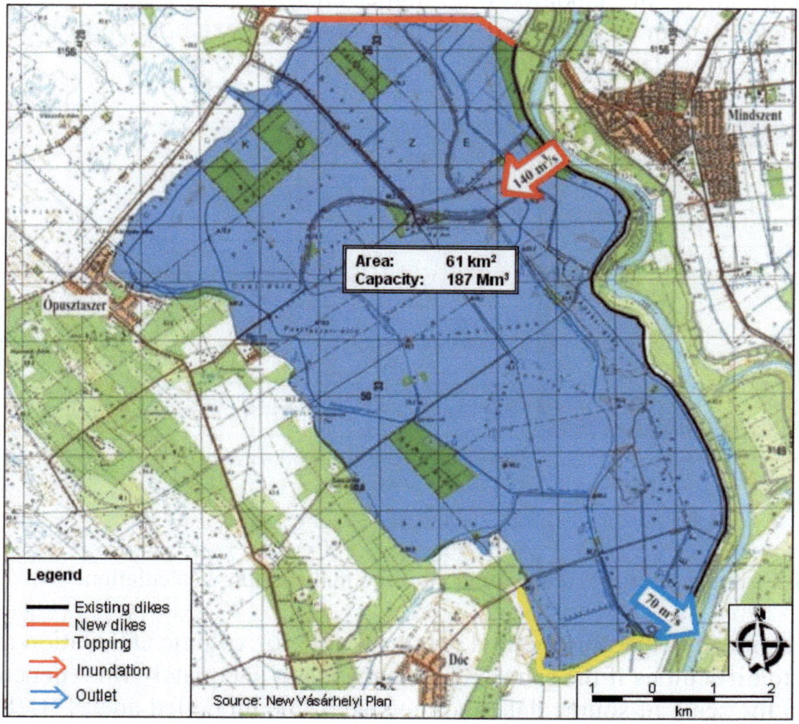

Fig. 4. The Szeged reservoir and the planned heightening earthworks at the north (in red) and the south (in yellow) border. Location is shown in Figure 3

According to the presented values and the information given by the water conservancy staff the reservoir capacity calculations in the original plan were carried out using a well-known profile area method that is suitable for water discharge calculation (knowing the velocity of the water!), but not for volume estimation in the case of a stationary reservoir. The applied method is the following: take the area of a few (i), arbitrary chosen parallel water profiles (A_{wpi}) and the distances (d_i) in between (Figure 5). Multiply the profile area by the distance to an adjacent profile. The presented volume (V) is the sum of their products:

$$V = \sum_{i=1}^{n} A_{wpi} d_i$$

The result of this method strongly depends on the sampling density (n) and the diversity of the surface. Since the source of the elevation data was the analogue topographic map (contour lines) of the area the calculations were time consuming and laboured (e.g. higher area, 'island' than the planned water level cannot be inundated). Figure 5 shows an example in the case of five, arbitrarily placed profiles.

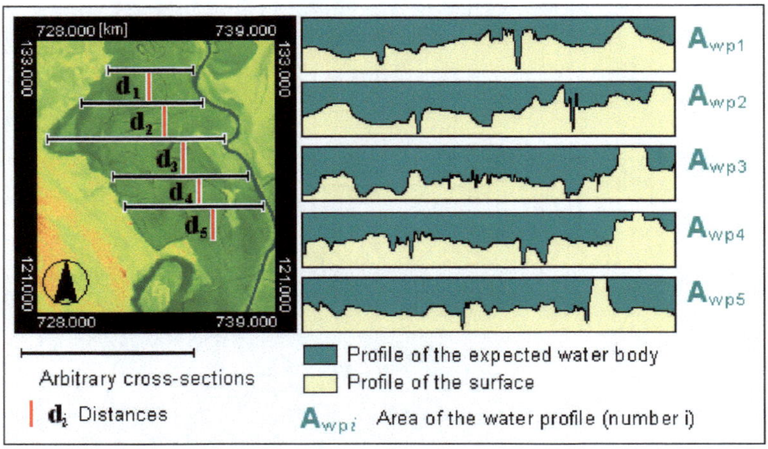

Fig. 5. Example of profile area method used to make volume calculation

As mentioned above this method is not suitable for volume calculation and leads to difficulties if the sampling distance is not adequately fine (depending on the scale of source data and the size of the modelled area, [7], [8]). Significant increasing of the division is almost impossible using analogue methods but we can easily apply a more appropriate technique using digital elevation models that was created, tested and suggested to the council of decision makers of the reservoir system.

As a first step the elevation data (contour lines and bench marks) were digitized from 1:10,000 scale topographic maps covering the study area (Figure 4). For farther modelling the line of natural streams and artificial channels were digitized, as well, together with their direction of runoff. The protective lines (road embankments and dikes) were digitized too. The data input operations were taken by on-screen digitizing (ESRI ArcView 3.2) using scanned (600dpi) geo-referenced and projected (ERDAS Imagine 8.4) topographic maps. The Digital Elevation Model was produced within ESRI Arc/Info Workstation environment. Aerial stereo-images were applied for visual correction and estimation the changing elevation of the existing dam-system (Bodis and Szatmari 2005).

Based on the created digital contour lines it was easy to demonstrate that even the previously applied traditional procedure using arbitrary topographic profiles did not give back the intended reservoir capacity in the case of both indicated inundation levels (80m, 83.4m) presented in the plans. Figure 6 shows theoretically possible settings of profiles. A few of them (number 1, 4, 7, 8) are probably not realistic in practice, but numerically still interesting in the capacity calculation point of view. Table 2 contains the referential results of storage capacity calculation based on the profiles presented on Figure 6. The average elevation was calculated (ESRI ArcView 3.2) by the mean value of all the intersection points of digital contour lines and the profiles within the area of the reservoir (61km^2).

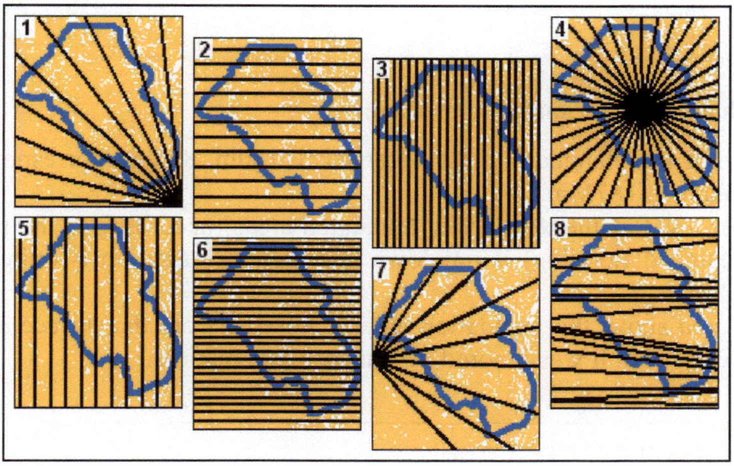

Fig. 6. Subjective settings of profile density and location. Table 2 shows calculations based on elevation of intersection points of digital contour lines (yellow lines) and the profiles (black lines) within the area of the reservoir (blue line)

Using a high resolution digital elevation model (DEM) derived from the same source (elevation data of topographic maps) we can get a more realistic impression about the surface conditions of the selected region. The reservoir volumes for both indicated extremes of inundation water levels (80m, 83.4m) have been calculated on the DEM (2m resolution). In the case of 80m the calculated capacity is 105Mm^3, in the case of 83.4m the capacity would be 306Mm^3 (Figure 7). Non of them corresponds to the reservoir capacity that was presented in the public plans [3] of the reservoir system. The GIS operations and capacity calculations were done within the Grid module of ESRI Arc/Info Workstation (Ver. 7.2.1).

Table 2. Capacity calculation based on the profiles presented on Figure 6. The values do not correspond to the planned reservoir capacity (187 Mm^3)

Code (Fig. 6)	Average elevation [m]	Average depth [m] (water level: 80m)	Average depth [m] (water level: 83.4m)	Capacity [Mm^3] (80 m)	Capacity [Mm^3] (83.4 m)
1	78.526	1.474	4.874	89.914	297.314
2	78.535	1.465	4.865	89.365	296.765
3	78.533	1.467	4.867	89.487	296.887
4	78.552	1.448	4.848	88.328	295.728
5	78.539	1.461	4.861	89.121	296.521
6	78.535	1.465	4.865	89.365	296.765
7	78.542	1.458	4.858	88.938	296.338
8	78.555	1.445	4.845	88.145	295.545

Evaluation of the analogue method:

- None of the planned highest water levels (80m, 83.4m) corresponds to the planned 187Mm^3,
- Water level of the temporally inundated areas (according to the natural or designed flooding) is unknown and their delimitation is quite difficult,
- The area above the inundation level is missing from the surface area of the reservoirs. It was ignored by the analogue method.

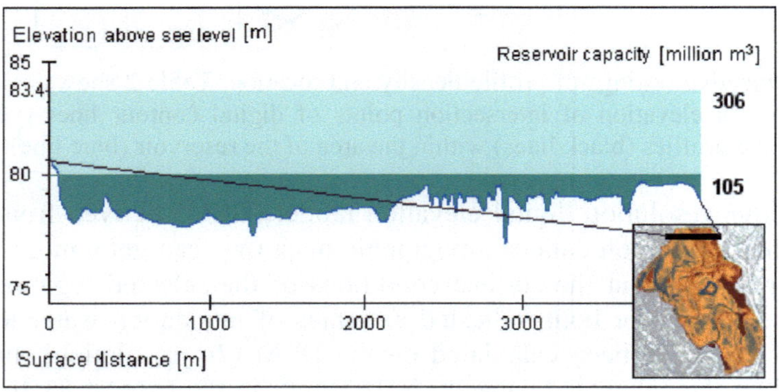

Fig. 7. Calculated reservoir capacity for the planned Szeged Flood Reservoir, in the case of 80 and 83.4m inundation. The black line in the inset at the lower right corner shows the location of the presented profile

3. The Digital Elevation Model of the Reservoir

3.1 Source of DEM

The main source of the DEM was provided by the elevation data of the National Topographic Map Series of Hungary in the scale of 1:10,000. Significant anthropogenic ground objects were also obtained by visual photogrammetric interpretation. The following digital data layers were taken into the database:

- Contour lines 0.5-1m contour interval, map scale = 1:10,000,
- Bench marks, spot heights, hill top elevations,
- Non-contributing areas (without surface outlet),
- Run-off direction of natural and regularized streams,
- Other ground objects which are important from the inundation point of view, such as dams, road embankments, elevated river bank.

3.2 Creation of DTM

The DTM was generated within the Topogrid module of ESRI Arc/Info Workstation. The applied deterministic interpolation method was specifically designed for the creation of hydrologically correct digital elevation models from comparatively small, but well selected elevation and stream layers. The complex iterative method is based on the ANUDEM model (Hutchinson 1988). ANUDEM calculates values on a regular grid of a discretised smooth surface fitted to large numbers of irregularly spaced elevation data points, contour lines and stream lines (Hutchinson 1997). The cells inherit the values from contour lines and elevation points if they fall into the centre of the cell, otherwise the model parameters of hydrological approach – drainage enforcement process that removes all unidentified sink points assuming small isolated areas without an outlet are generally rare in natural landscapes (Goodchild and Mark 1987), descending stream network in a dendritic pattern – and the iterative method ensure the smoothness of the intermediate values, avoiding the gross errors, steps in the surface and faults. The program uses a multi-resolution interpolation method, starting with a coarse grid and working toward the finer, user-specified resolution (ESRI 2004).

The protective lines (road embankments and dikes) in the area, the different surface conditions of the accreting flood plain and the protected sides give a special characteristic to the naturally flat area. To obtain a hydrologically sound DEM more GIS operations should have been used. Natural phenomena are mixed with artificial objects which could be misleading for the interpolation (Barton 2004). The rapid changes on both sides of dikes can result in over- and underestimated cell-values (Figure 8) or the two surfaces could 'flow' into each other without a distinct borderline.

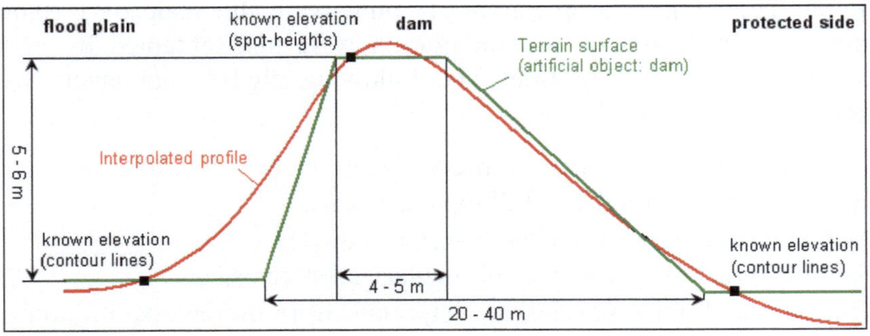

Fig. 8. Schematic example of rapidly changing terrain (flood plain, dam, protected side) and the occurring problem with interpolation (over- and underestimated surface)

To avoid this problem we divided the area into three parts along the dikes (flood plain and protected sides). The dikes and the road embankments were on an additional layer. The ANUDEM algorithm was applied separately on each part, then they were merged and completed with the elevation layer of the additional ground objects (Figure 9).

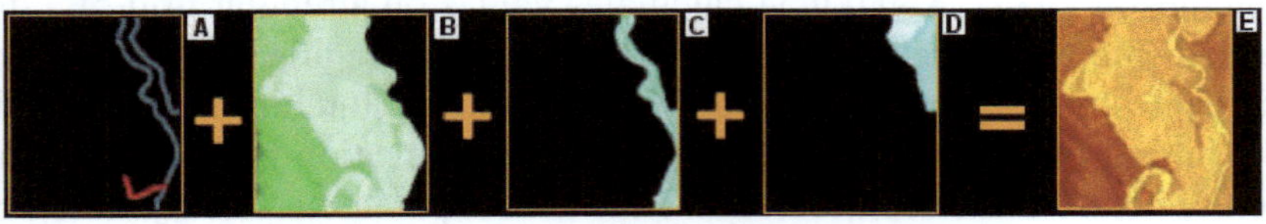

Fig. 9. Interpolation of coherent area and the merged result. A. Linear ground objects (dams). B. Protected side (right). C. Flood plain. D. Protected side (left). E. Merged result

To sum up the final DTM was generated by a complex, iterative method using deterministic interpolation that took the hydrological characteristics of the surface into the main focus.

3.3 Value Check of DTM

Assuming that the contour lines of the topographic map describe the surface adequately we should accept them as a kind of reference for the elevation information if the methodology is in the focus. The applied interpolation algorithm and the resulted elevation model can be tested using some simple procedures:

Overlaying the DEM with the topographic maps (input and reference data) we can compare the interpolated cell-values and the original discrete elevation data (Figure 10). The visual control was made by overlapping the 2m cell size raster elevation model with the original contour lines and elevation points. The deterministic interpolation and the proper settings of interpolation parameters (reflecting the accuracy and density of the elevation points, cause data smoothing, resulting in a more generalized output grid, which represents the amount of random, non-systematic error in elevation input) should result in cell values corresponding to the contour values above, furthermore the ascending and descending slopes in the model should follow the mapped topography (Szakal 2003).

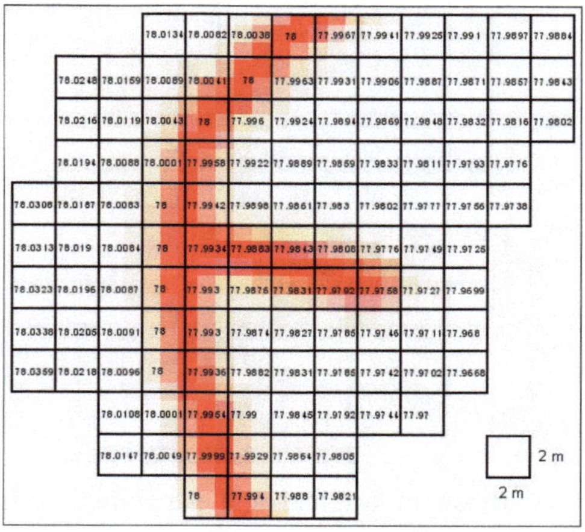

Fig. 10. Original contour line (red curve), a dropping-surface line (slope direction) and model-values. The interpolated cell values should correspond to the contour values

The digitized spot heights of hill tops and sink points of non-contributing areas were inputs for the interpolation. Their locations should appear as local maximum and local minimum on the DEM. This property can be tested visually overlapping the input layer of bench marks with the DEM. Map algebra operations can also be done in order to perform numerical evaluation (Tomlin 1990). Focal functions (focal maximum, focal minimum) produce an output grid in which the output value at each location is a function (maximum or minimum in our case) of the input value at that location and the values of the cells in a specified neighbourhood around the location. The differences of the grids resulted by the focal functions (FOCAL_MIN, FOCAL_MAX) and the raster maps of bench marks (LOCMIN_SINK, LOCMAX_PEAK) are supposed to be zero based on the deterministic approach. Calculations within the ArcInfo environment:

Grid: FOCAL_MIN = focalmin (DEM, rectangle, 5, 5)
Grid: FOCAL_MAX = focalmax (DEM, rectangle, 5, 5)
Grid: DIFFMIN = FOCAL_MIN - LOCMIN_SINK
Grid: DIFFMAX = FOCAL_MAX - LOCMAX_PEAK

Isolines generated at regular intervals from the DEM can be compared with the original input contour lines (Figure 11). The occurred deviation marks out those areas where the input surface information was not sufficient for realistic result of interpolation. In those cases it is necessary to densify the input data (intermediate contours, remote sensed data) in order to further modelling.

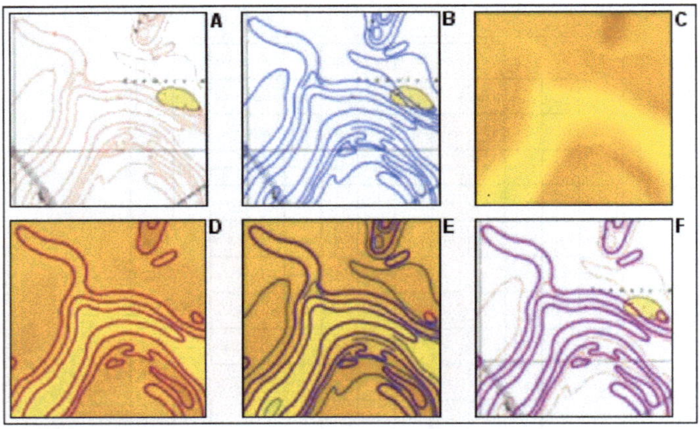

Fig. 11. Steps of comparison of original and generated contour lines (Szakal, 2003). A. Topographic map with original contour lines. B. Digitized contours. C. Elevation model. D. Generated isolines. E. Comparison between B and D. F. Comparison between A and D

On the Figure 11 the generated vectors were superposed over the original maps. The differences were less than the thickness of the scanned contour lines of the maps, so there was no need of additional inputs. The result DEM (Figure 12) is a 2 metre-resolution 5500x6500 sized grid of floating point data type with millimetre precision. This high precision does not indicate always high accuracy [5], but since the relative relief is very low (the minimum-cell-value is 75.288m, the maximum cell-value is 90.650m) it was worth keeping.

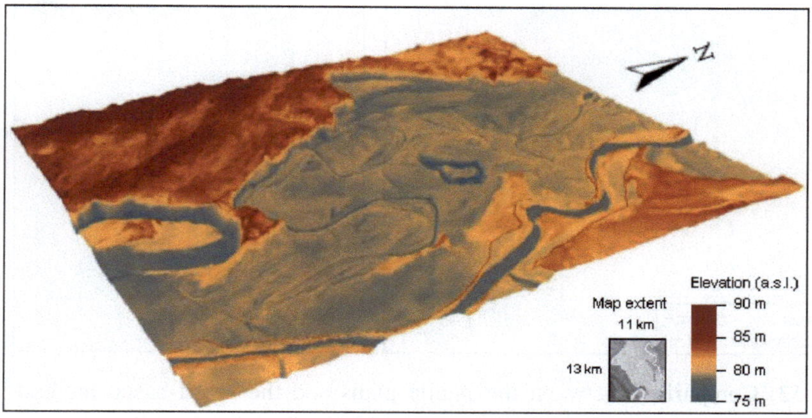

Fig. 12. 3D view of the Szeged Flood Emergency Reservoir based on the created DEM. The 3-dimensional, perspective visualization usually widely supports the design and the work of decision makers with the powerful presentation of plans too (e.g. for stakeholders)

4. Application of DEM to Flood Mitigation Plans

4.1 Quick Area and Reservoir Capacity Calculation

The need of quick, reliable and repeatable area and capacity calculation was the reason for the creation of the presented DEM. Simple GIS functions (overlay operations, map algebra, reclassification) on the digital dataset resulted in useful maps showing the presumably flooded area at a certain water level as well as the quantity of the stored water. We can compare the planned inundation area (Figure 13a) against the modelled possibilities (Figure 13b). The maps of Figure 13 show the inundation at 80m water level (the altitude of the old elevated river bank as the western border of the reservoir). It is visible that the capacity estimate in the plans (Figure 13a) did not take higher surfaces (islands) into consideration. It is one of the reasons for the over-estimated capacity in the plans. It meant more than $3Mm^3$ that would not have been possible to store according to the given circumstances in the public plans.

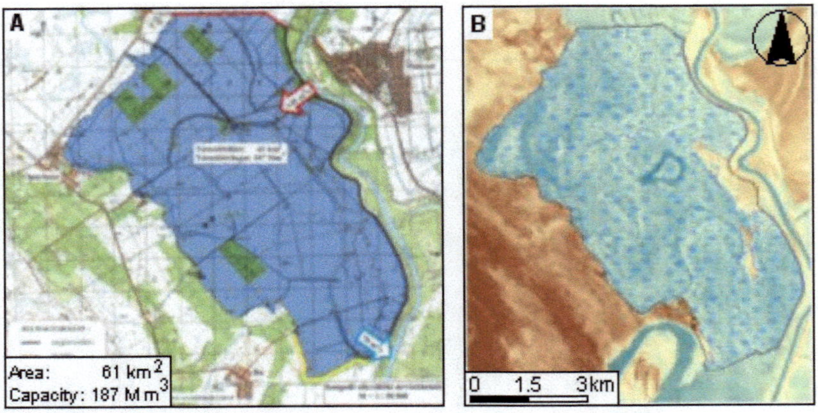

Fig. 13. Comparison between the public plans and the DEM-based method. A. Mapped reservoir area and capacity estimated by an unclear method (there is no known method to get back the presented volume). B. Wetted reservoir area below 80m by the reclassified elevation values on the DEM

Independently from the applied software the concept of calculation can be the following:

$$\text{Reclassification: SURFACE\_WATER}_h = \begin{Bmatrix} 1, \text{ where DEM\_RESERV} < h \\ 0, \text{ where DEM\_RESERV} \geq h \end{Bmatrix},$$

where SURFACE_WATER$_h$ is the grid containing altitude values under the given h inundation level, and DEM_RESERV is the digital elevation model of the limited area that can be used for water storage (only the reservoir, not the surroundings).

Storage capacity calculation:

$$\text{VOLUME}_h = \text{cellsize} * \text{SURFACE\_WATER}_h * \sum_{i=1}^{n} (\text{DEM\_RESERV}_i - h),$$

where VOLUME$_h$ is the calculated storage capacity relevant to the h inundation level, the SURFACE_WATER$_h$ is the Boolean grid of the available area to be inundated and i is the index of grid-cells running up to the total number of cells (n) under inundation.

The presented functions can be repeated on the DEM using any potential water level as input and the flooded area as well as the related storage capacity can be determined within minutes. The series of values (water level / capacity, in a table) builds up the calculated capacity curve.

4.2 Capacity Curve for Planning

It was presented in the previous section that the DEM provides a quick tool to define the volumetric capacity in the case of different inundation levels. The necessary water level that belongs to a desired volume can be easily assigned too. The capacity curve (Figure 14) shows the reservoir capacities of all the possible inundation levels.

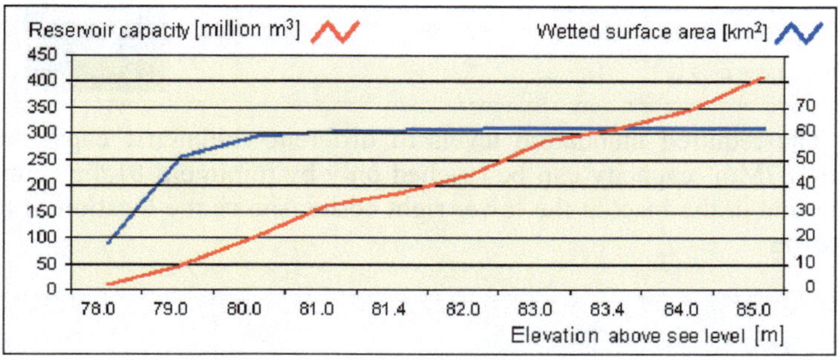

Fig. 14. Calculated capacity curve of the reservoir

Based on the numbers behind the graphs it is verifiable that the published $187Mm^3$ capacity is feasible only with about 81.4m inundation level (Figure 15). For this water level a new, app. 2-2.4-m-high and more than 18km-long dike must be built, which was not included in the operative and financial plans [3] (Figure 4).

4.3 Flow Direction, Inundation and Discharge Simulation, Running-off Modelling

The main operative tasks in the case of an emergency reservoir are the inundation and the running-off procedures. The hydrologically sound digital elevation model is applicable to the description of the run-off characteristics of the area. Based on the different morphometric models (Travis et al. 1975; Tarboton 1997; Costa-Cabral et al. 1994; Bauer et al. 1985; Zevenbergen et al. 1987; Haralick 1983) (Figure 16) and different flow direction algorithms (O'Callaghan and Mark 1984; Fairfield and Leymarie 1991; Greenlee 1987; Tarboton at al. 1991; Freeman 1991; Quinn et al. 1991) the elapsing time and the duration of inundation can be estimated by 2-dimensional hydro-dynamic model calculations and rainfall-runoff models for forecasting the necessary hydrological boundary conditions (discharge time-series) (Szabo 2006).

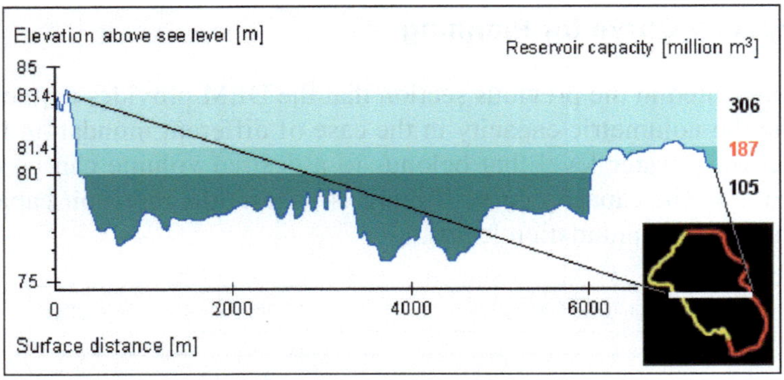

Fig. 15. The required inundation levels of different volumetric capacities. The published 187Mm^3 capacity can be reached only by minimum 81.4m inundation. The white line in the inset at the lower right corner shows the location of the presented profile

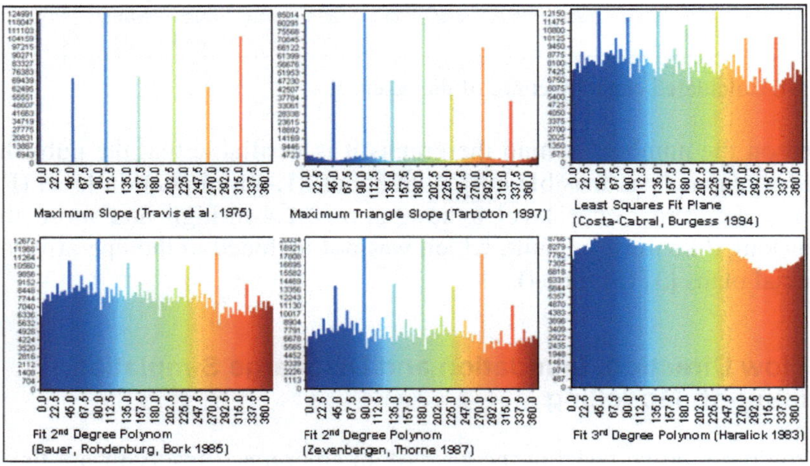

Fig. 16. Frequency histograms of slope directions (0-360 degree) based on different morphometric models (Danube-basin, SRTM30 dataset)

The running-off plans are even more important, since the kinetic energy from the gravity and the pressure of the flood wave are missing during the inverse operation. The DEM provides a preliminary overview about the necessary locations of temporal or permanent drainage systems that should be designed together with the protective water works. The DEM-based relative relief map (Figure 17a) shows the differences between the highest and lowest points on the surface within a predefined area. In the case of the studied reservoir the relief energy is very low. Relative relief indicates the relative velocity of vertical movements on the surface too. The map of Figure 17b shows flow direction from each cell to its steepest downslope neighbour (Jenson and Domingue 1988.). Its mosaic-like characteristics also indicates that there is a lack of natural drainage network on this very flat surface. The non-contributing areas (Figure 17c) can be localized by these maps. However the public maps of the plans (Figure 4) marks the outlet of the reservoir at the south-eastern corner of the reservoir with $70 m^3/s$ discharge it seems the release is impossible without artificial diversion of water.

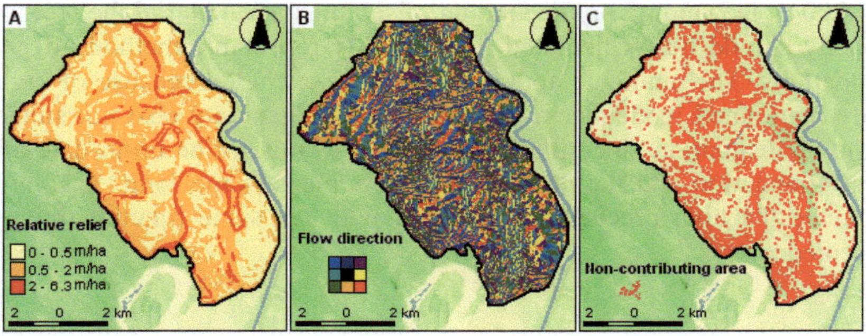

Fig. 17. Derivates of DEM. A. Relative relief map showing the maximum difference within 1 hectare. B. Flow direction map using the Rho8 model (Greenlee 1987; Fairfield and Leymarie 1991). C. Non-contributing area (without natural discharge)

The maps of Figure 17 are the result of the following functions within ArcInfo environment:

- Grid: RELREF = (focalmax (DEM_RESERV, rectangle, 50, 50) - focalmin (DEM_RESERV, rectangle, 50, 50)) (Figure 17a),
- Grid: FLOWDIR_RES = flowdirection (DEM_RESERV) (Figure 17b),
- Grid: SINK_RES = sink (FLOWDIR_RES) (Figure 17c).

Since the presented reservoir is only one (the last one on the Hungarian Tisza section) among the planned 12-14 reservoirs the timing of release of those reservoirs on the same river and nearly at the same time is a very difficult project. In order to avoid an artificial secondary flood wave a multicriteria optimization problem should be solved. One of the most useful tools to analyze the numerous alternative strategies of effective flood prevention is a Spatial Decision Support System (SDSS) that can be updated soon following the rapidly changing circumstances (Szabo 2006). The

DEM and its different derivatives are essential components of these systems.

4.4 Monitoring of Environmental Changes, Siltation

The surface characteristics will change after every flood event. The floated sediment of the Tisza River is relatively high (22 Mtons/year) (Yearbook of Hydrographical Service of Hungary 1985) and the highest densities (1-3 kg/m^3) can be measured during the rising periods of the flood waves (Mezosi 2004). Depending on the amount of sedimentary material the reservoir capacity should be recalculated by a quickly repeatable method after the inundations. The analogue approaches can be excluded because of the already mentioned inadequacy. The topographic maps are also not suitable in the new situation since their information would not be up-to-date after every inundation and the downtime of new maps is too long. A solution might be the high precision laser DEM with the same functionalities as presented.

5. Conclusion

In the last few years (1998-2006) extremely high flood waves have occurred successively on the Tisza River. The New Vásárhelyi Plan intends to mitigate the flood situation by a storage-drainage system on the Hungarian flood plains for temporal retardation of $1500Mm^3$ water in order to achieve gradual mitigation of flood peaks along the river (Vagas 2001).

A contour-based high resolution DEM of the planned Szeged emergency reservoir has been accomplished (Figure 12) and comparative calculations were performed using the DEM regarding the publicly available surface and volume parameters of the reservoirs. During the analyses some inconsistencies were discovered between the planned and the calculable possibilities: instead of the published $187Mm^3$ the maximum reservoir capacity is not more than $105Mm^3$ (Figure 7) regarding the same conditions (given area, possible inundation levels, existing and planned bordering dam-system). There are areas within the reservoirs that cannot be flooded as mapped in the public plans (Figure 4) because their topographical situation is higher than the possible water level. The major area of the reservoir has almost no natural discharge (Figure 17c), thus the running-off operation will require additional engineering for an artificial drainage system in order to mitigate the return flow.

The DEM and its derivations integrated within a Spatial Decision Support System could be an effective tool in some further phases of planning, like quick and repeatable capacity calculation of the reservoir, the elapsing time of inundation and the running-off operations, but the DEM provides essential basic data for hydrological and hydro-dynamic models (Szabo 2006) for forecasting the appropriate timing of opening and rhythm of the release too. The DEM and the derived capacity curve can be applied in planning new dikes and dike topping as well as making localizing plans. Therefore contour based high-resolution digital elevation models (including artificial objects, like dams) give important spatial information for design in the case of reliable elevation data sources.

Fortunately the approx. 4,100 topographic map sheets covering Hungary at the scale 1:10,000 in analogue form are known to be quite reliable (Winkler 2004). The estimated and overall quality-controlled accuracy of contour lines is ±0.5 - ±1.5m in absolute scale and is suitable to serve as a basis for creating a high-resolution DEM (Winkler 2004). The elaborated multiphase interpolation (Figure 9) ensures that the DEM follows the natural and artificial roughness of the rapidly changing surface. If the resolution is high enough the smaller ground objects are recognizable (Figure 18).

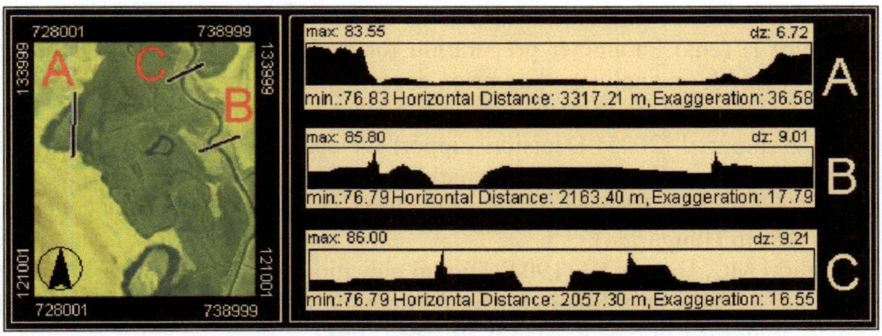

Fig. 18. Topographic profiles across the reservoir borders (dam, elevated river bank). Through the high resolution and the multiphase interpolation of the flood plain, the protected side and the artificial objects the DEM traces the rapidly changing surface

Based on the presented information regarding the inconsistencies in the original capacity plans, the National Directorate for Environment, Nature and Water (Hungary) proposed a workshop titled "Feasibilities and obstacles of DTM application" (31th March, 2004). The statement of this paper was one of the matters for discussion. The most relevant hydrological, geodetic and remote sensing institutes from Hungary, including two universities were involved in the decision support procedure. The result of the meeting was that the spatial characteristics were recalculated in the case of all the planned reservoirs using digital elevation models. However according to the current publications [4] the Szeged reservoir will not be realised in the first stage (2004-2007) of the New Vásárhelyi Plan.

Acknowledgements

Szabolcs Ede Szakál is thanked for his enthusiastic help in producing and testing the digital dataset.

References

Andom (2002) Hydro-geography of the Tisza river system. University of Szeged, Department of Physical Geography, Szeged (in Hungarian)

Barton G (2004) 20 metre-resolution Digital elevation model of Lake Balaton and its surroundings. Development, Improvement, Applications. Acta Geographica 38:11-22

Bauer J, Rohdenburg H, Bork H-R (1985) A Digital Elevation Model as requirement for deterministic water- and mass flow modelling, In: Bork H-R, Rohdenburg H (eds) Parameter preparation for deterministic areal water models, basic research on analysis of agrarian ecosystems, landscape genesis and landscape ecology. Agrar-Oekosystemen, H.10:1-15.

Bodis K, Szatmari J (2005) Decision support for design of lowland reservoirs using GIS). Proceedings of the Geospatial Information and Technology Association (GITA) Conference, 12-13 May 2005, Szeged, Hungary. Szeged, pp. 103-108 (in Hungarian)

Costa-Cabral MC, Burges SJ (1994) Digital elevation model networks (DEMON): A model of flow over hillslopes for computation of contributing and dispersal areas. Water Resources Research 30: 1681-1692.

ESRI (1994) Cell-based modeling with GRID. Environmental Systems Research Institute (ESRI) Inc., p. 481

Fairfield J, Leymarie P (1987) Drainage networks from grid digital elevation models. Water Resources Research 27:709–717

Freeman TG (1991) Calculating catchment area with divergent flow based on a regular grid, Computers and Geosciences, 17: 413–422

Goodchild MF, Mark DM (1987) The fractal nature of geographic phenomena. Annals of Association of American Geographers 77: 265-278

Greenlee DD (1987) Raster and vector processing for scanned linework. Photogrammetric Engineering and Remote Sensing 53:1383-1387

Gruber L, Geczi R (1996) Landscape changes in the Tisza river valley from the Turkish Rule to the 20th century. Acta Geographica 35:11-20

Haralick RM (1983) Ridge and valley on digital images. Computer Vision, Graphics and Image Processing 22:28-38

Hutchinson MF (1988) Calculation of hydrologically sound digital elevation models. Proceedings of Third International Symposium on Spatial Data Handling, August 17-19, Sydney. International Geographical Union, Columbus, Ohio

Hutchinson MF (1997) ANUDEM Version 4.6, User Guide, (Revision: 26 August 1997), The Australian National University, Centre for Resource and Environmental Studies, Canberra

Jenson SK, JO Domingue (1988) Extracting topographic structure from digital elevation data for geographic information system analysis. Photogrammetric Engineering and Remote Sensing 54:1593-1600

Kaan K (1939) Forests and waters of the Great Hungarian Plain. Stadium, Budapest (in Hungarian)

Mezosi G (2004) Physical geography of Hungary. JATE Press, Szeged (in Hungarian) (in press)

O'Callaghan JF, Mark DM (1984) The extraction of drainage networks from digital elevation data. Computer Vision, Graphics and Image Processing 28:323–44

Quinn PF, Beven KJ, Chevallier P, Planchon O (1991) The prediction of hillslope flow paths for distributed hydrological modelling using digital terrain models. Hydrological Processes 5:59–79

Rakonczai J (2002) Actual problems of floods in the Tisza Basin with a geographer's view. In: Rakonczai J (ed) Conference on Tisza Basin as complex region for analysis and development. Szeged, pp 107-114 (in Hungarian)

Süli-Zakar I (2001) The living Tisza. Historia 23:16-22 (in Hungarian)

Szabo JA (2006) Towards a spatial decision support system for integrated river basin flood control, based on distributed rainfall-runoff model. In: Peckham R, Jordan G (eds) Digital terrain modelling. Development and applications in a policy support environment. Springer Verlag, Berlin (in press)

Szakal SE (2003) Modelling a lowland reservoir. MSc Thesis, University of Szeged, Hungary (in Hungarian)

Tarboton DG, Bras RL, Rodrigues-Iturbe I (1991) On the extraction of channel networks from digital elevation data. Hydrological Processes 5:81-100

Tarboton DG (1997) A new method for the determination of flow directions and upslope areas in grid digital elevation models. Water Resources Research 33:309–319

Tomlin CD (1990) Geographic information systems and cartographic modelling. Prentice Hall, New York

Travis MR, Elsner GH, Iverson WD, Johnson CG (1975) VIEWIT: computation of seen areas, slope, and aspect for land-use planning. USDA F.S. Gen. Tech. Rep. PSW-11/1975, Berkeley, California, U.S.A.

Vagas I (2001) Flood waves on the Tisza River at the turn of the millennium. Magyar Tudomany 46:958-966 (in Hungarian)

Winkler P (2004) The national orthophoto program of Hungary completed under strict quality control. Proceedings of the 20th Congress of the International

Society for Photogrammetry and Remote Sensing, Turkey, 12-23 July 2004. (ISPRS 2004), Istanbul

Yearbook of Hydrographical Service of Hungary (1985), Volume 89, Budapest

Zevenbergen LW, Thorne CR (1987) Quantitative analysis of land surface topography. Earth Surface Processes and Landforms 12:47-56

Referenced Web Sites

[1] The EU Water Framework Directive - integrated river basin management for Europe: http://ec.europa.eu/environment/water/water-framework/index_en.html

[2] Environmental Protection and Infrastructure Operational Programme 2004-2006: http://akadalymentes.gkm.gov.hu/data/11334/KIOP_PC_eng040917.doc

[3] The new Vásárhelyi Plan (Conception Plan) (2002, March): http://www.vizugy.hu/vasarhelyi/koncepcio_1.htm

[4] The new Vásárhelyi Plan: http://www.vizugy.hu/vtt/altalanos_english.pdf

[5] Foote, Kenneth E., Huebner, Donald J. (2000): Error, Accuracy, and Precision, The Geographer's Craft Project, Department of Geography, The University of Colorado at Boulder: http://www.colorado.edu/geography/gcraft/notes/error/error_f.html

[6] Cell-Based Surface Modelling, Velence Mountains: http://www.geo.u-szeged.hu/~bodis/gis/velence/dem/

[7] The National Center for Geographic Information and Analysis (NCGIA) Core Curriculum in GIScience: http://www.ncgia.ucsb.edu/giscc/

[8] National Center for Geographic Information and Analysis (NCGIA) GIS Core Curriculum for Technical Programs: http://www.ncgia.ucsb.edu/cctp/

[9] Flood and inland water situation in Hungary, 20th April, 2006-05-10: http://www.kvvm.hu/dokumentum.php?content_id=1229

[10] http://www.vizugy.hu/vasarhelyi/abrak/20t.htm

[11] http://www.vizugy.hu/vasarhelyi/abrak/23.htm

A Quantitative Procedure for Building Physiographic Units for the European SOTER Database

Endre Dobos, Joël Daroussin, Luca Montanarella

1. Introduction

Soil data of various scales is needed for good management of agricultural and environmental resources. On the European level soil information is used for crop monitoring, yield forecasting, agricultural planning, feasibility studies for rural development, natural hazards forecasting, such as floods and landslides or slowly acting processes such as erosion, acidification and other types of chemical, biological and physical degradation of soils. However, no soil database for the European Union to support these goals had existed before the late eighties. The strong need for policy support has speeded up the database compilation and resulted in the first version of the Soil Geographical Database of Eurasia at scale 1:1 million (SGDBE1M). Despite its limitations, SGDBE1M is still among the few databases which serve as a flagship in the development of the small scale spatial databases in Europe. The version 2.0 of SGDBE1M was published recently, in 2004. The refinement of the database and the extension of its geographic coverage to Eurasia and the Mediterranean Africa are in preparation (ESB 2004)

A similar soil database would be needed for global scale yield forecasting, modelling and research. However, the only available soil map with a global coverage is the 1 to 5 million scale FAO (Food and Agriculture Organization, United Nations) soil map of the world, which has been compiled from data collected up to the late seventies. Since the completion of the FAO soil map much new data have been documented and new approaches of mapping and database development have been developed. The lack of a standardized, compatible, reliable soils database at appropriate scale is a major constraint to global environmental and agricultural modelling. Therefore, the SOTER (World SOil and TERrain Digital Database) project was initiated by the International Society of Soil Science (ISSS) in 1986 (ISSS 1986). SOTER is intended to have a global coverage at 1:1 million scale (Batjes 1990; ISRIC 1993), which goal was later degraded to 1:5 million scale due to the lack of financial means. Other international organizations, such as the United Nations Environmental Programme (UNEP), FAO of the United Nations and the International Soil Reference and Information Centre (ISRIC) joined this project and supported the idea of having a global scale soil and terrain database useful for a series of applications. An international committee was appointed to develop a 'universal map legend system' and to define a minimum necessary set of soil and terrain attributes suitable for compilation of a small-scale soil resources map. The database can provide information for a wide range of applications such as 'crop suitability, soil degradation, forest productivity, global soil change, irrigation suitability, agro-ecological zonation, and risk of drought' (ISRIC, 1993). It is not feasible to delineate soil polygons on regional-to-global scale soil databases. Only homogeneous terrain units defined by their physiographic and parent material information can be defined to represent homogeneous units of the soil forming environment. Soil information appears only on the attribute level as assigned soil associations.

The central and eastern part of Europe is completed based on the SOVEUR project (Batjes and Bridges 1997). The database is currently operational and has been used for assessing different land and soil degradation processes acting in the area. However, it still shows the most typical limitations of data inconsistency listed below. The database was compiled from national databases provided and translated to the 'SOTER language' by the national soil survey institutions. Differences in the interpretation of soil parameters are evident from the acidification map (Figure 1). Almost all political borders are visible on the thematic map. Besides the variation in resolution and quality of the incorporated data, this artefact is most likely due to the differences in interpretation of numerous soil terms, and to the misplacement of the national soil variability on the global variability range. More appropriate quantification procedure for soil property characterization is needed to solve this problem.

The work has been started to prepare the SOTER database for the western part of Europe as well. The European Commission has agreed to complete the SOTER database for the EU countries at scale 1:5 million. The soil information for the EU SOTER map is taken from the SGDBE1M with an expert knowledge and decision rule controlled procedure developed by INRA Orleans (King et al. 2002). A preliminary version of the database has been completed by aggregating the existing SGDBE1M polygons. Unfortunately, the SGDBE1M polygon delineation does not follow the SOTER methodology. Thus the SOTER Unit delineation based on the aggregation of SGDBE1M polygons does not necessarily match SOTER criteria.

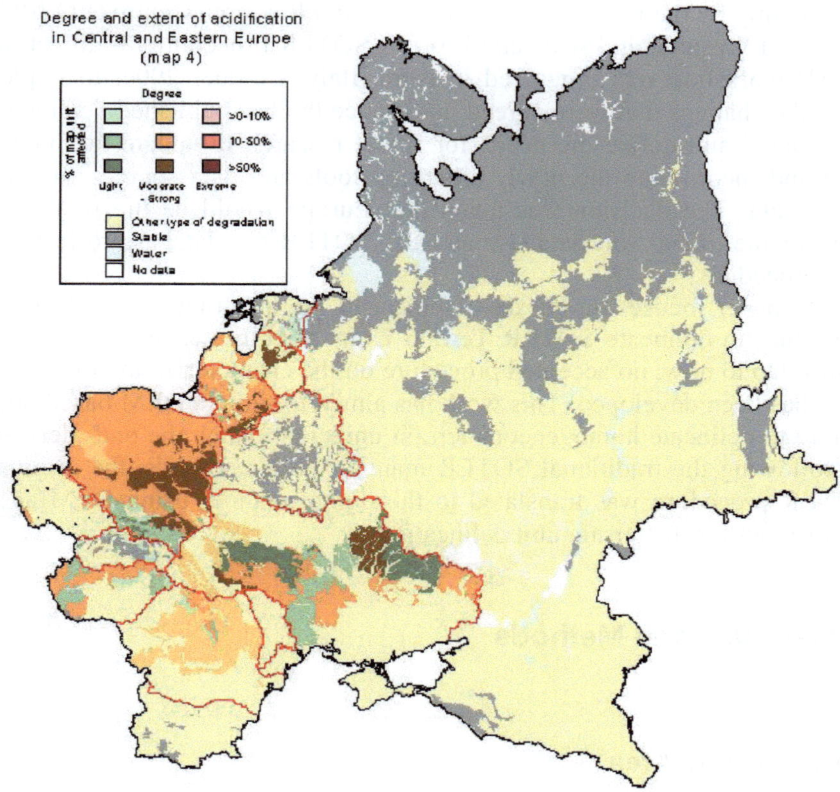

Fig. 1. Degree and extent of acidification risk in central and Eastern Europe derived from the SOVEUR database

Emerging digital technology and high resolution digital terrain information, such as the digital elevation model (DEM) obtained by the Shuttle Radar Topography Mission (SRTM) (Farr and Kolbrick 2000), represent a great resource and potential for developing a quantitative procedure to replace the existing SOTER procedure. The main advantage of a quantitative procedure lies in the spatial and thematic consistency of the final product.

The existing SOTER procedure needs to be modified slightly, partly due to its natural evolution and partly because of the quantification of the criteria that were originally defined qualitatively. Dobos et al. (2004) did preliminary studies on the use of digital terrain and remotely sensed data for the SOTER Unit delineation. The results were promising and a program was launched to update and modify the SOTER procedure to incorporate digital elevation data in the delineation of the Terrain Units. This work is done jointly by the European Commission Joint Research Centre (EC JRC) - Soil and Waste Unit, ISRIC and FAO. A SOTER Procedure Modification (SPM) workshop was organized in Ispra, Italy, October 2004, to implement the changes that were agreed upon since the last published version of the manual and define the needs for future research to update the procedure and incorporate the newly emerging tools and data sources such as DEM data. The decision was taken that Europe would be the first pilot area for the global soil database and EU SOTER the database to test the new procedure.

This paper focuses on the development of the quantitative DEM-based procedure to delineate SOTER Terrain Units at both 1:1 and 1:5 million scales. Up to now, no accepted procedure on the characterization of terrain units has been developed. This work has aimed to setup a DEM based procedure to delineate homogeneous terrain units that match the ones derived by following the traditional SOTER manual. In other words, the original SOTER procedure was translated to the digital manner using DEM as a major input for the terrain unit delineation.

2. Materials and Methods

2.1 The Study Area

The study area covers the countries of the European Union. However, in this paper as well as for the procedure development, a smaller pilot area was selected, namely the eastern half of the Carpathian basin (Figure 2).

2.2 The Data

The Shuttle Radar Topography Mission (SRTM) global elevation data covers almost 80 percent of the globe, almost all terrestrial land surfaces. Its coverage extends between 60° north and 56° south latitudes. SRTM is a joint project between the National Aeronautics and Space Administration (NASA) and the Department of Defence's National Geospatial - Intelligence Agency (NGA) to produce a near global digital elevation data coverage at a relatively high spatial resolution. Farr and Kolbrick (2000) describe the data capturing and processing procedure. The data is handled and distributed by the United States Geological Survey and can be downloaded from their ftp site [1].

Fig. 2. The study areas. The big window (No. 1) shows the full extent of pilot area, while the windows No. 2 (Bükk Mts.) and No. 3 (Great Plain) are the smaller areas to show specific characteristics of the procedure

The basic product has a 1 arc second resolution, but it is publicly available for the US territories only or under agreement with the NGA. This data was resampled by the data provider to 3-arc seconds resolution by averaging a 3 by 3 cells area. This 3-arc seconds resolution data is used and suggested for SOTER studies to keep the consistency of the procedure over all regions of the world. The metric, projected, resolution of this data at the equator is approximately 90m. The data has gaps, i.e. areas with no information, where water bodies or mountains prevented the backscattering of the radar beams, thus providing insufficient signal to process. These voids can limit its use, e.g. for hydrological modelling. The Consultative Group for International Agricultural Research - Consortium for Spatial Information (CGIAR-CSI) has produced a post-processed version of the 3 arc second data using a hole-filling algorithm to provide continuous surfaces. The procedure is described by Jarvis et al. (2004). The algorithm vectorizes the data to produce contours with 10m vertical intervals and then uses the ArcInfo® TOPOGRID tool (Hutchinson 1988, 1989) to produce a hydrologically sound DEM. Data from this DEM is taken to fill the '*no data*' area of the original SRTM data. Jarvis has reported that most macro-scale features have been captured by the algorithm and no significant difference which can alter the use of the data was found between the processed SRTM and the DEM derived from cartographic map. The hole-filled data used in this project can be downloaded from the CGIAR-CSI website [2].

The SRTM data was then projected from WGS84 geographical latitude/longitude coordinates to the standard Lambert Azimuthal Equal Area projection to make the metric calculations possible.

Before the SRTM data could be used, the input DEM had to be hydrologically corrected. Sinks that are due to errors in the data and the micro-scale natural sinks, like the sinkholes – which, at that scale, appear as noise – had to be filled, while the meso- and macro-scale natural variability of the area were kept. Filling all sinks may result in removing all the natural variability, including real geomorphologic units such as depressions or basins, which need to be kept for meaningful landscape characterisation on flat areas. Figure 3 shows an example of such situations occurring on the northern half of the Hungarian Great Plain. The image on the left shows the original, unfilled SRTM data, representing elevations between 87 and 172m. The image to the right shows the same area after using the FILL command within the ArcInfo® GRID module. Notice how all variability has been eliminated and the entire Great Plain was filled up to the elevation of 90m. Using this second image would be useless for terrain characterization of the area. According to our experience in using SRTM data, a limit of 20m has to be set as a maximum sink depth to be filled. This limit was suggested in this study as well.

2.3 Methods

The classifications for the four terrain layers are described in the four sections below. Figure 4 shows the flowchart of the analysis. ArcInfo® geographical information system software and its GRID raster analysis module were used to achieve the work.

2.3.1 SOTER Terrain Unit Delineation

SOTER Unit delineation is based on two primary soil formation phenomena: terrain and lithology. Each SOTER Unit represents a unique combination of terrain and soil characteristics. The two major differentiating criteria are applied in a step-by-step manner, leading to a more detailed identification of the land area under consideration. Physiography is the first differentiating criterion to be used to characterize a SOTER Unit. The term physiography is used in this context as the description of the landforms on the Earth's surface. It can be best described as identifying and quantifying the major landforms, on the basis of the dominant gradient of their slopes and their relief intensity. The use of these variables, in combination with a hypsometric (absolute elevation above sea level) classification and a factor characterizing the degree of dissection, can make a broad subdivision of an area and delineate it on the map. (Refer to the 'Calculation of Potential Drainage Density Index (PDD)' chapter of this book to get more information on the role of dissection on the terrain parameterization.) Further subdivision of the SOTER Unit according to the lithology (parent material) needs to be done to complete the delineation procedure.

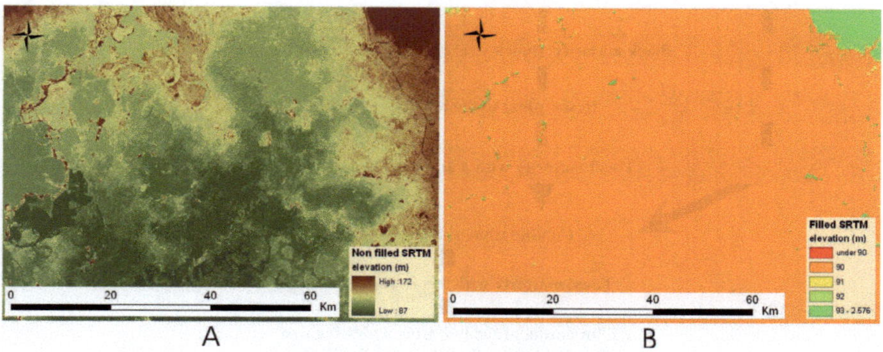

Fig. 3. SRTM images of the northern half of the Hungarian Great Plain, pilot window No.3. The left image shows the area before filling the sinks, while the right one shows the same DEM once filled

Until recently, manual methods were used to delineate SOTER Units, the geometric objects of the SOTER database. The availability of DEM makes it feasible to use a quantitative approach. Worstell (2000) has proposed an approach using DEM for SOTER Terrain Units delineation based on Edwin Hammond's (1954) landform classification methods. Worstell has adapted and modified his methods to create a quantitative procedure to classify landforms on a regional scale.

A new, quantitative method for creating a SOTER database has been suggested by Dobos et al. (2004) using the 1km resolution SRTM30 data as the base DEM. Although the procedures were promising, they suggested more research and quality check on the results.

The aim of the present study was to develop a quantitative method to derive terrain classes that match the criteria of the SOTER Manual of Procedure (ISRIC 1993). According to the manual, four terrain attributes are used to define the SOTER Terrain Unit: hypsometry (elevation), slope percentage, relief intensity and dissection. The GIS layers of these attributes were derived from the digital elevation model by translating and reformatting the terrain class characteristics given by the SOTER manual. These four layers are combined to produce the complex landform classification. This combined layer was then vectorized, and finally generalized to achieve the polygon size limit appropriate for the 1:1 and 1:5 million scales of the database to be produced.

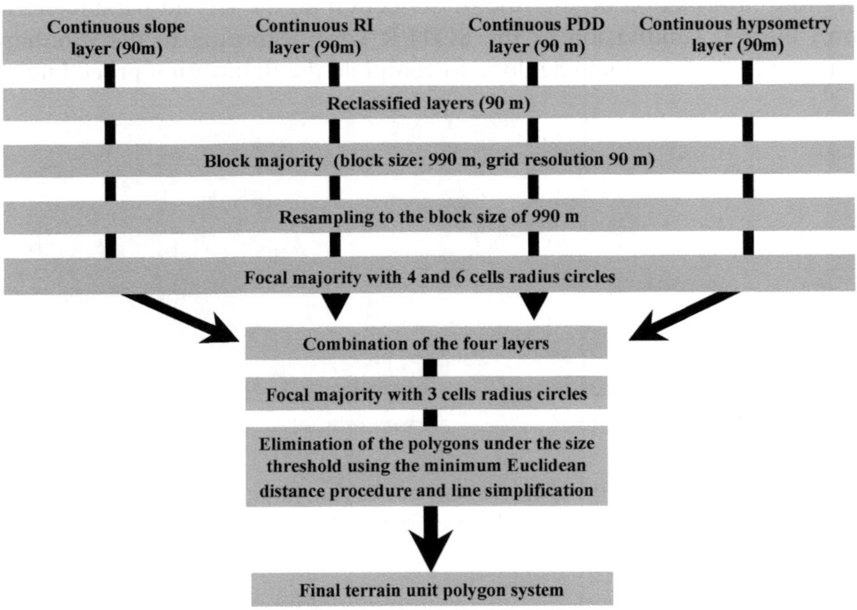

Fig. 4. Flowchart of the Terrain Unit delineation. RI stands for relief intensity while the PDD means potential drainage density

The class limits of these attributes are defined more or less quantitatively in the Manual, ('Attribute coding'), except for the dissection for which only qualitative definitions are given. Changes in the class borders were implemented as well, which have been proposed and agreed upon in the SOTER Procedure Modification workshop. The dissection class limits were derived from the Potential Drainage Density (PDD) layers (Dobos et al. 2000) through an empirical approach.

A coarse resolution DEM tends to generalize the land surface and eliminate the micro- and meso-scale features of the surface, drastically decreasing the slope and relief values of the area (Wilson et al. 2000). Therefore, a relatively fine resolution DEM (SRTM) was used to maintain the higher scale landscape elements and to derive the terrain descriptor values for the area. The resulting variables show much more detail than what the targeted scales are capable of handling. Thus, a generalization and aggregation procedure was used to obtain the appropriate resolution.

2.3.2 Creation of the Four Thematic Layers

Slope

The slope layer was derived from the SRTM data using the slope function available from the ArcInfo® GRID module. This function uses the average maximum technique (Burrough 1986). It calculates the slope for a fitted plane surface on a 3 by 3 cells neighbourhood centred on the processing cell (marked as 'e' in Figure 5) as follows:

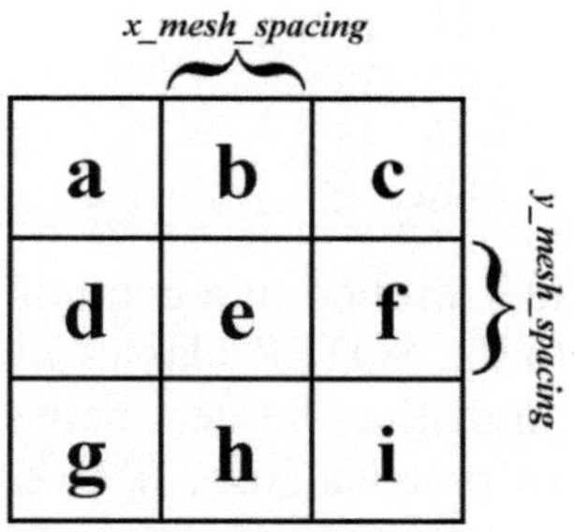

Fig. 5. Example of a 3 by 3 cells moving window. Cell 'e' is the centre cell. Other cells are the neighbouring cells

$$rise\_run = \sqrt{\left(\frac{dz}{dx}\right)^2 + \left(\frac{dz}{dy}\right)^2}$$

$$degree\_slope = \arctan(rise\_run) * 57.29578$$

where the deltas (dx, dy and dz) are calculated using a 3 by 3 cells moving window as follows:

$$\frac{dz}{dx} = \frac{(a + 2d + g) - (c + 2f + i)}{8 * x\_mesh\_spacing}$$

$$\frac{dz}{dy} = \frac{(a + 2b + c) - (g + 2h + i)}{8 * y\_mesh\_spacing}$$

where [a] to [i] are the altitude values of cells a to i, and [x_mesh_spacing] and [y_mesh_spacing] are the distances between cell centres in the x and y directions respectively. Altitudes and distances are expressed in the same measurement unit.

The SOTER modified classification scheme for the slopes is shown Table 1.

Table 1. The modified slope classification scheme of SOTER

Slope class	Range of slope percentages
1	0 - 2
2	>2 - 5
3	>5 - 10
4	> 10 - 15
5	> 15 - 30
6	> 30 - 45
7	Above 45

The use of SLOPE function resulted in a continuous slope layer, which was reclassified according to the SOTER classes in Table 1. This kind of classification rarely results in distinct borders between the classes, necessary for defining polygons of practical size. It gives too much spatial detail, which cannot be represented at the target scale ('salt and pepper' effect). Therefore, adequate filters have to be used to derive the slope layer with homogenous pattern of slope classes and the resolution has to be degraded to reach the spatial detail needed for the targeted 1:1 million and 1:5 million scales.

Raster-based Generalization Procedure

The main steps of the generalization procedure are shown in Figure 6 and Figure 7. The appropriate spatial resolution of the grid for a 1:1 million scale target database is around 1x1 km (McBratney et al. 2004). Two options were considered for degrading the resolution, (i) averaging the cell values within 1km^2 area or, (ii) taking the majority class and assigning it to the spatially degraded cell (blocks). The majority class of the area characterises the landscape better than the average value.

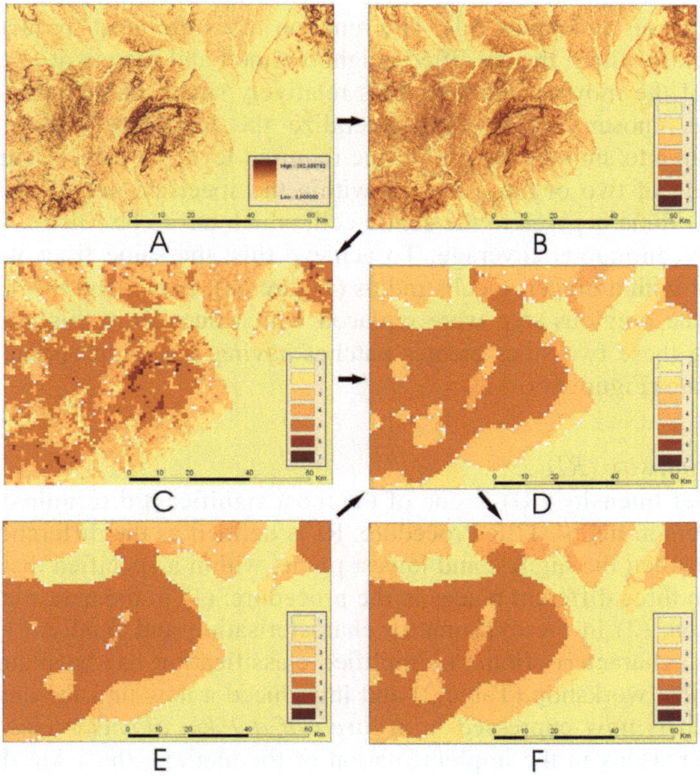

Fig. 6. Flowchart explaining the creation of the slope grid illustrated with an example from the Bükk-mountain, pilot area No. 2. Table 1 provides the corresponding slope classes. A. Continuous slope layer. B. Classified slope layer. C. Resampled, classified slope layer. D. Filtered with majority filter, 4 cells radius. E. Filtered with majority filters 6 cells radius. F. The final slope layer

Therefore, the blocking approach was applied to define the majority class within a square shape area with the size of 11 by 11 cells (990m by 990m). The resulting grid remains with the original 90m resolution. Therefore it was then resampled to the target resolution of 990m, which was decided by the authors to be appropriate for the target 1:1 million scale (Figure 7c).

The resulting grid still had some salt and pepper effect, having a mixture of stand-alone cells or small contiguous areas, especially on the transition zones between the classes (Figure 6c). This phenomenon represents a significant problem when representative polygons are to be drawn with a minimum polygon-size requirement. To overcome this problem, a filter was applied to that grid layer by using a majority value function with a 4 cells radius circle (Figure 6d). The function takes the most frequently occurring class within the specified neighbourhood and assigns it to the centre cell of the moving window. This relatively small neighbourhood of 4 cells was chosen not to over-generalize the landscape, while having enough area to smooth the grid to the required level. In case of equal representation of two or more classes within the specified neighbourhood, a '*no data*' value is given to the centre cell, which has to be filled to achieve complete/continuous coverage. To achieve this, the same filter was used again but with a bigger, 6 cells radius (Figure 6e), circle, and the '*no data*' cells in the previous step were replaced with values from this grid layer. Applying these two steps creates patches having sizes appropriate to that target scale (Figure 6f).

Relief Intensity (RI)

The Relief Intensity (RI) is one of the most significant discriminating terrain factors in the SOTER Procedure. RI is defined as the difference in altitude between the highest and lowest points within a specified distance. It is used in three different places in the procedure: (1) in the major landform description, (2) in the hypsometry characterisation and finally, (3) in the dissection characterisation. A simplified classification has been suggested by the SPM workshop (Table 2) and introduced a new unit for measuring RI, which is now expressed in *m/[area of a 1 km diameter circle]*. For practical reasons in the implementation of the method, the 1 km diameter is approximated with a 990m diameter because, when using 90m resolution projected SRTM data, it can thus be expressed simply with a 5 cells radius circle (5 cells radius = 11 cells diameter = 990m diameter circle).

The algorithm defines a circle shape neighbourhood with a radius of 5 cells and identifies the highest and the lowest points within that area. The difference between these two points is assigned to the centre cell of the moving processing window as an RI value.

The rest of the procedure was basically the same as in the one used for deriving the slope layer. High resolution, 90m SRTM data was used to create the original RI layer, which was later classified according to the classes listed in Table 2. Then this classified image was generalized following the generalization procedure described above (Figure 7).

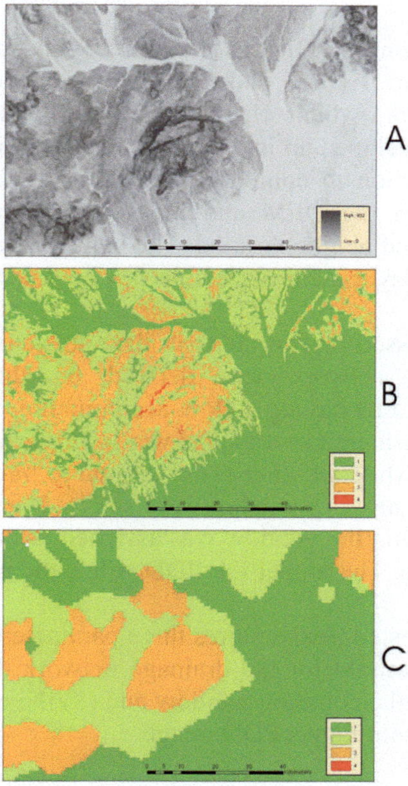

Fig. 7. The Relief Intensity (RI) layers illustrated with an example from the Bükk-mountain, Hungary, pilot area No. 2. Table 2 provides the corresponding RI classes

Table 2. The Relief Intensity classification as suggested by the SOTER Procedure Modification workshop

Relief Intensity class	Altitude range in meters within a 990 m diameter circle
1	0 - 50
2	> 50 - 100
3	> 100 - 300
4	Above 300

Dissection (PDD)

The degree of dissection is difficult to quantify with traditional methods (ISRIC 1993). The use of DEM makes it feasible to derive an artificial drainage/valley network, which characterizes the landscape dissection. Dobos et al. (2000) developed an index called the Potential Drainage Density (PDD) and a function to compute the PDD. The function derives a drainage network from the DEM and measures the network's density within a predefined sized neighbourhood. The nature of (and the procedure for creating) the PDD layer are described in this book among the technical notes.

The data was processed in two steps in this study. In the first step, a DEM-based drainage network was derived by thresholding flow-accumulation values. Cells having a flow-accumulation value higher than this threshold were considered as drainage ways. These drainage cells were assigned a value of 1, while all other cells were set to '*no data*'. In the second step, a size for a moving window was selected, and a count of the drainage-way cells within the window was assigned to the centre cell. The result is the PDD value. The higher the PDD value, the more dissected the terrain.

The procedure suggested here requires three parameters to be set: (1) the flow accumulation threshold for the drainage network derivation, (2) the radius of the circle for the counting window and (3) the class limits for the reclassification of the continuous PDD image.

Choosing a Flow Accumulation Threshold for the Drainage Network

The threshold value was set to 100. In the case of the SRTM data, it translates to approximately a $1 km^2$ catchment's area for a drainage line to start. Lower values create a very dense network with too many details, while higher thresholds decrease the pattern density, thus missing some necessary details.

Choosing a Radius for the Circle to Count Drainage cells

The radius was set to 20 pixels/cells. Several radii were tested. The rule of thumb is that a radius being too small is not able to deliver meaningful information about the general landscape. The resulting images show buffer-like zones along the drainage lines with relatively big portions of the image having '*no data*'. Choosing a radius that is too large tends to over-generalize the image, there again masking general landscape characteristics. The optimal choice is the smallest circle, which is still big enough to pick up at least one drainage cell. After several trials and errors, the radius value of 20 was found to be appropriate for the drainage network derived from the SRTM data using the flow-accumulation threshold value of 100.

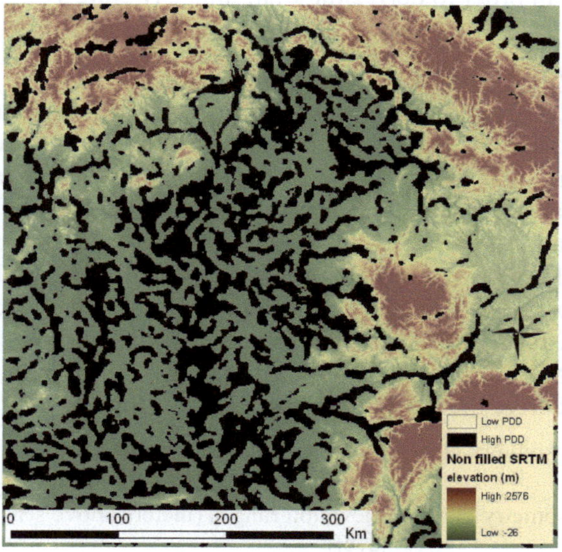

Fig. 8. The PDD layer derived from the SRTM elevation data in the Hungarian Great Plain and part of the Carpathians. The black areas represent the high PDD value class

Class Limits for the Reclassification of the Continuous PDD Image

The dissection measurement unit as defined in the 'Manual of Procedure' (ISRIC 1993) is the length of permanent and seasonal streams and rivers within a $1km^2$ neighbourhood. Three dissection classes are distinguished: 0-10, 10-25, and above $25km/km^2$. In this study, using these class limits would have been meaningless because the approach is different in the sense that we use 'potential' drainage lines *versus* actual ones. In the digital procedure, only two classes were defined with value ranges as given in Table 3.

Table 3. The PDD class ranges

Class	PDD value range
1: less dissected areas, convex surfaces	0-90
2: more dissected areas or depressions, concave surfaces	Above 90

The procedure then followed the same line as for the other data/model/map layers by applying the generalization and filtering procedure described above for slope in order to derive the final image (Figure 10).

Hypsometry (Elevation)

The original SOTER Procedure Manual suggests a two steps procedure for elevation classification. The first step divides the area into three general relief types, namely (i) the level lands, (ii) the sloping lands and (iii) the steep sloping lands. The second step further divides each of these three types into elevation subclasses but using a different classification scheme for each type. These two steps classification system was simplified by the SPM workshop and a new one was introduced (Table 4) which was used here as well. After the SRTM image was reclassified using the classes in Table 4, it followed the same generalization and filtering procedure as the other layers to derive the final hypsometry image (Figure 11d).

Table 4. The hypsometry classes suggested by the SPM workshop

Hypsometry class	Elevation ranges (meters above sea level)
1	Up to 10
2	> 10 - 50
3	> 50 - 100
4	> 100 - 200
5	> 200 - 300
6	> 300 - 600
7	> 600 - 1500
8	> 1500 - 3000
9	> 3000 - 5000
10	Above 5000

Database Compilation

The next step in the database development was to combine the information from the above four SRTM-derived terrain thematic layers (slope, RI, PDD and hypsometry) into a single grid constituted of Terrain Units. Expectedly this grid had many small and meaningless patterns. It was therefore filtered using a majority function with a 3 cells radius circle shape neighbourhood to create a more homogeneous appearance of the Terrain Units. This grid was then vectorized to create polygon coverage.

Vector-based Generalization of the Raw SOTER Terrain Units

Figure 9 shows a close view on part of the raw SOTER Terrain Units polygon system (black thick lines and red thin lines). It is evident that even after the many steps of filtering, there are still a lot of small polygons, which are below the size limit specified in the SOTER Procedure Manual (ISRIC 1993) (hatched polygons on Figure 9). According to the Manual, the minimum area of a polygon must be 25km^2 if represented on a 1:1 million scale map and 625km^2 on a 1:5 million scale map. Any polygon smaller than these thresholds, depending on the target scale, has to be aggregated with one of its neighbouring polygons, preferably with the one that has the most similar terrain characteristics. To evaluate this similarity in a quantitative manner, the Euclidean distance between each polygon pair is calculated using the same four thematic terrain layers as produced above. These distances (red numbers on Figure 9) are then used to select the neighbouring polygon that has the minimum Euclidean distance from the polygon to aggregate.

In the first step, the four continuous landform parameter grids (slope, RI, PDD and hypsometry) are normalized to a range of 0 to 1000 using equation (1). This step is important to give equal weights to all the four landform parameters.

$$Xnorm = \frac{X - X\min}{Xrange} * 1000 \qquad (1)$$

where Xnorm is the normalized value for parameter X in the processing cell, X is the original value for parameter X in the processing cell, Xmin is the minimum value of parameter X calculated from all cells in the grid layer, and Xrange is the value range of parameter X (Xmax - Xmin) calculated from all cells in the grid layer.

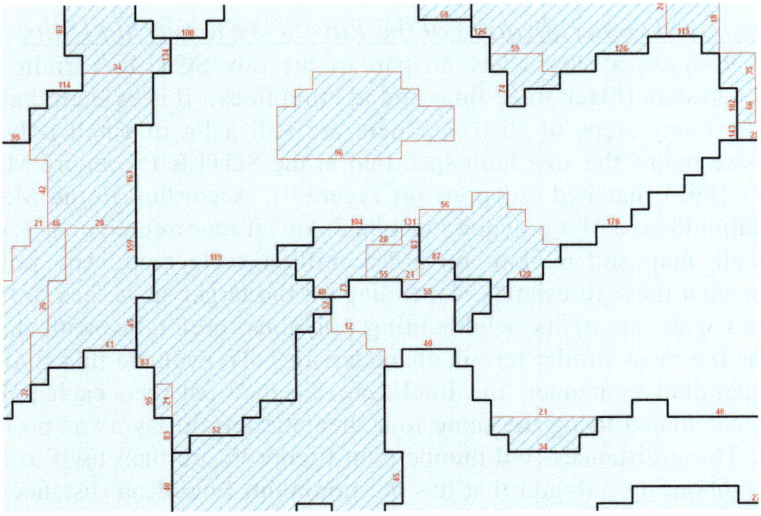

Fig. 9. The minimum Euclidean distance-based aggregation procedure. The red numbers are the Euclidean distances calculated between one polygon and its neighbour. The hatched polygons are those that have a surface area below the threshold (to be aggregated to one of their neighbours). The red thin lines are the polygon borders removed by the algorithm for aggregating

At this stage, each cell is characterised by the four normalized values $SLOPE_{norm}$, RI_{norm}, PDD_{norm} and $HYPSO_{norm}$. In the second step, the mean values of these four normalized landform parameters are computed for and assigned to each polygon. Thus, each polygon is characterised by the four normalized mean values $SLOPE_{Poly}$, RI_{Poly}, PDD_{Poly} and $HYPSO_{Poly}$. In the third step, the Euclidean distance (*Ed*) is calculated as a measure of similarity for each neighbouring polygon pair using equation (2) and assigned to the arc that divides that pair.

$$Ed = \sqrt{D_1^2 + D_2^2 + D_3^2 + D_4^2} \qquad (2)$$

where

$D_1 = SLOPE_{PolyLeft} - SLOPE_{PolyRight}$

$D_2 = RI_{PolyLeft} - RI_{PolyRight}$

$D_3 = PDD_{PolyLeft} - PDD_{PolyRight}$

$D_4 = HYPSO_{PolyLeft} - HYPSO_{PolyRight}$

and where $SLOPE_{PolyLeft}$ is the mean normalised slope value of the polygon standing to the left of the arc, $SLOPE_{PolyRight}$ is the mean normalised slope value of the polygon standing to the right of the arc and so forth for each of the four parameters.

At this stage, each arc holds a quantitative estimation of the similarity between the two polygons that it separates in terms of landform characteristics (red numbers on Figure 9). In the fourth and final step, the algorithm can now select – for each of the polygons that are candidates for aggregation – the arc to delete that has the lowest Euclidean distance value of all arcs making up the polygon. Deleting that arc results in aggregating the small polygon to its most similar neighbour.

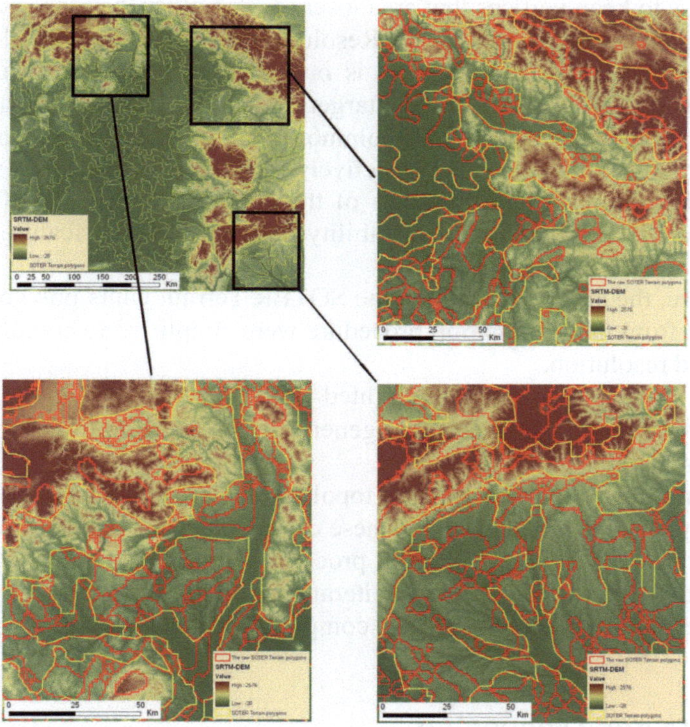

Fig. 10. The results of the aggregation and line simplification procedure to the target scale of 1:5 Million in the Hungarian Great Plain and part of the Carpathians

Aggregating several small polygons together may result in a new polygon that still has an area below the specified threshold. Therefore, the procedure is iterative. It starts again from the second step, i.e. calculating the mean values of the 4 normalized landform parameters for each newly aggregated polygon, calculating the Euclidean distances between the pairs

and eliminating the arc with the smallest Euclidean distance value. The procedure is repeated until the resulting polygon system remains stable.
The procedure was implemented as a standalone AML tool, which can either be downloaded from the JRC's Soil and Waste Unit homepage or provided by the authors on request.

Line Smoothing and Simplification
In a vector GIS data form, polygons are structured into a set of arcs, which in turn are structured as a set of vertices (X and Y coordinate pairs). It is meaningless to keep vertices that are closer to each other than the specified resolution of a geographic database. Resolution is related to nominal scale, i.e. the scale at which the database is meant to be used. As mentioned above, the SOTER Europe project targets both the 1:1 and 1:5 million scales. At these target scales, it is commonly admitted that the resolution can be set to 200 and 1000m respectively. This is because these values match with 0.2mm on map printouts of the corresponding scales, 0.2mm being the commonly admitted separability distance between features on a map.

Therefore, the arcs that make up the SOTER Terrain Units polygon system output from the aggregation procedure were simplified to comply with the targeted resolution.

Figure 10 shows, on three highlighted areas, the results of applying the small polygon aggregation and line generalization procedure displayed at the target scale of 1:5 million.

Line simplification alters polygon topology and sometimes results in dividing polygons into smaller ones. These can then become smaller than the specified threshold area. The overall procedure, both aggregation and line simplification, therefore needs to reiterate. There again iterations repeat until the resulting coverage reaches a completely stable state.

3. Results and Discussion

Figure 11 shows the four terrain variables derived from the SRTM DEM grid for the eastern half of the Carpathian basin. The resulting data sets were found to be realistically characterizing the terrain. Although, the centre and south-western part is a low relief area (Hungarian Great Plain) with little terrain variability, expressed as only a few meters range in elevation, this variability is still the most important factor controlling the soil formation processes. In contrast, it is evident from all four derived terrain variable images that the ring of the Carpathians surrounding the plain shows strong variations in the terrain. The challenge of this study was to capture all the significant natural physiographic variation on both plain and hilly areas by applying a quantitative, consistent procedure using the 3-arc second resolution SRTM DEM.

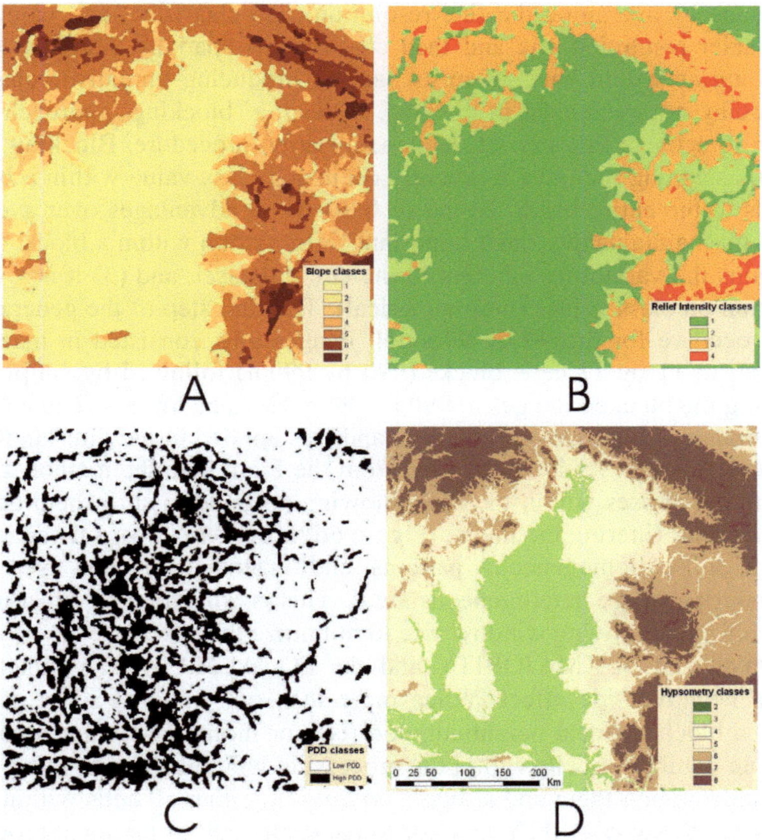

Fig. 11. Maps of the four classified landform parameters in the pilot area. Refer to the method section for the class meanings

Dobos et al. (2004) used a coarser resolution 30 arc seconds SRTM DEM (~1km) for delineating SOTER Terrain Units. They concluded that the spatial detail needed for deriving a 1:1 million scale database is well achieved with such a 1 by 1km resolution dataset. However, a coarse spatial resolution DEM tends to 'over generalize' the terrain features by decreasing their value ranges.

The approach developed here is an attempt to take advantage of the availability of a fine spatial resolution DEM. The thematic terrain variable layers were derived from the 90m resolution DEM to keep as much details of the landscape information as possible. These resulting images were classified based on the SOTER criteria. The classification was followed by the generalization procedure. In the first stage of the generalisation procedure, the parameter grids were resampled to coarser resolution. This could have been done by averaging the fine resolution (non-classified) cell values to the coarser resolution one and then classify this spatially degraded grid. But doing so would have taken us back to producing unrealistic classes, which may not even exist in the area. Instead, a 'blocking' approach was chosen to achieve this step of the generalisation procedure. Blocking consists in assigning the most frequently occurring class value within a block of cells to the entire block. Doing so has several advantages over averaging: (1) it maintains the 'most important' information within a block, (2) it maintains the variability inherent to the entire dataset, and (3) it is a valid operation on class values (non-numerical). The first step of the generalisation procedure for the 90 by 90m cell images thus consisted in majority blocking to 11 by 11 cells blocks (990 by 990m) followed by simply resampling the blocked images to 990 by 990m cell size images. The method is illustrated in Figure 6 and Figure 7 and was applied to the four landform parameter images. As it can be seen from the Figure 6c, the method keeps the original classes within the area, showing no trend to shift them downward. Further filtering the image (Figure 6d) however shows a tendency to enlarge solid, homogeneous patterns and extend their area over the neighbouring, more heterogeneous areas. This is why filtering must use a relatively small moving window size to minimize this effect.

Both the BLOCKMAJORITY and the FOCALMAJORITY functions have a negative side effect. When more than one majority class occurs within a block or focal neighbourhood (two or more classes are found in the same number of cells within the processing window), the function cannot decide among them and assigns '*no data*' to either all cells within that block (BLOCKMAJORITY case) or to the centre cell of the focal processing window (FOCALMAJORITY case) (see the white spots in Figure 6d). Introducing a second filtering step with a larger window is a turnaround 'trick' to get rid of almost all these '*no data*' cells by simply picking up for them a value that is found within a larger neighbourhood. The insignificant portion of '*no data*' cells that still remain even after applying this method is treated in the later stages of the generalisation procedure, namely with using the polygon aggregation algorithm. It is a more 'intelligent' and target oriented procedure and it uses the cell properties directly, instead of estimating the cell class based on its neighbourhood characteristics. However, running the algorithm on a database having a lot of tiny neighbouring polygons with often missing bigger, dominant adjacent polygons to join, the product of the aggregation would be influenced much more by order of polygon processing than by the semantic characteristics of the polygon and would result in a random like aggregation of these polygons. That is why filtering was necessary first to decrease the number of polygons to an acceptable level. The polygon aggregation function was then used to deal with the rest of the '*no data*' polygons. Thus, a raster based filtering and a vector based aggregation was found to be the optimal combination for generalization.

New classification schemes were introduced by the SPM workshop for the four landform parameters derived from DEM data. This paper does not aim to discuss their benefits and impacts. The only conclusion made from the visual interpretation of the polygon system is that the new scheme follows well the geomorphologic units of the landscape and creates meaningful delineation. The changes in the class limits for slope and hypsometry were found to be a great improvement in this study. The new slope classes make a smoother and better discrimination of the land. The classes are grouped around the most frequently occurring values, therefore making a more balanced distribution among them. This phenomenon is even more evident with the hypsometry classes. New classes were introduced to improve differentiation among the landscape units on the low-lying plain areas, where the elevation above the sea level is almost the only significant terrain factor besides the PDD.

Three of the four terrain variables, namely the slope, RI and hypsometry, have not any contribution to the plain area characterization (Figure 11). The hypsometry classes contribute a little, but just 'accidentally' here because the Hungarian Great Plain lies along the 100m altitude class limit. The only parameter, which contributes to the landscape unit delineation on a plain area is the PDD. The slope and RI parameters, complemented with the hypsometry, are very efficient in characterizing the landscape of the hilly and mountainous lands. The combination of these three elements creates a very detailed physiographic characterization even without considering the PDD parameter. A large portion of the information provided by the PDD for these higher relief regions is already delivered by the other three parameters. However it is evident that the PDD carries much additional information over the other three parameters, but this appears mainly on a higher scale, not appropriate to the target scales of this project. As opposed to the high relief areas, only the PDD can provide meaningful information for unit delineation within the plain areas by highlighting the depressions and low lying areas where wetness potentially occurs. Using the three dissection classes suggested in the Manual (ISRIC 1993) would result in a lot of details and small patterns not adequate at the target scales. Instead, only two classes were created and adjusted to delineate both the low lying areas and valley bottoms in hilly and mountainous regions, and the depressions in plain areas. The class boundaries were defined empirically by testing different setups and matching them to real physiographic features.

The pilot area selected for the study was quite challenging due to its very complex natural and anthropogenic geomorphologic patterns. Starting from the second half of the nineteenth century an extensive dike system was built along the major rivers of the Hungarian Great Plain to prevent a huge area from annual floods and to expand agricultural land. At that time, it was often difficult to identify the major watercourse because huge areas were completely covered by ponding or flooding water appearing as temporal lakes. Nowadays the geomorphologic setup of the Plain still resembles the one from before the dike system. Huge low lying areas, depressions, narrow sand barriers, sand dunes, loess plateaus and old, abandoned river beds create a mosaic of geomorphologic patterns (Figure 12). This picture was further diversified by the man made dike and channel structure. Dikes along the Tisza River are captured in the SRTM DEM data and are visible on the DEM image as well (see the brown linear pattern along the Tisza River, pointed with two arrows in Figure 12). This of course has a significant impact on the drainage network determination necessary for deriving the PDD. The PDD is generally successful at detecting depressions and local heights. However, the area along the Tisza river is quite problematic. Dikes, especially where the surrounding area is relatively high as well are often taken as heights, preventing the water from flowing through to the river and are thus classified as local elevated areas. This phenomenon extends the area of the natural heights over the low laying ones (see the polygon peninsula extending along the Tisza and pointed by the two arrows). In other cases, where the low lying character of the surrounding area is strongly expressed, thus collecting a lot of drainage lines, these lines are trapped along the dikes on both sides of the river. This creates a high drainage density for those areas, despite the existence of the dikes (see the lower left corner of Figure 12). In contrast, these are classified as low lying areas. Although this effect hampers the continuous delineation of the Tisza River channel and shifts the balance towards the low PDD area over the high one, it is not as erroneous decision as it may appear. Indeed, dikes have a great impact on the surface water flow, which impact is reflected on the image as well.

The RI classification has changed a lot after the SPM workshop, but mainly in a formal way. The various RI units that were used in different contexts within the previous methodology were replaced with one common unit: m/[area of a 1 km diameter circle]. This new unit still maintains a link to the original SOTER RI units while it is easier to handle within a GIS and meets the definition used by geographers for that measure.

The final step in the procedure aims at generalizing the polygon system so that it meets the requirements of the SOTER Manual regarding the target scales: polygons that are below the minimum size limits of 25 and 625km^2 at the 1:1 and 1:5 million scales, respectively, must be eliminated and polygon borders must be simplified. Figure 13 shows the resulting images for the pilot area at these two target scales. Figure 14 is a zoom into the lower right corner of the pilot area. It shows all three polygon systems together to give an idea of the effect of the procedure: the raw polygon system as derived directly from the grids (blue lines), the polygon systems produced for the 1:1 million (red lines) and 1:5 million (yellow lines) target scales are displayed.

The procedure eliminates small polygons by aggregating each of them with its most similar neighbour. Similarity between a small polygon and its neighbours is measured by considering the Euclidean distance between their respective landform characteristics. The most similar neighbour, namely the one that is at the smallest Euclidean distance, is selected for aggregation.

On our dataset, despite reducing the number of polygons by 82% at the 1:1 million scale (see the Appendix notes at the end of the paper), aggregation still moderately altered the original polygon system. Only small, few cell-sized polygons were eliminated, while the overall look of the polygon system was well-kept. At that scale, most of the landscape units that are visible on the SRTM image have been successfully delineated. With hardly any exception, all the blue polygons in Figure 14 that were aggregated to form the red ones were correctly assigned, followed the landform units of the area.

As expected, aggregation to the 1:5 million scale resulted in a more drastic change of the polygon system, reducing the number of polygons by 98%. Characteristic – sometimes very different – landscape units had to be combined in a meaningful way. This required a lot of compromise, especially when a small intrusion had to be eliminated but none of its neighbours was really similar. Nevertheless, the algorithm generally performed successfully and no major problem was identified on the resulting polygon system. Many of the linear features such as valley bottoms were kept in the output and representative (to the geomorphologic units) Terrain Units were formed, appropriate for the target scale.

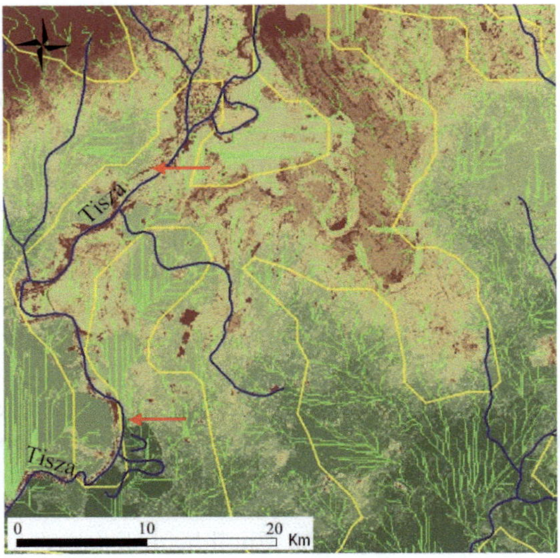

Fig. 12. A 50 by 50 km zoom into the Tisza Valley on the Hungarian Great Plain. The 15 m elevation range (81 to 96m above sea level) is represented with green to brown colours. Yellow lines show the resulting 1:5 million SOTER polygon system borders. Blue lines show the actual drainage system (rivers, streams and channels) whereas the green ones show the drainage system derived from the SRTM DEM as part of the PDD procedure. The red arrows highlight some of the dikes along the Tisza River

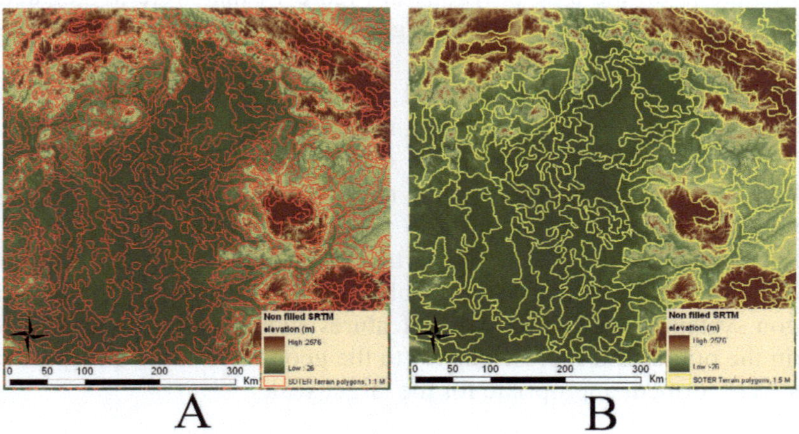

Fig. 13. The generalized polygon system derived from SRTM DEM data to produce the 1:1million (A) and 1:5 million (B) scales SOTER Terrain Units for the Hungarian Great Plain and for part of the Carpathians

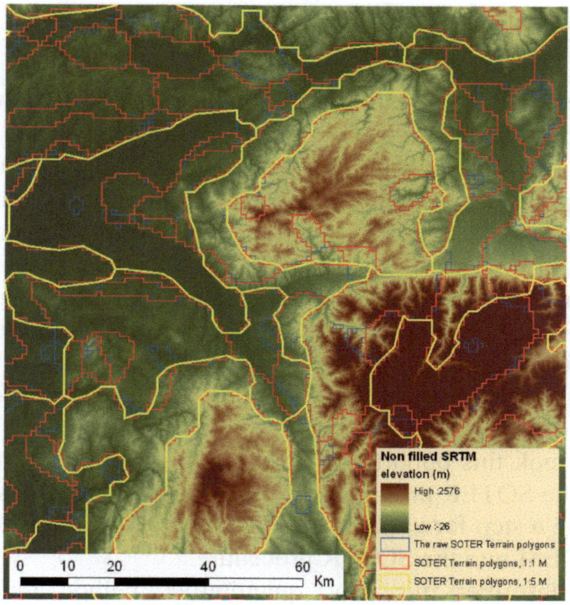

Fig. 14. The SOTER Terrain Unit structure aggregated to 1:1 and 1:5 million scales

Generalizing polygons means also generalizing the borders of the polygons that remain after aggregation so that their resolution meets the target scale. The corresponding tolerances (200 and 1000m, respectively) were chosen according to cartographic common sense (0.2mm on a map at both scales). Generalizing the arcs is achieved through line simplification. Among the two algorithms provided within the ArcInfo® GENERALIZE command, in our case and only by visual inspection of the results, the Douglas-Peucker algorithm (1973) appeared to perform better than the shape recognition one. A 'cosmetic' side effect of the line simplification is that the stair-like appearance of the arcs due to the vectorization of the original 990 by 990m grid cells is smoothed out. Applying both polygon aggregation and line simplification reduced the number of vertices (coordinate pairs) in our dataset by 53 and 89% at the 1:1 and 1:5 million scales, respectively, thus drastically reducing the database volume. Regardless of the scales, the aggregation method produces more or less homogenous Terrain Units giving satisfying results.

4. Conclusions

The development of a quantitative procedure for compiling a harmonized, consistent database is seen as a promising way to speed up the completion process of the global SOil and TERrain project. Many segments of the digital soil mapping technology have been made available since the late 1990s. Numerous studies were carried out to test the usefulness of digital elevation data for soil survey and characterization and much knowledge has been accumulated on this topic. The Shuttle Radar Topography Mission (SRTM) project developed a global digital elevation model coverage, which is now freely available and easily accessible for use. This emerging set of data and technology can help creating a common platform for the global SOTER database development.

The authors took this opportunity to develop a procedure and test it in the context of the SOTER project for the European Union. This pilot study aimed at making a step forward on this road by creating a methodology for incorporating DEM into the SOTER procedure.

This paper describes this new, quantitative method for creating a SOTER Terrain Unit polygon system. The method is designed for mapping large areas of the world quickly and cost effectively. The resulting SOTER database will have the advantages of quantitatively derived databases, namely consistency, homogeneity, limited data generalization problems, and it will avoid edge-matching and harmonisation problems. The procedure is based on the SOTER Manual specifications and is meant to be compatible with the datasets formally developed using the traditional way. But it also incorporates the procedural changes, which have occurred since 1995, when the last revision of the Manual was published.

The procedure has been tested on a representative pilot area covering the eastern half of the Carpathian Basin. The delineation of the terrain features is appropriate to the targeted scales. Meaningful and homogeneous geomorphologic units were identified at the 1:1 million scale in the test area. More complex but still uniform units were identified at the 1:5 million scale as well.

The method is used to develop the SOTER database for the Member States of the European Union. Further refinement and characterisation of the Terrain Units will be done using the Soil Geographical Database of Eurasia at scale 1:1,000,000 (King et al. 2002).

Acknowledgement

This study was supported by the European Commission, the Hungarian National Science Foundation (OTKA 34210), the Bolyai Foundation and the French National Agronomic Research Institute (INRA). The authors wish to express their thanks to Vincent van Engelen, Freddy Nachtergaele, Norm Bliss, Koos Dijkshoorn and Jan Nemecek for their scientific and technical support.

References

Batjes NH (1990) Macro-scale land evaluation using the 1:1 M World Soils and Terrain Digital Database. Working paper and preprint 90/9. Int. Soil Reference and Information Centre, Wageningen, The Netherlands

Batjes NH, Bridges EM (1997) Implementation of a soil degradation and vulnerability database for Central and Eastern Europe. Proceedings of ISRIC Workshop. Wageningen, 1-3 October 1997. ISRIC Wageningen, The Netherlands

Burrough PA (1986) Principles of geographical information systems for land resources assessment. Oxford University Press, New York

Dobos E, Montanarella L (2004) The development of a quantitative procedure for soilscape delineation using digital elevation data for Europe. Digital Soil Mapping Workshop, 14-17 September, Montpellier, France

Dobos E, Micheli E, Baumgardner MF, Biehl L, Helt T (2000) Use of combined digital elevation model and satellite radiometric data for regional soil mapping. Geoderma 97:367-391

Dobos E, Montanarella L, Negre T, Micheli E (2001) A regional scale soil mapping approach using integrated AVHRR and DEM data. International Journal of Applied Earth Observations and Geoinformation 3:XX

Douglas DH, Peucker TK (1973) Algorithms for the eeduction of the number of points required to represent a digitized line or its caricature. Canadian Cartographer 10:XX

ESB (2004) The European Soil Database. CD Version 2. European Commission, Joint Research Centre, Ispra, Italy

ESRI Inc. (1996) Automation of map generalization: the cutting-edge technology. Technical paper. White Papers section of ArcOnline [3]

FAO (1995) Global and national soil and terrain digital databases (SOTER). Procedures Manual. World Soil Resources Reports, 74 REV. 1

Farr TG, Kolbrick M (2000) Shuttle radar topography missions produces a wealth of data. American Geophysical Union, EOS 81:583-585

Hammond EH (1954) Small scale continental landform maps. Annals of Association of American Geographers 44:33-42

Hutchinson M (1988) Calculation of hydrologically sound digital elevation models. Third International Symposium on Spatial Data Handling. International Geographical Union, Columbus, Ohio

Hutchinson M (1989) A new procedure for gridding elevation and stream line data with automatic removal of spurious pits. Journal of Hydrology 106:211-232

Jarvis A, Rubiano J, Nelson A, Farrow A, Mulligan M (2004) Practical use of SRTM data in the tropics: comparisons with digital elevation models generated from cartographic data. Working Document No. 198. International Centre for Tropical Agriculture, Cali

ISRIC (1993) Global and National Soils and Terrain Databases (SOTER): Procedures Manual. UNEP-ISSS-ISRIC-FAO, Int. Soil Reference and Information Centre, Wageningen, The Netherlands

ISSS (1986) Project proposal World soils and terrain digital database at a scale 1:1 M (SOTER). Baumgardner MF (ed) Int. Society of Soil Science, Wageningen, The Netherlands

King D, Saby N, Le Bas C, Nachtergaele F, van Engelen V, Eimberck M, Jamagne M, Lambert JJ, Bridges M, Reinhard R, Montanarella L (2002) A method for generalization of a soil geographical database: the example of a transfer of the European database EUSIS at 1:1M to the world SOTER program at 1:5M. 17th World Congress of Soil Science, Bangkok, Thailand

Lambert JJ, Daroussin J, Eimberck M, Le Bas C, Jamagne M, King D, Montanarella L (2002) Soil geographical database for Eurasia and the Mediterranean. Instructions guide for elaboration at a scale 1:1,000,000 version 4.0. European Commission, Joint Research Centre, Ispra, Italy

Wilson JP, Repetto PL, Snyder RD (2000) Effect of data source, grid resolution, and flow-routing method on computed topographic attributes. In: Wilson JP, Gallant JC (ed) Terrain analysis. Principles and applications. John Wiley and Sons, New York

Worstell B (2000) Development of soil terrain (SOTER) map units using digital elevation models (DEM) and ancillary digital data. MSc Thesis, Purdue University, Indiana, U.S.A.

Referenced Web Sites

[1] ftp://edcsgs9.cr.usgs.gov/pub/data/srtm/

[2] http://srtm.csi.cgiar.org/SELECTION/inputCoord.asp

[3] http://arconline.esri.com

Appendix

The ArcInfo® command used for filling sinks:
FILL dem demfill SINK 20

The ArcInfo® commands used for the slope section:
slope = SLOPE (demfill, PERCENTRISE)
slopeclass = RECLASS (slope, slopeclass.txt)

The ArcInfo® commands used for the generalization section:
/* BLOCKING AND RESAMPLING:
slopeblock990 = BLOCKMAJORITY (slopeclass, RECTANGLE, 11, 11)
slopeclass990 = RESAMPLE (slopeblock990, 990)

/* FILTERING:
slope990filt4 = FOCALMAJORITY (slopeclass990, CIRCLE, 4)
slope990filt6 = FOCALMAJORITY (slopeclass990, CIRCLE, 6)
slope990 = CON (ISNULL (slope990filt4), slope990filt6, slope990filt4)

The ArcInfo® commands used for the Relief Intensity section:
ri = FOCALRANGE(demfill, CIRCLE, 5)
riclass = RECLASS (ri, riclass.txt)

The ArcInfo® commands used for the dissection (PDD) section:
flowdir = FLOWDIRECTION(demfill)
flowacc = FLOWACCUMULATION(flowdir)
drainage = CON(flowacc > 100, 1)
pdd = FOCALSUM(drainage, CIRCLE, 20)
pddclass = RECLASS(pdd, pddclass.txt)

The ArcInfo® command used for the hypsometry section:
hypsclass = RECLASS (demfill, hypsclass.txt)

Generalization results for the 1:1 million:
Arc: generalize_polys mlf mlf mlf1m MLF.DIS1 mlf_distances 1000000 .2 25

After 3 iterations one moves
from coverage MLF to coverage MLF1M
from 6589 to 1167 polygons (-82%)
from 18151 to 3370 arcs (-81%)
from 69495 to 32684 segments (-53%)

```
 Proportion of the area in resulting MLF1M polygons effectively
 concerned by the predominant value of MLF (key item):
  minimum:   23.636363983154297%
  mean:      86.632505730944672%
  maximum:   100.000000000000000%
  std:       17.280993168451701%

 Program generalize_polys.aml
   started on 31 Jan 05 15:33:36 Monday
   ended on   31 Jan 05 16:06:53 Monday
 Elapsed time: 0.55444444444

Generalization results for the 1:5 million:
Arc: generalize_polys mlf mlf mlf5m MLF.DIS1 mlf_distances 5000000
.2 25
 After 3 iterations one moves
  from coverage MLF to coverage MLF5M
  from 6589 to 120 polygons (-98%)
  from 18151 to 350 arcs (-98%)
  from 69495 to 7383 segments (-89%)

 Proportion of the area in resulting MLF5M polygons effectively
 concerned by the predominant value of MLF (key item):
  minimum:   13.505257606506348%
  mean:      55.283419232408541%
  maximum:   98.905235290527344%
  std:       26.534571550618274%

 Program generalize_polys.aml
   started on 28 Jan 05 17:01:47 Friday
   ended on   28 Jan 05 17:42:37 Friday
 Elapsed time: 0.68055555556
```

Solar Resource Modelling for Energy Applications

Marcel Šúri, Thomas Huld, Ewan D. Dunlop and Jaroslav Hofierka

1. Introduction

Solar energy is the main driver of natural processes on the Earth surface. It is an important input parameter into environmental, ecological and risk-simulation models as the energy budget at the land surface and depends on the terrain. Vegetation biodiversity, and biomass production are also related to the radiation input.

The Sun is an abundant and inexhaustible source of energy that is available everywhere on the planet and still is exploited in a minimum way. Around the turn of the century, the energy technologies exploiting solar radiation began to receive high political attention as they were recognised as possible alternatives to the conventional energy systems based on fossil fuels (see e.g. the conference 'Renewables 2004', [1]). It is generally accepted that solar energy systems can help to mitigate global problems, such as climate change, poverty in the developing world and insufficient security of energy supply in most of the world economies.

Like most new innovative technologies, there are factors blocking the full expansion in the world markets, such as low awareness about the technology options, higher initial investment costs, and conservatism of society to accept necessity of fundamental changes in the mode of energy generation, distribution and consumption. To overcome these barriers, countries may set up mid-term support policies (up to 10-20 years), defining appropriate legal and financial instruments. In this respect, in the European Union, the most important is the directive on the promotion of electricity produced from renewable energy sources (Directive 77/2001/EC) and directive promoting renewable energy for heating and cooling that is under preparation.

Solar energy impinging on the Earth surface is spatially and temporally variable with a seasonal dynamics (given by astronomic factors) that is modulated by stochastic variability of the weather. It influences the performance, reliability and economics of solar energy systems. Photovoltaics is a technology that experiences rapid growth (Jäger-Waldau 2005), and more detailed knowledge of the primary solar energy resource is needed. The improved knowledge can significantly contribute to better siting and economic assessment of the new installations, monitoring of their performance and forecasting of the delivered energy. Improved understanding is also needed for integration of solar energy technologies into existing energy and economic structures. The geographical dependency and distributed nature of renewable energy generation impose questions that require specific location-dependent answers also in policy making. Therefore map-based assessment of the availability of solar resource greatly contributes to setting up effective policies at the level of the European Union, its member states and regions.

A new generation of global digital elevation models (DEM), derived from the SRTM satellite mission [2] plays an important role in improving our solar energy resource knowledge. Therefore the objective of this chapter is to show the role of DEMs in solar radiation modelling within the GIS. We then present a case study where we have used the DEM data and developed GIS models – in solar energy resource assessment and estimation of electricity generation from photovoltaic (PV) systems in Europe.

2. Solar Radiation Modelling

According to the accepted terminology (Page 1986), two terms for solar (i.e. short-wave) radiation are used in this chapter. The term *irradiance* denotes the solar power (instantaneous energy flux) falling on a unit area per unit time ($W.m^{-2}$). The term *irradiation* denotes the amount of solar energy falling on a unit area over a stated time interval ($Wh.m^{-2}$). Solar radiation, incident on the Earth's surface, is a result of complex interactions of energy between the atmosphere and surface, determined by three groups of factors (see more in Šúri and Hofierka 2004):

- Earth's geometry, revolution and rotation (declination, latitude, solar hour angle),
- Terrain (elevation, surface inclination and orientation, shadows),
- Atmospheric attenuation (scattering, absorption) by:
  - Gases (air molecules, ozone, CO_2 and O_2),
  - Solid particles (aerosols) and non-condensed water,

- Clouds (condensed water).

The first group of factors determines the fluxes of extraterrestrial radiation that are dependent on the position of the Sun above the horizon; it can be precisely calculated using astronomical and geometry formulae.

The radiation input to the Earth's surface is then modified by terrain (elevation above sea level, slope inclination and aspect), and by shadowing effects of neighbouring terrain features. The second group of factors can be also modelled at a high level of accuracy using DEMs.

Solar radiation traversing from the top of the Earth's atmosphere to the surface is attenuated by atmospheric constituents – gas, liquid and solid particles (third group of factors). The path length of the solar beam through the atmosphere is also critical. The attenuation by gas constituents (factor 3.1) can be determined with a good level of precision. The attenuation by particles and water vapour (factor 3.2) is described by the Linke atmospheric turbidity factor and due to its dynamic nature and uncertainty, calculation and subsequent averaging leads to an inevitable generalisation. Clouds (factor 3.3) are the strongest attenuators, but due to their dynamics and complexity, most often only simplified empirical techniques are applied in the estimation of their effects. In recent years a lot of effort has been made (e.g. within the Heliosat-3 project, [3]) to improve estimates of the influence of atmospheric turbidity and cloudiness from satellite data.

The radiation that is not reflected or scattered and reaches the surface directly is known as *direct* (beam) radiation. The scattered radiation that reaches the ground is called *diffuse* radiation. The small part of radiation that is reflected from the ground onto an inclined receiver is described as *reflected* radiation. These three components together create *global* radiation.

Taking into account all three factors (3.1-3.3) of atmospheric attenuation in a calculation scheme gives *real-sky* (cloudy) irradiance values. Omitting the cloud attenuation (factor 3.3) results in *clear-sky* (cloudless) irradiance values; these are often used in satellite models such as Heliosat-2 (Rigollier et al. 2004).

3. Spatially Distributed Solar Databases

A number of solar energy resource databases and dedicated tools for the assessment of solar energy systems are available on the Internet, such as NASA SSE [4], European Solar Radiation Atlas (ESRA 2000, [5]), Meteonorm (Remund et al. 1999, [6]), Satellight, (Hammer et al. 1998, [7]), RETScreen [8] or PVGIS (Šúri et al. 2005, [9]). Many of them are interconnected and accessible via the SoDa web portal (Wald 2000, [10]). These databases were developed from ground measurements and/or satellite images.

Throughout Europe there are several hundred ground meteorological stations where solar radiation is directly measured or estimated from other climatologic parameters (sunshine, cloudiness, etc.). To prepare spatially-distributed databases from these measurements, interpolation techniques are used, such as splines, weighted averages or kriging (see Hutchinson et al. 1984; Hulme et al. 1995; Zelenka et al. 1992). The meteorological data are heterogeneously located around the continent and in most cases they do not represent spatial and vertical variability of solar radiation adequately. Therefore the use of additional information from DEMs and satellite images is used to improve the quality of interpolation, mainly by a co-kriging approach (D'Agostino and Zelenka 1992; Beyer et al. 1997) or by multivariate splines (Šúri and Hofierka 2004).

Since the mid 1980s, the polar-orbiting and geostationary satellites (such as NOAA, Meteosat, GOES, GMS) provide an alternative to monitor solar irradiance regularly from space. The advantage of satellites is continuous data coverage over vast territories (Noia et al. 1993a,b) in fast revisit time interval. For example, a new generation satellite, Meteosat-8 provides data with spatial resolution up to 1km, in 12 spectral bands every 15 minutes. To derive high resolution data in near real-time regime from satellite missions new algorithms based on deterministic approaches are under development (see e.g. Girodo et al. 2006). These algorithms can benefit from incorporation of high resolution DEMs as particularly in mountains varying elevation gradients combined with terrain shadowing cause significant local irradiance fluctuations and thus determine the lower accuracy of the model outputs.

To account for spatial variations of solar radiation in areas with mountainous terrain, models integrated within geographical information systems (GIS) were developed. Solar radiation models use ground-measured data and DEMs as inputs into physically-based and empirical equations to provide rapid and accurate estimates of radiation over large regions, while considering surface inclination, orientation and potentially also shadowing effects. In our previous work (Šúri and Hofierka 2004) we have analyzed GIS-based solar radiation models, such as SolarFlux (Dubayah and Rich 1995, Hetrick et al. 1993), Solei (Mészároš 1998), Solar Analyst (Fu and Rich 2000, [11]) and SRAD (Wilson and Gallant 2000). The analysis showed that these models have some restrictions in terms of their applicability for large territories and handling all necessary input parameters as spatially-distributed data. Therefore we decided to develop a new model, denoted as *r.sun* that is discussed later.

In the approaches used for developing solar radiation databases, the DEMs significantly contribute to improved knowledge of geographical and time variability of the solar energy resource. GIS plays an important role as a tool for integration of data from various sources and modelling algorithms. On the other hand, combination of solar radiation with other geographical data in GIS provides opportunities for customised analyses that are needed by decision-makers.

4. Solar Radiation Model *r.sun* and Terrain Parameters

Šúri and Hofierka (2004) have developed a GIS-based methodology for computation of solar irradiance/irradiation for any geographical region and for any time moment or interval. This approach has been implemented in the GIS software GRASS and it is based on use of the solar radiation model *r.sun* and the spatial interpolation techniques *s.surf.rst* and *s.vol.rst*. The model *r.sun* has the following features:

- It is integrated within GRASS GIS from version 5.0. the modular structure and open source code is available for further development (Neteler and Mitasova 2002).
- Inputs can be set-up flexibly as spatially-distributed data or constants (where appropriate).
- Calculation can be performed assuming solar or civil time.
- Calculation procedures use data either from ground meteorological stations or from satellites.
- Input and output data can be processed by other programs within the GRASS GIS.
- Memory management and code optimization allows to use high resolution data from local scale up to global scale.

The *r.sun* algorithm uses equations published in the European Solar Radiation Atlas (ESRA 2000). It estimates beam, diffuse and reflected components of the clear-sky and real-sky global irradiance/irradiation on horizontal and inclined surfaces. The model works in two modes. In *mode 1* for the instant time (second), it calculates grid maps of solar irradiance ($W.m^{-2}$) and solar incident angle (degrees). In *mode 2*, the grid maps of daily sum of solar irradiation ($Wh.m^{-2}$) and duration of the beam irradiation (minutes) are computed from the integration of irradiance values that are calculated at a user-selected time step from sunrise to sunset. If selected, for each time step the computation can account for sky obstruction (shadowing) by local terrain features.

By shell-scripting, these two modes can be used separately or in combination to provide estimates for any desired time step or interval. The theoretical background and applications are described in papers by Šúri and Hofierka (2004) and Šúri et al. (2005). Synopsis, description and notes to the command can be consulted at the *r.sun* manual page [12].

The *r.sun* model was used as a basis for developing another couple of GRASS modules. For example *r.sunoptangle* and *r.sunyear*, were developed to calculate the optimum inclination of a southwards-oriented surface with given orientation. These modules are used for calculation of optimum inclination of PV modules at which the daily or yearly irradiation input is at maximum. The program *r.sunoptangle* calculates the optimum angle for a given day in the year while *r.sunyear* calculates the optimum angle for the entire year (assuming a PV system with a fixed inclination angle). The programs output the optimum angle as well as the corresponding global irradiation values. In the model calculation scheme, the DEM parameters play an important role.

4.1 Elevation Above Sea Level

The elevation above sea level determines the attenuation of radiation by optical thickness of the atmosphere (air mass). At high elevation, due to lower air mass, the influence of atmospheric gases and turbidity is lower and the incident irradiance is thus less attenuated. As demonstrated below (Figure 2) the DEM resolution significantly improves the spatial pattern of the modelled irradiation.

4.2 Inclination and Aspect

The inclination and aspect (orientation to the cardinal points) of the surface modify the incident beam, diffuse and ground-reflected irradiance components. These two parameters influence the spatial-temporal pattern of radiation fields. The influence of slope and aspect is more pronounced under clear-sky (sunny) weather, compared to the cloudy conditions when diffuse radiation dominates.

In environmental and ecological studies, the information on spatial and temporal pattern of solar energy impinging on the terrain surface is needed for study of the geographical and ecological factors and processes (e.g. air and soil temperature, moisture, snow melting, photosynthesis and evapotranspiration).

In technical sciences the information about solar energy incident onto artificial surfaces (e.g. solar photovoltaic or thermal modules, and walls of buildings) is needed.

On of examples the western part of the Alpine region (Figure 1), the two maps show the yearly average of global irradiation (in kWh.m^{-2}) incident on a horizontal surface (Figure 1a) and on the real terrain (Figure 1b). While the first map is used as a basis for energy generation assessment from a solar system (e.g. for calculation of an energy input onto a PV module), the second one (where slope and aspect of the ground surface are taken into account) is more used in hydrological and ecological modelling.

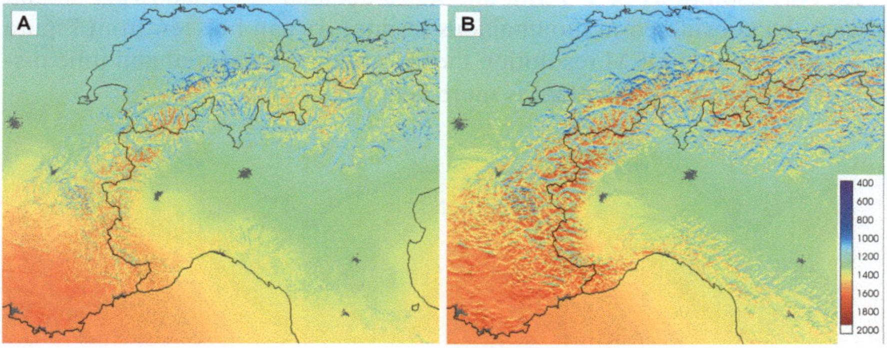

Fig. 1. Yearly sum of global irradiation (kWh.m^{-2}) received (A) by a horizontal surface, e.g. PV module and (B) by the real terrain (simulated by the 1-km grid DEM). The figure shows an example of the western part of the Alps. Large cities are mapped by grey colour (source: CORINE Land Cover)

4.3 Shadowing

The effects of neighbouring terrain features can considerably change the radiation in the mountains. Therefore the *r.sun* model has incorporated routines to deal with shadowing by the local terrain.

The visibility of a grid cell is evaluated in the direction of solar beam towards the Sun for each time step from sunrise to sunset (in our case the time step is 15 minutes). The sampling path from the cell under consideration towards the Sun is adjustable and can be changed to offer either fast or more precise estimations. A slower, but more precise algorithm uses short sampling distances to calculate the visibility. A faster algorithm with longer sampling distances uses the maximum altitude found in 4 nearest surrounding cells. The curvature of the Earth is taken into consideration.

The calculation of the shadowing effect is a computationally intensive task and therefore a separate GRASS module *r.shadow* was developed to speed-up the calculation. For each time step the terrain shadowing is calculated for each cell and the results are stored as binary values (shadowed – yes/no) in separate grid files that are read afterwards by *r.sun*.

Considering the DEM influence, the scale issues are important. From the above-mentioned factors, the shadowing is the most sensitive to the DEM grid resolution as can be seen in Figure 2. For applications at global scale, solar radiation calculated with grid resolution 10-40 km or even rougher provides an overview information and no shadowing is considered. For resolutions of 1-5 km, and more detailed, the shadowing should already be considered in mountainous terrain. As can be seen from Figure 2, with increasing DEM resolution the information on spatial and time distribution of shadowing pattern is more detailed.

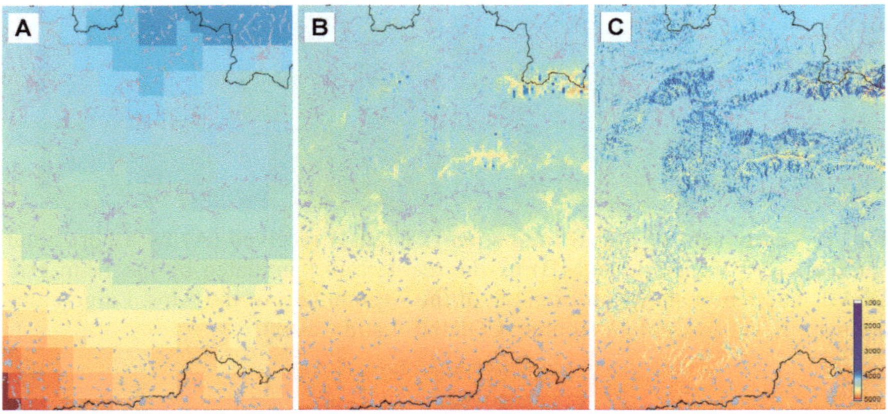

Fig. 2. Monthly average of daily sums of global irradiation (Wh.m^{-2}) on a horizontal surface (August, average for the period 1981-1990). A. Dataset with 10-km grid resolution without terrain shadowing (source: ESRA). B. Data calculated by the *r.sun* model with 1-km grid resolution assuming terrain shadowing (source: PVGIS). C. data calculated by *r.sun*, using DEM with 100-m grid resolution assuming terrain shadowing

In detailed studies with DEM at resolutions of hundreds to tenths of meters, the character of the data strongly affects the visibility. In such case, the DEMs which preserve terrain extremes (tops of the mountains, saddles, ridges and bottoms of valleys) should be preferred to DEMs with smoothed data.

5. PVGIS: Application of solar Radiation Model in an Assessment of Photovoltaic Power generation

The *r.sun* model was applied in the Photovoltaic Geographic Information System (PVGIS), which combines the long-term expertise from laboratory research, monitoring and testing of PV devices at the JRC, Renewable Energies Unit with geographical knowledge in order to serve three purposes:

- As a research tool, it improves understanding of performance of PV technology in regions of Europe,
- The system assists the policy-making process in the European Union,
- The developed interactive web interface provides an access to the basic data, maps and tools to decision-makers, professionals from manufacturing industry, installation and maintenance companies, the PV systems owners as well as to interested individuals.

The spatial extent of the GIS database is 5000 x 4500km, the grid resolution is 1x1 km and the map projection is Lambert azimuthal, equal-area on WGS84 ellipsoid. The main inputs used in the computation were as follows:

- Monthly averages of daily sums of global and diffuse irradiation for 566 ground meteorological stations (these data cover a period 1981-1990; source: ESRA 2000, [5]),
- Linke atmospheric turbidity (source: Remund, 2003, available also from [10]),
- Digital elevation model with a grid resolution of 1x1km, derived from the USGS SRTM-30 data (source: [2]).

The primary grid database represents the period 1981-1990 and it contains 12 monthly averages and a yearly average of the following climatic parameters:

- Daily global irradiation on a horizontal surface,
- Ratio of diffuse to global horizontal irradiation,
- Clear-sky index (describes the average cloudiness of the sky).

The global irradiation was calculated by the model *r.sun* using the input grid layers of DEM, Linke turbidity, clear-sky index and diffuse to global radiation ratio. The last two parameters were processed by interpolation from the monthly average values available for the meteorological stations using the multivariate regularized spline with tension, implemented in GRASS GIS as the module *s.vol.rst* (Hofierka et al. 2002; Neteler and Mitasova 2002). The outline of the calculation scheme is shown in Figure 3.

Methodology in details can be consulted in the papers by Šúri and Hofierka (2004) and Šúri et al. (2005).

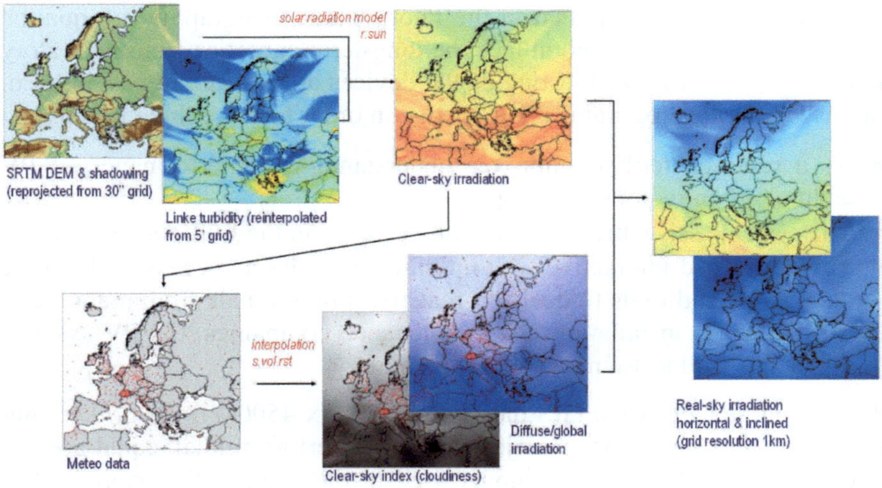

Fig. 3. PVGIS solar radiation database: the calculation scheme

The primary database, in combination with developed tools, can be used for calculation of derived data that are related to solar electricity generation:

- Global irradiation for horizontal and inclined surfaces,
- Clear-sky and average real-sky daily profile of irradiances (considering also terrain shadowing),
- Average beam, diffuse and reflected components of the global radiation,
- Average electricity generation from fixed and tracking PV systems,
- Optimum inclination and orientation of fixed PV modules to maximize energy yields.

The current PVGIS database, covering the European continent, is based on the archive of climatic data from ground stations and provides only averages over the 10-years period determined by the available measurements from meteorological stations. The level of spatial detail of the data is determined by the DEM with a grid resolution of 1km and therefore the structure of the terrain features (elevation and shadowing) is represented only approximately. Besides solar radiation and derived outputs, the PVGIS database contains also other geographical data (temperatures, administrative boundaries, cities, land cover). These are used in spatial analyses of the technical and economic performance of PV systems (Dunlop et al. 2005; Huld et al. 2005).

The GIS database in itself is useful only for a limited number of experts. To provide access to the general public, we decided to link the data with web-based interactive applications. The web interface provides access to solar radiation and PV performance assessments, in the form of maps and graphs/tables for any location that can be chosen by clicking on a map, by choosing a country and a city from a list, or by setting latitude/longitude. All the information can be accessed from [9].

6. Conclusions

The solar radiation model *r.sun* integrated into the PVGIS decision-support system, improves knowledge of the geographical diversity of the solar energy resource and the technical and socioeconomic aspects of distributed energy generation from solar systems. This knowledge improves decision making at political level, in the industry and business, as well as in the daily life.

Currently, the PVGIS development continues and with an integration of the Meteosat satellite data to provide probability statistics at higher accuracy, especially in regions that are not well represented by the ground measurements.

Better understanding of solar radiation is needed to satisfy the growing needs of the solar energy sector as well as in environmental modelling. The new fields of research and applications are emerging that develop methods to provide historical, actual and predicted data with high spatial and temporal resolutions. These data are needed for the solar energy industry, utilities, policy-makers and users for planning, siting, monitoring, and performance forecasting of the renewable-energy technologies. Similarly to the solar energy industry, environmental modelling needs data to improve knowledge of short-term fluctuations of solar energy. The long-term averages (monthly, yearly) derived from satellite and climatologic archives are needed for improved statistical assessments.

GIS is extensively used in the PVGIS project, as it offers an environment for:

- Data integration: data fluxes from satellites and other sources need a unified infrastructure for maintenance and processing of the spatially-distributed data with an option for spatially-defined retrieval of time series,

- Implementing of new modelling algorithms: the new algorithms need input parameters that are measured and calculated at distant facilities, Proper networking and spatial-temporal data processing are needed,
- Hierarchical data processing: various applications need products at different levels of complexity and spatial and temporal resolutions,
- Links to other data and algorithms: solar radiation products are generally used as input to other models or assessments,
- Data visualisation: map-based visualisation and animation is needed for understanding the dynamics of the phenomena and the related effects in regions,
- It provides a link to the web interface and the interactive tools developed using Open Source code.

DEMs provide important input parameters into the modelling algorithms. The availability of high-resolution data from the SRTM missions (DEMs with spatial resolution of 1" and 3" in geographical coordinates, see [2]) opens new horizons for research in mountainous environments where the spatial pattern of solar radiation fields is complex and dynamic. DEMs can be integrated directly into new satellite image processing algorithms in order to better cope with problems of terrain shadowing and cloudiness.

This chapter provides only overview information; the reader is invited to consult more details in our referenced publications and also at the PVGIS web site [9].

Acknowledgement

This work has been supported by the European Commission Joint Research Centre, the Action No. 2324 'SOLAREC', and by the Scientific Grant Agency of the Slovak Republic, project No. 1/3049/06.

References

Beyer HG, Czeplak G, Terzenbach U, Wald L (1997) Assessment of the method used to construct clearness index maps for the new European solar radiation atlas (ESRA). Solar Energy 61:389-397

D'Agostino V, Zelenka A (1992) Supplementing solar radiation network data by co-kriging with satellite images. International Journal of Climatology 12:749-761

Directive 2001/77/EC of the European Parliament and of the Council of 27 September (2001) on the promotion of electricity from renewable energy sources in the internal electricity market. Official Journal L 283 of 27.10.2001

Dubayah R and Rich PM 1995 Topographic solar radiation models for GIS. International Journal of Geographical Information Systems 9:405-19

Dunlop ED, Huld TA, Šúri M, Ossenbrink HA (2005) Is solar electricity expensive? Proceedings of the 20th European Photovoltaic Solar Energy Conference and Exhibition, 6-10 June 2005, Barcelona, Spain

European Solar Radiation Atlas (ESRA) (2000) Greif J, Scharmer K. Scientific advisors: Dogniaux R, Page JK. (eds). Published for the Commission of the European Communities by Presses de l'Ecole des Mines de Paris, Paris, France (web site: [5])

Fu P, Rich PM (2000) The Solar Analyst 1.0 User Manual. Helios Environmental Modeling Institute (web site: [11])

Girodo M, Mueller RW, Heinemann D (2005) Influence of three-dimensional cloud effects on satellite derived solar irradiance estimation – First approaches to improve the Heliosat method. Solar Energy 80:1145-1159

Hammer A, Heinemann D, Westerhellweg A, Ineichen P, Olseth JA, Skartveit A, Dumortier D, Fontoynont M, Wald L, Beyer HG, Reise Ch, Roche L, Page J (1998) Derivation of daylight and solar irradiance data from satellite observations. Proceedings of the 9th Conference on satellite meteorology and oceanography, Paris, May 1998, pp. 747-750

Hetrick WA, Rich PM, Barnes FJ, Weiss SB (1993) GIS-based solar radiation flux models. American Society for Photogrammetry and Remote Sensing Technical papers. GIS, Photogrammetry and Modeling 3:132-43

Hofierka J, Parajka P, Mitasova M, Mitas L (2002) Multivariate interpolation of precipitation using regularized spline with tension. Transactions in GIS 6, 135-50

Hutchinson MF, Booth TH, McMahon LP, Nin HA (1984) Estimating monthly mean values of daily total solar radiation for Australia. Solar Energy 32:277-290

Huld TA, Šúri M, Kenny RP, Dunlop ED (2005) Estimating PV performance over large geographical regions. Roceedings of the 31st IEEE Photovoltaic Specialists Conference and Exhibition, 3-7 January 2005, Coronado Springs Resort, Lake Buena Vista, Florida, USA, pp. 1679-1682

Hulme M, Conway D, Jones PD, Jiang T, Barrow EM and Turney CA (1995) 1961-1990 climatology for Europe for climate change modelling and impact applications. International Journal of Climatology 15:1333-1364

Jäger-Waldau A (2005) European photovoltaics in world wide comparison. Proceedings of the 21st International Conference on Amorphous and Nanocrystalline Semiconductors, September 4-9 2005, Lisbon, Portugal

Mészároš I (1998) Modelovanie príkonu slnečnej energie na horské povodie. Acta Hydrologica Slovaca 1:68-75

Neteler M and Mitasova H (2002) Open Source GIS: A GRASS GIS Approach. Kluwer Academic Publishers, Boston

Noia M, Ratto CF, Festa R (1993a) Solar irradiance estimation from geostationary satellite data: I. Statistical models. Solar Energy 51:449-456

Noia M, Ratto CF, Festa R (1993b) Solar irradiance estimation from geostationary satellite data: II. Physical models. Solar Energy 51:457-465

Page JK (ed) (1986) Prediction of solar radiation on inclined surfaces. Solar Energy R&D in the European Community, Series F: Solar Radiation Data, Vol. 3, D. Reidel Publishing Company, Dordrecht

Remund J, Kunz S, Lang R (1999) METEONORM: Global meteorological database for solar energy and applied climatology. Solar Engineering Handbook, version 4.0, Meteotest, Bern (web site: [6])

Remund J, Wald L, Lefèvre M, Ranchin T, Page J (2003) Worldwide Linke Turbidity Information. ISES Solar World Congress on Solar Energy for a Sustainable Future, Göteborg, Sweden, 14-20 June 2003

Renewables (2004) International Conference for Renewable Eenergies, Bonn, Germany, 1-4 June 2004 (web site: [5])

Rigollier C, Lefèvre M, Wald L (2004) The method Heliosat-2 for deriving shortwave solar radiation from satellite images. Solar Energy 77:159–169

Šúri M, Hofierka J (2004) A New GIS-based Solar Radiation Model and Its Application to Photovoltaic Assessments. Transactions in GIS 8:175-190

Šúri M, Huld TA, Dunlop ED (2005) PVGIS: a web-based solar radiation database for the calculation of PV potential in Europe. International Journal of Sustainable Energy 24:55-67

Wald L (2000) SODA: a project for the integration and exploitation of networked solar radiation databases. European Geophysical Society Meeting, XXV General Assembly, Nice, France, 25-29 April 2000

Wilson JP, Gallant JC (2000) Secondary topographic attributes. In: Wilson JP, Gallant JC (eds) Terrain analysis. Principles and applications. Wiley, New York, pp. 87-132

Zelenka A, Czeplak G, D'Agostino V, Josefson W, Maxwell E, Perez R (1992) Techniques for supplementing solar radiation network data, Technical Report, International Energy Agency, # IEA-SHCP-9D-1, Swiss Meteorological Institute, Switzerland

Referenced Web Sites

[1] http://www.renewables2004.de/
[2] http://srtm.usgs.gov/
[3] http://www.heliosat3.de/
[4] http://eosweb.larc.nasa.gov/sse/
[5] http://www.helioclim.net/esra/
[6] http://www.meteotest.ch/
[7] http://www.satellight.com/
[8] http://www.retscreen.net/

[9] http://re.jrc.ec.europa.eu/pvgis/pv/
[10] http://www.soda-is.com/
[11] http://www.fs.fed.us/informs/download.php
[12] http://grass.itc.it/gdp/html_grass5/html/r.sun.html

GRASS and R - Advanced GIS and Statistical Tools for DEM Analysis

Radoslav Bonk

1 Introduction

The aim of this paper is to present two Open Source applications - *R* (programming environment for data analysis and graphics) and *GRASS* (geographic resources and analysis support system) Open Source Free GIS. Even-though there is a plethora of GIS and statistical applications on the market, the two applications are among the leaders in Open Source statistical and GIS solutions(1). Both are well-suited for research, educational purposes and for studies where non-trivial analyses are required. Moreover, sophisticated algorithms make them the right choice for scientists, and access to the source code makes them the right choice for those who can alter the code and suit it for specific purposes.

Geographic Resources Analysis Support System, commonly referred to as *GRASS* GIS(2), is used for data management, image processing, graphics production, spatial modelling, and visualisation of many types of data (Neteler and Mitasova 2002). It is free software released under GNU General Public License(3). *GRASS* is widely used not only in universities and research institutions, but in the business area as well. *R* is a language and environment for statistical computing and graphics(4). It is a GNU project which is similar to the S language and environment which was developed at Bell Laboratories by John Chambers and colleagues (Chambers and Hastie 1992). *R* project is the results of *R* core development team, and Internet volunteers. However, initial work was done by Ross Ihaka and Robert Gentleman (Ihaka and Gentleman 1996). *R* provides a wide variety of statistical (linear and nonlinear modelling, classical statistical tests, time-series analysis, classification, cluster analysis) and graphical techniques, and is highly extensible. One of *R*'s strengths is the ease with which well-designed publication-quality plots can be produced, including mathematical symbols and formulae where needed. Both applications run on a variety of UNIX platforms and similar systems (including FreeBSD and Linux). It also compiles and runs on Windows 9x/NT/2000 and MacOS. The real analytical power arises from the fact, that the two applications can be linked together. In such a case *GRASS* serves as the GIS subsystem and *R* serves as a statistical subsystem and publication-quality graphics production engine (Figure 1). Even-though there are graphical user interfaces available, both applications offer the command line approach. The objective of this

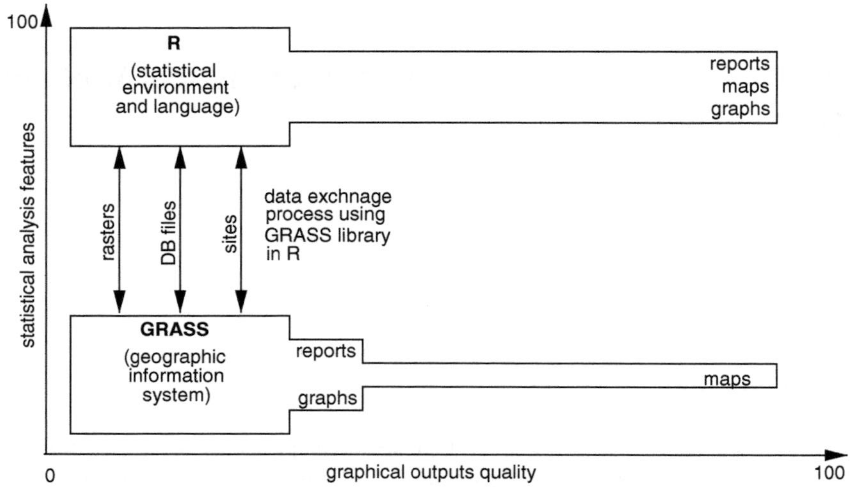

Fig. 1. *GRASS* GIS and *R* environment can be linked together for data exchange. While *GRASS* is strong in GIS analysis, *R* can produce publication-quality graphics and is a powerful statistical analysis tool. They both form a powerful application for modern geomorphological research

article is to show how *GRASS* and *R* can be linked together. Focus is given on how to exchange the data, perform the statistical analysis and produce high-quality graphics, while the full methodology and illustrative results are provided by Bonk (2003). These features will be presented on a case study in Slovakia, Devínska Kobyla ridge. It is expected that the user has a *GRASS* and *R* installation on the system (the installation won't be described here). The objective for the case study is: Determine the degree of accuracy of DEM-derived morphometric parameters (elevation – Z, slope angle – γ_N, profile curvature – $(K_N)_n$, tangential curvature – K_r), when compared with field observations, using original DEM, and smoothed DEM.

2 Case Study

There were 75 geodetic points (Figure 2), and four marginal points for each geodetic point, measured at the region of Devínska Kobyla (Minar et al. 2001). Marginal points values were used to compute four morphometric properties for the central point (elevation – Z, slope angle – γ_N, profile curvature – $(K_N)_n$, tangential curvature – K_r). Original and smoothed DEM, were produced to compare them with field-measured morphometric parameters (Bonk 2003). The original DEM (Z_{040}) passes directly through topographic maps contour lines, thus producing no residuals. This was achieved by specifying the parameters of tension ($\varphi = 40$) and smoothing ($B_a =$

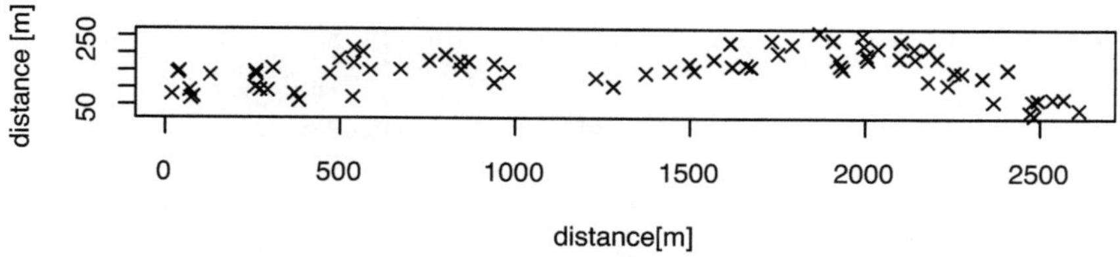

Fig. 2. Location of geodetic points measured at Devínska Kobyla region (Western Slovakia)

0) in regularised spline with tension interpolation function (Mitasova and Hofierka 1993). The smoothed DEM (Z_{0820}) doesn't pass exactly through the contour lines, and thus produces residuals from contour lines (Bonk 2003). *R* session starts from within the *GRASS* prompt, and it can look like this one:

```
GRASS:> R
> library(GRASS)
> G <- gmeta()
> kzdevi0820 <- sites.get(G, slist=c("kzdevi0820"))
> ls()
> [1] "kzdevi0820"
> summary(kzdevi0820\$var1)
Min. 1st Qu. Median Mean 3rd Qu. Max.
-3.74100 -0.37070 -0.04152 -0.03549 0.31060 5.27500
```

In this example the *R* session was started from the *GRASS* prompt, and the *GRASS* library was loaded into *R* environment (since it is not automatically loaded when *R* starts). Then *GRASS* region info was stored into *G* object and desired *GRASS* site file was imported into *R* and finally listed. Then basic statistics on site file residuals were obtained. Consequently, newly computed DEMs (Figure 3) were subtracted from the reference elevation data resulting in regular DEM elevation residuals (R^Z_{040}), and smoothed DEM elevation residuals (R^Z_{0820}). Other residuals of slope angle ($R^{\gamma_N}_{040}, R^{\gamma_N}_{0820}$), profile curvature ($R^{(K_N)_n}_{040}, R^{(K_N)_n}_{0820}$), and tangential curvature ($R^{K_r}_{040}, R^{K_r}_{0820}$) were computed in the same manner ("040" subscript refers to original DEM where smoothing was set to 0 and tension was set to 40, "0820" subscript refers to smoothed DEM where smoothing was set to 0.8 and tension to 0.2). Exploratory data analysis (EDA) graphics were plotted to see the nature of these residuals. It is quite easy to produce sets of EDA graphics for elevation and slope angle residuals to study their nature (Figure 4). How to produce simple EDA graphics and how to export it the graphical output into high-quality encapsulated postscript file is covered in the following example:

```
> library(Simple)
> postscript("KGZ.kz040.eps",height=2.3,width=4)
> simple.eda(KGZ.kz040$var1)
> dev.off(2)
```

Fig. 3. Comparison of original and smoothed morphometric surfaces. Morphometric parameters surfaces are presented as follows: **a.** elevation based on the original DEM Z_{040}, **b.** elevation based on the smoothed DEM Z_{0820}, **c.** slope angle based on the original DEM $(\gamma_N)_{040}$, **d.** slope angle based on the smoothed DEM $(\gamma_N)_{0820}$, **e.** profile curvature based on the original DEM $((K_N)_n)_{040}$, **f.** profile curvature based on the smoothed DEM $((K_N)_n)_{0820}$, **g.** tangential curvature based on the original DEM $(K_r)_{040}$, **h.** tangential curvature based on the smoothed DEM $(K_r)_{0820}$

In this session *Simple* library was loaded at the beginning. Then postscript driver was initialised together with filename and dimensions of the graphic. By using *simple.eda* command, with R^Z_{040} as an argument, histogram, boxplot and normal Q-Q plot was written into specified filename. Consequently, it is necessary to close the postscript device. To study the independence of estimation errors it is necessary to compare estimation errors (e.g. original DEM residuals – R^Z_{040}) against actual values of field-measured morphometric parameters (KGZ). This can be done in *R* by plotting simple scatterplot, in this example enhanced by the histograms (Figure 5):

```
> str(KGZ)
'data.frame':   69 obs. of  4 variables:
 $ east : num  40.4 35.8 70.1 79.0 19.3 ...
 $ north: num  147.9 142.6  91.1  73.0  80.2 ...
 $ var1 : int  1 2 3 5 6 8 9 11 12 13 ...
 $ var2 : num  139 139 138 137 136 ...
> str(KGZ.kz040)
'data.frame':   69 obs. of  3 variables:
```

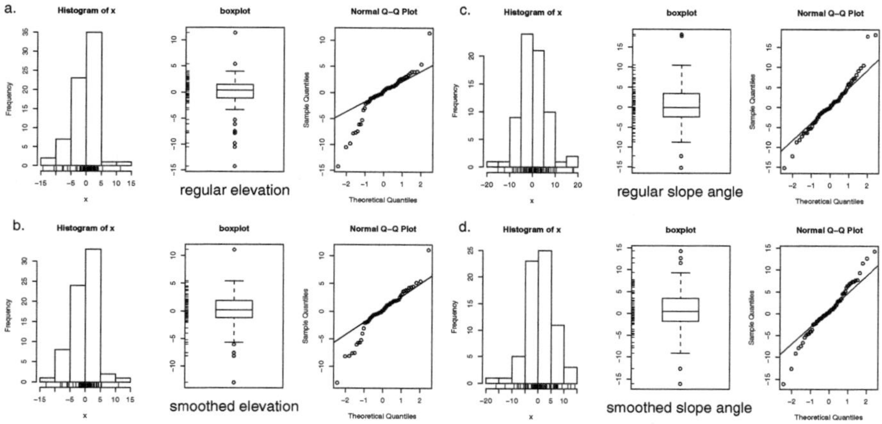

Fig. 4. Exploratory data analysis for elevation residuals R^Z and slope angle residuals R^{γ_N} at Devínska Kobyla region. Residuals are presented as follows: **a.** residuals for elevation based on the original DEM R^Z_{040}, **b.** residuals for elevation based on the smoothed DEM R^Z_{0820}, **c.** slope angle residuals based on the original DEM $R^{\gamma_N}_{040}$ **d.** slope angle residuals based on the smoothed DEM $R^{\gamma_N}_{0820}$

```
 $ east : num  1868 1992 1908 1734 2102 ...
 $ north: num  260 252 240 238 236 ...
 $ var1 : num  -2.102  1.203 -0.724 -0.333 -1.851 ...
> simple.scatterplot(KGZ$var2,KGZ.kz040$var1)
```

In *R* one can have several fields (columns) of data, in one *R* object. To specify the desired one for analysis, *$* sign have to be used followed by the name of the column (e.g. *$var1*). Here, column names of *R* objects were listed (*str*), and then used the desired one in *simple.scatterplot* command. This example is visually very helpful.

To study the degree of error it is helpful to plot two boxplots next to each other (Figure 6) using the same scale. Here the boxplots of morphometric parameters derived from the regular DEM were plotted against the boxplots of its residual (*R* code is presented for *Z* only, other morphological parameters can be plotted in the same way). Rugs, representing the actual observations, are plotted on both sides of the graphic:

```
> boxplot(KGZmask$var2, KGZ.kz040$var1, notch=T,col=
"grey")
> rug(KGZmask$var2, side=2)
> rug(KGZ.kz040$var1, side=4)
```

The second set of analyses is represented by the spatial analysis performed by plotting residual values at their locations (Figure 7), to learn whether there is any spatial dependence. The size of the circles in the map increases with the magnitude of the residual by the step defined by value of the deciles of the particular dataset:

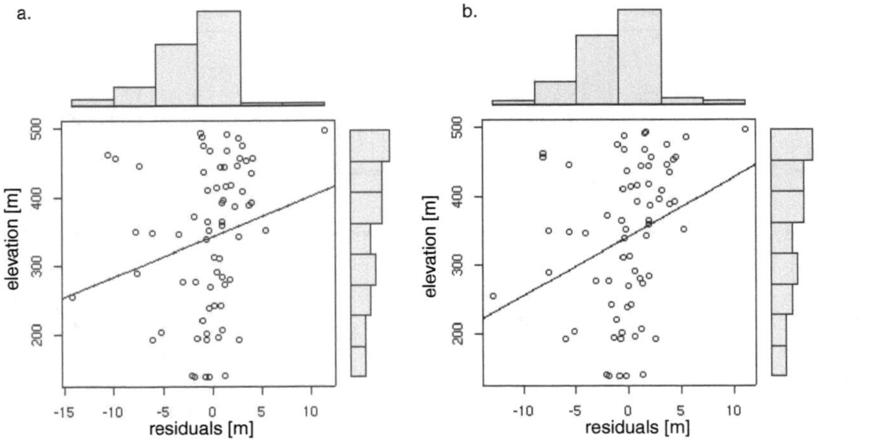

Fig. 5. Scatter plots representing independence of estimation accuracy errors and field-measured data for elevation Z. Field-measured values are on "Y" axis, and residuals on "X" axis. Residuals and field-measured elevation are presented as follows: **a.** elevation and residuals based on the original DEM Z_{040}, **b.** elevation and residuals based on the smoothed DEM Z_{0820}

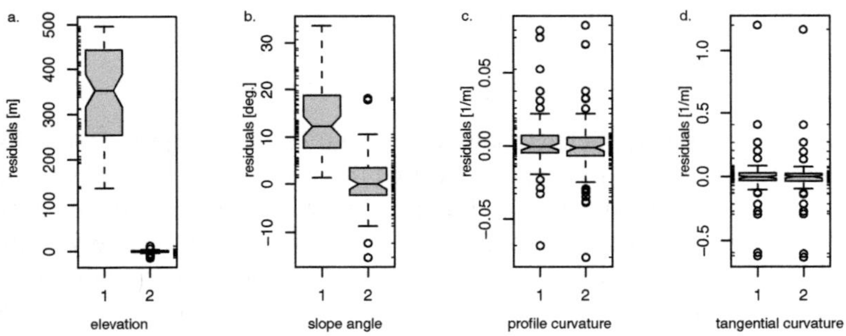

Fig. 6. Relative boxplots representing both, morphometric parameters (left) and its residuals (right) for Devínska Kobyla region. The datasets are presented as follows: **a.** elevation based on the original DEM Z_{040}, elevation residuals based on the original DEM R^{Z}_{040}; **b.** slope angle based on the original DEM $(\gamma_N)_{040}$, slope angle residuals based on the original DEM $R^{\gamma_N}_{040}$; **c.** profile curvature based on original DEM $((K_N)_n)_{040}$, profile curvature residuals based on the original DEM $R^{(K_N)_n}_{040}$; **d.** tangential curvature based on the original DEM $(K_r)_{040}$, tangential curvature residuals based on the original DEM $R^{K_r}_{040}$

```
> library(geoR)
> library(GRASS)
> rgeo1 <- as.geodata(KGZ.kz040)
```

```
> G2 <- gmeta()
> plot.grassmeta(G2)
> points.geodata(rgeo1, add.to.plot=TRUE, pt.divide=\
"data.prop", col=gray(seq(1, 0.1,l=100)), cex.max=2,\
 cex.min=0.1)
> title(xlab="distance [m]", ylab="distance [m]")
```

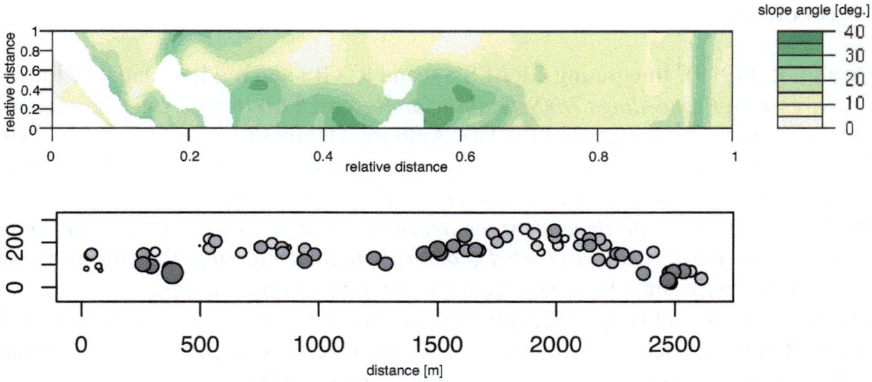

Fig. 7. Spatial location and degree of elevation residuals based on the original DEM (R^Z) at Devínska Kobyla region (upper part of the figure is to illustrate terrain properties and is not the output of the code)

First, *geoR* and *GRASS* libraries were loaded in *R* session. Then the regular data frame was transformed to "geodata" and plotted using cartographic entities with variable radius. A title for the map was added at the end. This way, many types of spatial phenomena can be plotted in their actual location using *R*.

3 Conclusions

In this short introduction to *GRASS* and *R* main focus was given to presenting efficient statistical approaches and publication-quality graphics production. However simple, but helpful examples were shown to provide users with the look and feel of *R* language in the GIS and terrain modelling environment. It might be stated, that *R* provides many helpful tools users need for modern geomorphological data analysis, and if they are not included in *R* directly, they can be programmed using the existing commands. In this way the functionality of *R* can be easily extended. The main advantage comes from using both applications in one modular system. Its strength comes from the fact that *GRASS* is powerful in the GIS field, and thus it does not have to include advanced statistical functions. Although *R* has some mapping functions, its main strength is in statistics and spatial statistics. Thus, the combination of both tools makes it an extremely powerful system for the researcher. Both tools are being used in JRC Ispra where European scale digital elevation models are being produced and tested for flood forecasting and flood hazard mapping. More in depth information and examples can be found at work of Roger Bivand (Bivand 1999, 2000). Also books on *GRASS* (Neteler and Mitasova 2002) and *R* (Venables and Ripley 2002) exist.

References

Bivand, R. (1999). Integrating GRASS 5.0 and R: GIS and modern statistics for data analysis. In *Proceedings 7th Scandinavian Research Conference on Geographical Information Science*, pages 111–127, Aalborg, Denmark

Bivand, R. (2000). Using the R statistical data analysis language on GRASS 5.0 GIS database files. *Computers & Geosciences*, (26):1043–1052

Bonk, R. (2003). *Scale-dependent impact of selected factors on morphometric parameters accuracy and automated geomorphological mapping*. PhD thesis, Comenius University Bratislava, Slovakia, Faculty of Natural Sciences

Chambers, J. M. and Hastie, T. J. (1992). *Statistical models in S.* Chapman & Hall

Ihaka, R. and Gentleman, R. (1996). R: A language for data analysis and graphics. *Journal of Computational and Graphical Statistics*, 5(3):299–314

Minar, J., Barka, I., Bonk, R., Bizubova, M., Cernansky, J., Faltan, V., Gasparek, J., Koleny, M., Kozuch, M., Kusendova, D., Machova, Z., Mician, L., Micietova, E., Michalka, R., Novotny, J., Ruzek, I., Svec, P., Trembos, P., Trizna, M., and Zatko, M. (2001). Geoecological (Complex Physical-Geographical) Research and Mapping in Large Scales. In *Geographical Spectrum*, number 3, page 209. Faculty of Sciences, Comenius University, Bratislava. in Slovak

Mitasova, H. and Hofierka, J. (1993). Interpolation by regularized spline with tension: II. Application to terrain modeling and surface geometry analysis. *Mathematical Geology*, 25(6):657–669

Neteler, M. and Mitasova, H. (2002). *Open Source GIS: A GRASS GIS Approach.* Kluver Academic Publishers. 1st.edition

Venables, W. N. and Ripley, B. D. (2002). *Modern Applied Statistics with S. Fourth Edition*. Springer. URL http://www.stats.ox.ac.uk/pub/MASS4. ISBN 0-387-95457-0

Referenced Web Sites

[1] `http://www.opensource.org`

[2] `http://grass.itc.it`

[3] `http://www.gnu.org`

[4] `http://www.r-project.org`

Calculation of Potential Drainage Density Index (PDD)

Endre Dobos and Joël Daroussin

1. Introduction

This paper focuses only on the technical details of the usage and creation of PDD. More details on the theory can be found in the papers written by Dobos (1998) and Dobos et al. (2000). The Potential Drainage Density index, abbreviated as PDD, can be used for geomorphologic, pedologic and geologic characterisation of the landscape.

PDD is a measure of landscape dissection. The degree of surface dissection is determined by the surface runoff and the permeability of the soils and of the underlying rock strata. The less the water infiltrates into the soil, the higher is the water runoff on the surface, causing erosion and dissecting the land surface. Thus, the surface runoff is inversely proportional to permeability. A sandy parent material absorbs the majority of the surface water income, so there is basically no, or only limited runoff. A clayey parent material produces the most dissected surface, due to the very limited permeability and high surface runoff and erosion. Therefore, the landscape dissection is an indirect indicator of the soils and bedrocks.

Dissection is difficult to measure. A potential approach to characterize the degree of landscape dissection is to measure the total lengths of valleys or drainage lines. Drainage density (D) is the total length of the permanent and seasonal streams and rivers divided by a unit size of area (A):

$$D = \frac{\sum_{1}^{n} L}{A}$$

where 'L' is the length and 'n' is the number of rivers. Unfortunately, real drainage density is also difficult to measure, because it requires good topographic maps in an appropriate scale. Instead, the quantitative approach of PDD is presented here. PDD is a terrain attribute, which can be derived from Digital Elevation Models (DEM). The word 'Potential' was added to the name 'Drainage Density' to emphasize the difference between the actual drainage density as could be measured in the field and the DEM-derived one. Any software equipped with hydrological modelling tools can be used to derive the PDD index.

In practice, PDD is used to delineate ridges, backslopes and valley bottoms on high relief areas, i.e. hilly and mountainous regions (Figure 1). Low PDD values represent local heights, elevated areas, ridge tops. Intermediate values represent backslopes, with higher values for the steeper sloping and lower for the more gently sloping areas. High PDD values refer to low-laying areas, depressions, valley bottoms and basins (Figure 2). On flat areas, where the relief is very low, PDD highlights depressions and local lows, where the surface drainage system forms a convergent flow channel pattern. Such a pattern often results in extremely high PDD values.

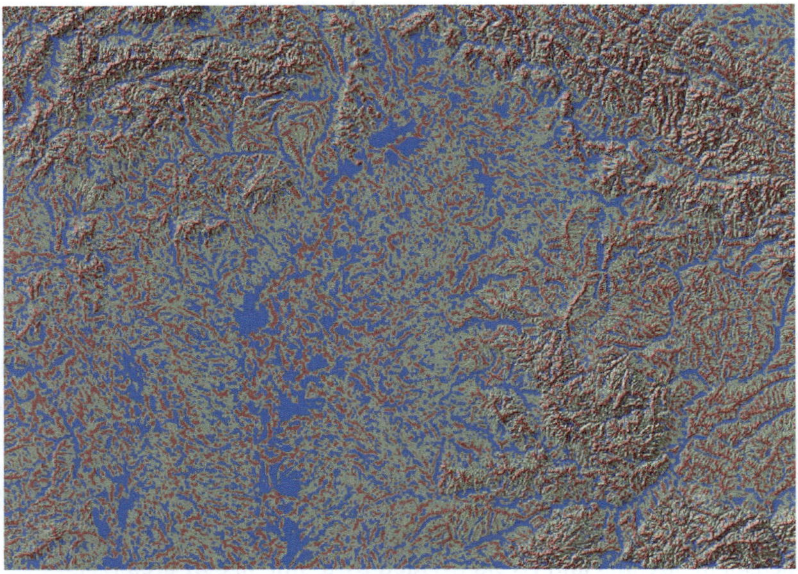

Fig. 1. A classified PDD image of the north-eastern part of the Carpathian-basin covering the major part of the Great Hungarian Plain and sections of the Carpathians. Brown colour represents ridges and local heights, greenish colour stands for backslopes and transitional areas, and blue for depressions and valley bottoms

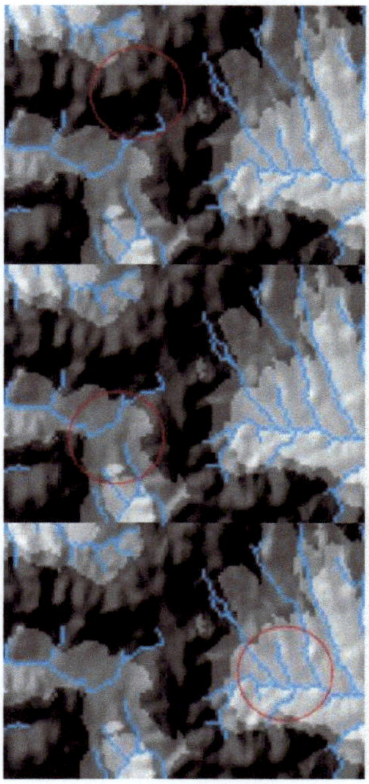

Fig. 2. Low, medium and high drainage densities within a circle-shape neighbourhood window

2. Derivation of the PDD Layer

Most raster based Geographic Information System (GIS) software packages have built-in tools that can be used for this purpose. In this example, ArcInfo GRID commands and functions are used to describe the procedure. For their detailed descriptions, please refer to ArcInfo manuals.

2.1 Input DEM

Running the PDD procedure requires a depressionless or hydrologically corrected DEM. Sinks due to errors in the data have to be removed from the DEM. This can be done using the FILL command of the ArcInfo® GRID module. However, it is often difficult to distinguish the sinks, which naturally occur in the data (e.g. sinkholes in karst regions) from those that are erroneous. Filling all sinks regardless of this characteristic may remove all natural depressions. In our data this has the effect of leveling up to the same elevation huge natural plain areas. By doing so, we loose all chances of differentiation on the plain areas, where few meters or even centimetres of elevation range represent a great variety of natural conditions. Instead of filling all sinks, it is therefore more appropriate to introduce a sink depth limit. Any sink deeper than the specified limit will be kept untouched. Only the sinks less deep than the threshold will be filled up to the level of their pour point. Our experience with the 90 m resolution Shuttle Radar Topographic Mission (SRTM) DEM data showed that a threshold value of 20 metre helps to remove the majority of the small sinks, while keeping the major elevation patterns untouched on the Pannonian plain. Of course this is in view of our objective to obtain a picture of the dissection of the area but not necessarily to obtain a hydrologically correct DEM.

2.2 Step 1. Flow Direction

Within the ArcInfo® GRID environment, the FLOWDIRECTION function can be used to create a grid/image of flow direction. The flow direction is the direction from each cell to its steepest downslope neighbour. The flow direction (Jenson and Domingue, 1988) is based on the highest drop value calculated for each of the eight neighbouring pixels as:

$$\text{drop} = \frac{(\text{change\_in\_elevation\_value})}{(\text{cell\_center\_distance}) * 100}$$

If the descent to all adjacent cells is the same, the neighbourhood is enlarged until the steepest descent is found.

2.3 Step 2. Flow Accumulation/Contributing Area/Catchment Area

Based on the flow direction grid, the size of the contributing area to each grid cell can be calculated with the FLOWACCUMULATION function. The output grid represents the catchment area size of each grid cell expressed as cell counts.

2.4 Step 3. Drainage Network

Applying a threshold value to the flow accumulation grid can create a stream network. All cells, which have a contributing area higher than a certain threshold value, are assigned a value of 1, representing the drainage path. All other cells, which have a flow accumulation value below the threshold are assigned '*no data*' and become background cells (Figure 3).

The threshold value is set by the user. There is no certain quantitative rule to define it. For a 100-meter resolution DEM a threshold value of 100 cells (1 km^2 area) has been found to provide a realistic density of drainage lines characterizing well the real landscape. As the resolution becomes coarser, the representation of topography gets less detailed, while the area represented by one pixel is increased. Keeping the original minimum size of the catchment area would increase the drainage density unrealistically, thus the minimum catchment has to be increased as well. We found that taking the 100-meters pixel size as a basic case, the incremental rate in pixel size can be used as a multiplication factor for the catchment area size. For example, as we move from 100 to 500-meters pixel size, the size of the catchment area should be increased from 1 to 5 square kilometres, while the number of threshold pixels is decreased from 100 to 20 accordingly.

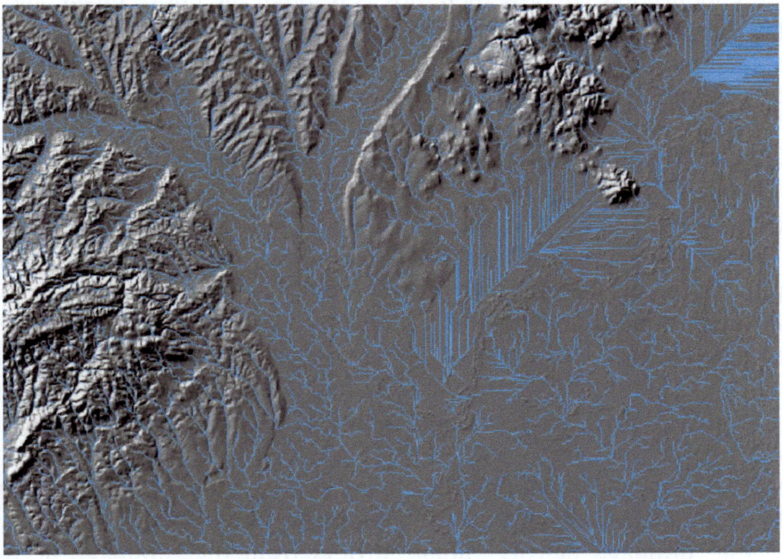

Fig. 3. The drainage network derived from the 90 m resolution SRTM DEM

From Figure 3 the reader can identify some artefacts in the form of parallel, densely located drainage lines. This pattern occurs on areas where the original image had sinks that were filled up and levelled to large flat areas by the FILL command of ArcInfo® GRID to make the drainage network creation possible. On such flat areas, the program has difficulties with identifying the flow direction, because the descent to all adjacent cells is the same. In this case the neighbourhood is enlarged until the steepest descent is found. This procedure results in straight parallel lines running towards the pour points, which often produces a very high drainage density. Although it is an artefact, the interpretation is still valid and useful because this pattern highlights depressions within the landscape, which are generally representative of valley bottoms.

2.5 Step 4. Potential Drainage Density (PDD)

The final step of creating the PDD image is to run the FOCALSUM function on the drainage network layer. The function sums up the cell values that fall within a predefined shaped and sized neighbourhood and assigns the sum to the centre cell. In our case, summing is equivalent to a simple count of the cells representing drainage lines as they all have a value of 1. The neighbourhood window then moves through the entire image and calculates the PDD value for every pixel.

For computing the PDD, a circle shape neighbourhood is chosen for which a radius expressed as a number of cells has to be specified. The radius size can vary based on the need of the user. However, in order to produce a full, continuous coverage of the area, one should make sure that the size of the radius is big enough for the function to always ‘catch’ at least one drainage cell so as to avoid having empty neighbourhoods. A large radius tends to generalize the image, while a smaller one helps to maintain the physiographic patterns. Moreover, enlarging the neighbourhood increases computing time, which is a problem when a large DEM is used. As an example the image of Figure 4 was produced from a 90 m SRTM DEM, with a drainage threshold of 100 cells and a FOCALSUM radius of 15 cells. Figure 5 presents the same image after classifying it into three classes.

The PDD is an artificial image with cells having a value referring to the length of drainage ways in a specified neighbourhood of each cell. In this sense, the ’drainage density’ image differs from the ‘drainage network’ image. It has to be clarified here that the term "drainage density" is not identical to the one used by geographers and geomorphologists. It is not based on the real drainage network; it is only derived from a digital terrain model. Therefore it does not take into consideration the loss of surface water due to infiltration into the soil. The potential drainage density is always higher or equal to the real drainage density of an area.

Fig. 4. PDD image derived from a 90 m resolution SRTM DEM

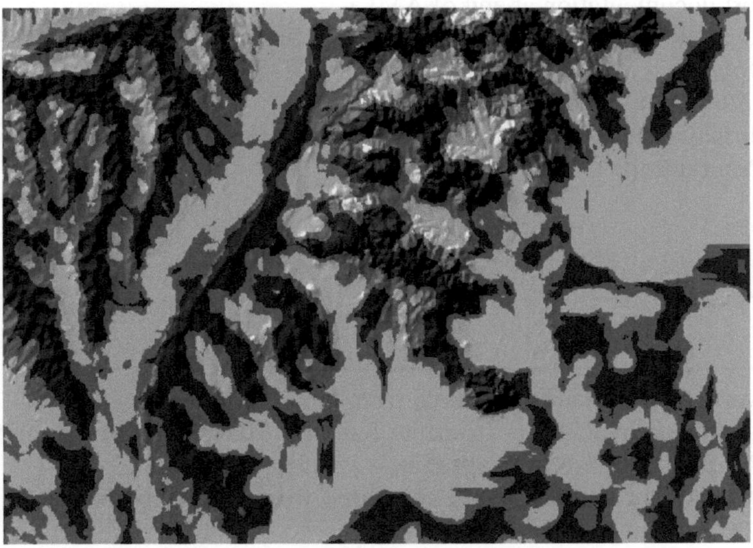

Fig. 5. Classified PDD. Light colour refers to depressions and valley bottoms, intermediate colour to backslopes, and dark colour to ridges and local heights

References

Dobos E, Micheli E, Baumgardner MF, Biehl L, Helt T (2000) Use of combined digital elevation model and satellite radiometric data for regional soil mapping. Geoderma 97:367-391

Dobos E (1998) Quantitative analysis and evaluation of AVHRR and terrain data for small scale soil pattern recognition. Ph.D. thesis, Purdue University, West Lafayette, Indiana USA

Jenson SK, Domingue JO (1988) Extracting topographic structure from Digital Elevation Data for Geographic Information System Analysis. Photogrammetric Engineering and Remote Sensing 54:1593-1600

Appendix: An Arc/Info® AML file to derive a PDD layer from a filled DEM

(This AML can be created by copy/pasting the following text into a text editor and saving it to pdd.aml or can be downloaded either from http://support.esri.com, Downloads/ArcScripts or from http://eusoils.jrc.it) (see short documentation at end of AML)

```
/* PDD.AML
&echo &off save_echo
&args InAltGrid CatchmentAreaKm2 PddRadiusKm OutPddGrid rest:rest
&call ProgramInit

/* ******
/* ******
/* Main:

/* -------------------------------------------------------------------
/* Convert catchment area threshold from km to number of cells:
&sv CatchmentAreaCells = [truncate [calc ( ( %CatchmentAreaKm2% *
1000 ) ** 2 )/( %grd$dx% * %grd$dx% ) + .5]]
&type Catchment area threshold converted from %CatchmentAreaKm2%
km2 to %CatchmentAreaCells% cells

/* Create drainage network layer:
pddtmp1 = con(flowaccumulation(flowdirection(%InAltGrid%)) >=
%CatchmentAreaCells%, 1, 0)

/* Convert radius of the PDD moving circle from km to number of cells:
```

```
&sv PddRadiusCells = [truncate [calc %PddRadiusKm% * 1000 /
%grd$dx% + .5]]
&type Radius of the PDD moving circle converted from %PddRadi-
usKm% km to %PddRadiusCells% cells

/* Create PDD layer (as number of drainage cells within the %PddRadius-
Cells% moving circle):
&type Computing PDD...
%OutPddGrid% = focalsum(pddtmp1, circle, %PddRadiusCells%)
&messages &off
kill pddtmp1
&messages &on

/* Convert PDD from number of cells to km/km2:
&type Computing PDD in km/km2...
/** PDD holds the number of drainage cells within the %PddRadiusKm%
circle.
/** The circle is (Pi * %PddRadiusKm% ** 2) km2:
&sv area = 3.141592 * %PddRadiusKm% ** 2
/** Each cell represents a drainage length of either the grid's resolution
/**  or its diagonal. Let's set it to an average of the two:
&describe %OutPddGrid%
&sv length = ( %grd$dx% + %grd$dx% * 2 ** .5 ) / 2 / 1000
/** Each PDD value thus represents:
%OutPddGrid%km = %OutPddGrid% * %length% / %area%    /* km/km2

/* ----------------------------------------------------------------------
/* End of main:

&type Resulting grids:
&type  %OutPddGrid%: PDD expressed as cell counts
&type  %OutPddGrid%km: PDD expressed as km/km2
&call reset
&return


/* **********
/* **********
/* Standard program management subroutines:

/* ----------------------------------------------------------------------
/* Program initialisation:
```

```
&routine ProgramInit
&severity &error &routine error
&sv save_messages = [show &messages]
&messages &on

/* Debugging tool:
&sv debug off  /* set to on or off.
&echo &%debug%
&if %debug% = on &then &messages &on

&call ArgumentsControl
&return

/* -----------------------------------------------------------------------
/* Arguments control:
&routine ArgumentsControl

&if [show program] <> GRID &then &do
 &type This AML works only from Grid.
 &call error
 &end

&if [null %OutPddGrid%] or not [null %rest%] &then &call usage

&if not [exists %InAltGrid% -grid] &then &do
 &type Grid %InAltGrid% not found.
 &call error
 &end
&describe %InAltGrid%
&if %prj$units% <> METERS &then &do
 &type This AML works only on projected grids with units METERS.
 &call error
 &end

&if [type %CatchmentAreaKm2%] > 0 &then &call usage
&if [type %PddRadiusKm%] > 0 &then &call usage

&if [exists %OutPddGrid% -file] or [exists %OutPddGrid% -directory]
&then &do
 &type File or directory %OutPddGrid% already exists.
 &call error
 &end
```

```
&if [exists %OutPddGrid%km -file] or [exists %OutPddGrid%km -
directory] &then &do
 &type File or directory %OutPddGrid%km already exists.
 &call error
 &end

&if [exists pddtmp1 -file] or [exists pddtmp1 -directory] &then &do
 &type Temporary file or directory pddtmp1 already exists.
 &call error
 &end

&if [exists pddtmp2 -file] or [exists pddtmp2 -directory] &then &do
 &type Temporary file or directory pddtmp2 already exists.
 &call error
 &end

&return

/* ---------------------------------------------------------------------
/* Error management:

/* General error handling routine:
&routine error
&severity &error &ignore
&type Abnormal termination of program [upcase %aml$file%].□
&call reset
&return &error

/* Usage:
&routine usage
&type Syntax error.□
&type Usage: &RUN PDD <input elevation grid> <drainage area thresh-
old (km2)>
&type              <PDD moving circle radius (km)> <output PDD grids
prefix>
&type More help available from pdd.txt and pdd.doc.
&call reset
&return &error

/* ---------------------------------------------------------------------
/* Restore environment:
&routine reset
```

```
&messages %save_messages%
&echo %save_echo%
&return

/* ************
/* ************
/* Program and user documentation:
```

Joel Daroussin - JRC Ispra - 07/10/2004
Updated 20/12/2004: documentation.

AML built under ArcInfo 9.0 on PC.

ASSISTANCE-COMPUTES THE POTENTIAL DRAINAGE DENSITY (PDD) INDEX FROM AN ELEVATION GRID.

Usage: &RUN PDD <input elevation grid> <drainage area threshold (km2)>
<PDD moving circle radius (km)> <output PDD grids prefix>

Arguments:

<input elevation grid> - Name of the elevation grid input to the procedure.
Grid must be projected with units METERS.

<drainage area threshold (km2)> - Threshold value in km2 used to initiate delineation of the drainage network.

<PDD moving circle radius (km)> - Radius in km of the moving circle used to count drainage network pixels within the centre cell's neighbourhood.

<output PDD grids prefix> - Name of the output PDD grid. This first grid will hold counts of drainage network cells. A second grid named <output PDD grids prefix>km will be output to hold km of drainage network by km2.

For more details on the PDD index, please read pdd.pdf. Notes:

- for better results consider using a FILLed -hydrologically corrected- input,
- altitude grid (see ArcInfo Grid command FILL for more information),
- Computing km of drainage network by km2 is a simple unit conversion process,
- neglecting diagonal versus horizontal (or vertical) lengths of drainage lines,
- The length used is averaged from the straight and the diagonal lengths of cells.

About the Authors

José I. Barredo is a geographer with experience in spatial analysis, GIS, multi-criteria evaluation techniques and cellular automata. He works in the fields of natural hazards and urban and regional planning. In 1995 he received a PhD in GIS and remote sensing from University of Alcalá, Spain. Currently he works for the DG Joint Research Centre, European Commission in Ispra, Italy. His research focuses on flood and drought risk and damage assessment, development and application of dynamic spatial models for land-use simulation and natural hazard scenarios resulting from climate change. Contact: *jose.barredo@jrc.it*.

Katalin Bódis completed her studies on mathematics and geography at the University of Szeged and on surveying and geoinformatics engineering at Budapest University of Technology and Economics, Hungary. As a scholarship holder she spent a semester at the Karlsruhe University of Applied Sciences and at the Johann Wolfgang Goethe-Universität in Frankfurt am Main, Germany. She also took part in the PhD program in 'Geographical and Geological Analysis of Regional Processes' at the University of Szeged. She has been involved in the scientific work of the Applied Geoinformatics Laboratory and the Department of Physical Geography and Geoinformatics of Szeged University since 1994. According to the locality and the flood-affected history of Szeged, the most researches dealt with the neighbouring Tisza River and with landscape analysis with GIS. As a junior lecturer Katalin has developed a significant part of the curricula for students of geography specialized in GIS and was responsible for several wide-ranging GIS courses. Katalin Bódis joined the Institute of Environment and Sustainability of the DG Joint Research Centre, European Commission in Ispra, Italy, as a GIS analyst in the Floods and Other Weather Driven Natural Hazards Action in 2003. Her main research interest is focusing on DEM applications in hydrological modelling, accuracy and reliability of algorithms providing digital elevation models (DEM) and flow networks, interpolation and (fuzzy) classification of surface and landscape elements. Contact: *katalin.bodis@jrc.it*.

Rado Bonk graduated in the field of Physical Geography and Landscape Planning at Comenius University, Bratislava, Slovakia in 1998. He proceeded with his Master's studies at the University of Nebraska at Omaha in the field of GIS and remote sensing during 1999-2001. He obtained his PhD degree in 2003 at Comenius University with the thesis on digital elevation model accuracy and automated geomorphological mapping. During the period 2003-2006 he was working at the Environment Institute of DG Joint Research Centre, European Commission in Ispra, Italy in the Natural Hazards group, where he was designing hydrological databases and studying application of DEMs for flood forecasting. He currently works as Assistant Professor at the Department of Physical Geography and Geoecology at the Faculty of Natural Sciences, Comenius University, Bratislava, Slovakia. His primary interest is in application of high-resolution DEMs in natural hazards assessment, especially in flood mapping and flood risk assessment. He also works as a UNIX and database administrator and as a GRASS GIS consultant. He specializes in Open Source software implementation for both research and education purposes. Contact: *rado.bonk@gmail.com*.

Tomas Cebecauer is working as a researcher in the fields of geography and geoinformatics at the Institute of Geography, Slovak Academy of Sciences. He received his MSc in Geography and Cartography in 1994 and PhD in Cartography and Geoinformatics in 2002 from the Comenius University, Bratislava, Slovakia. His research activities are devoted to processing and analysis of GIS and remote sensing data, development and implementation of new GIS methods for data processing and visualisation. With respect to digital terrain modeling, he is focused mainly on data interpolation, utilisation of DTM in technical and environmental applications and digital cartography. Since 2006 he has been working for the DG Joint Research Centre, European Commission in Ispra, Italy, Institute for Environment and Sustainability, Renewable Energy Unit in the field of solar energy resource mapping. Contact: *tomas.cebecauer@jrc.it*.

Roberto Colombo graduated in Geology at the Department of Earth Sciences, University of Milan, Italy. He wrote his MSc thesis on the use of remotely sensed data and GIS for geomorphological mapping of the inland delta of the Niger River in Mali. From 1993 he was in contact with the Istituto Agronomico per l'Oltremare in Florence and he was involved in the hydrological modeling of the Mareb River in Eritrea on a two years research fellowship. From 1996 to 2000 he worked at the Remote Sensing Department of the CNR in Milan in research regarding the use of remotely sensed data for the determination of energy balance components. At the same time he also conducted research at the DG Joint Research Centre, European Commission in Ispra, Italy into developing methodologies for extracting river networks and catchment boundaries by combining DEMs and environmental parameters at pan European scale. Roberto Colombo is now a senior researcher at the Università degli Studi di Milano-Bicocca and is associated with the Laboratorio di Telerilevamento delle Dinamiche Ambientali of DISAT where he is investigating the capability of hyperspectral observations for the retrieval of biophysical parameters of terrestrial vegetation. Contact: *roberto.colombo@jrc.it*.

Joël Daroussin is a research engineer at the French National Institute for Agronomic Research (INRA). He works for the Soil Science Department of Orléans (FR). He spent 16 months in 2004 and 2005 at the DG Joint Research Centre, European Commission in Ispra, Italy where he worked for the European Soil Bureau based in Ispra. His field of experience relates to GIS technology applied to soil science and pedology. His research projects concern the characterisation of the spatial organisation of the pedological landscape from digital elevation models, the implementation of various models into GIS such as erosion, carbon flux and crop growth models, and the problem of scale transfer. He participates in the development of the European Soil Information System and teaches GIS technology at the University of Orléans. Contact: *joel.daroussin@orleans.inra.fr*.

Endre Dobos made a PhD in Soil Mapping with GIS and Remote Sensing at Purdue University, Agronomy Department, Indiana, USA in 1998 and a MSc in GIS and Environmental Survey at the Technical University of Budapest, Faculty of Civil Engineering in 1996. Endre Dobos is an Associate Professor at the University of Miskolc, Department of Physical Geography and Environmental Sciences. His educational activity covers soil science and GIS related courses. Endre's research activity focuses on digital soil mapping and the development of global, small scale soil databases using remotely sensed data, digital elevation models and other ancillary data. He is a chair of the Digital Soil Mapping Working Group of the European Soil Bureau Network. Contact: *ecodobos@uni-miskolc.hu*.

Ewan D. Dunlop holds a PhD in Applied Physics specialising in surface engineering of thin-film material. He joined the European Solar Test Installation (ESTI) of the Institute for Environment and Sustainability, Renewable Energy Unit at the DG Joint Research Centre, European Commission in Ispra, Italy in 1993 as a scientific consultant leading the group's research in thin-film photovoltaic device measurement. Since 1997, he is the group leader for the Photovoltaic Device Technology activities and has since been appointed as Technical Manager of the ESTI Laboratory. Contact: *ewan.dunlop@cec.eu.int.*

Jaroslav Hofierka is an Associate Professor of Physical Geography and Geoecology and head of GIS Laboratory in the Department of Geography and Regional Development at the University of Presov, Slovakia. He received his PhD in Cartography and Geoinformatics at the Comenius University, Bratislava, Slovakia in 1998. His main research activites have been focused on digital terrain modelling and applications, spatial interpolation and modelling of landscape processes (solar radiation, water erosion) using GIS. He has been participating in the development of Open Source GRASS GIS since 1992. His further research areas include renewable energies, spatial and temporal landscape changes and municipal information systems. Contact: *hofierka@fhpv.unipo.sk.*

Thomas Huld has more than 15 years experience in mathematical modelling, numerical simulation and scientific software development. He holds a M.Eng. in electrical engineering and a PhD in plasma physics from the Technical University of Denmark in Copenhagen. He joined the Renewable Energies Unit in the Institute for Environment and Sustainability at the DG Joint Research Centre, European Commission in Ispra, Italy in 2001, where he has been working mainly on modelling of the geographical distribution of the solar energy resource over large geographical areas and on developing the interactive interface to the PVGIS web system. Contact: *thomas.huld@jrc.it.*

Alfred de Jager started his professional career with Logica CMG in which he developed various computer systems with a geographical processing component for the Dutch government. For the last ten years he has worked for the DG Joint Research Centre, European Commission in Ispra, Italy. As a physical geographer and informatician he has been working in projects where the use of detailed and large geographic databases for the whole of Europe was important. Amongst others he was involved in the redesigning of the GISCO (GIS for the Commission) spatial database system, the exploitation of the meteorological component of the Crop Growth Modeling System (CGMS) and the development of the Catchment Characterization and Modeling system (CCM). For the latter project various algorithms were developed to correct elevation models like the SRTM model and to extract water presence from satellite imagery. Currently he is in charge of the Water Information System for Europe in which geographic and environmental quality status data are reported to the Commission by the 27 Member States and Norway. Contact: *alfred.de-jager@jrc.it.*

Ad de Roo is Action Leader of the Natural Hazards project at the Institute for Environment and Sustainability, DG Joint Research Centre, European Commission in Ispra, Italy, dealing with floods, droughts, land use and climate change. He is a senior scientist with a PhD in Physical Geography on the topic of hydrological and soil erosion modelling. Both his MSc and PhD were completed at Utrecht University, the Netherlands, where he has also worked as lecturer and researcher until 1997, before coming to JRC. He developed the physically-based catchment models LISEM for soil erosion and LISFLOOD for flood simulation and has 20 years professional experience with hydrological simulation models linked to GIS. His many publications have appeared in international journals such as Hydrological Processes, Journal of Hydrology, CATENA, Geomorphology, European Journal of Soil Science, Journal of River Basin Management, Hydrology and Earth Systems Sciences, Physics and Chemistry of the Earth and Earth Surface Processes and Landforms, and in several books. His main expertise is in hydrology, hydrological modelling, GIS, soil physics and soil erosion. Contact: *ad.de-roo@jrc.it*.

Gyozo Jordan is a senior researcher at the Hydrogeology Division of the Geological Institute of Hungary. His research includes environmental modelling with GIS, digital terrain modelling applications, sediment and run-off modelling in catchments and 3D digital geological modelling for hydrogeological applications. Gyozo Jordan completed his undergraduate and MSc studies at the Eötvös University of Sciences, Hungary and at the Queen's University of Belfast, UK, and received an MSc degree in Geology and Environmental Geology from the Eötvös University in 1993. He obtained an MSc degree in Environmental Geochemistry from the Uppsala University, Sweden in 1995. He received his PhD on digital terrain modeling at the Uppsala University, Sweden in 2004. He was detached to the International Institute for Geo-Information Science and Earth Observation (ITC), The Netherlands, as visiting scientist to collaborate in sediment transport modelling and digital terrain modeling in 2000. More recently, he worked for the University of Natural Resources and Applied Life Sciences (BOKU), Austria and at the United States Geological Survey, U.S.A. as visiting scientist in 2006. He joined DG Joint Research Centre, European Commission in Ispra, Italy in the period 2001-2004 to work on GIS applications for a European environmental inventory and risk assessment project. Currently he acts as the National Representative for Association of European Geological Surveys. Contact: *jordan@mafi.hu*.

Carlo Lavalle after graduating in Physics for the University of Bologna, and achieving specialization in applied Geophysics, he spent the initial part of his professional career as researcher in the Italian CNR in the fields of Atmospheric Dynamics and environmental impact of global changes. Prior to joining DG Joint Research Centre, European Commission in Ispra, Italy in 1990, he worked for the European Space Agency. He is with the Land Management and Natural Hazards Unit of the Institute for Environment and Sustainability at JRC. Currently, he is working in the following areas of applications: urban and regional planning, environmental impact of human activities and of extreme weather events at local and regional scale, environmental protection and impact assessment of regional development programmes. Contact: *carlo.lavalle@jrc.it*.

Luca Montanarella studied Agricultural Sciences and Agricultural Engineering at the Faculty of Agriculture, University of Perugia, Italy, and graduated in 1984. This was followed by post-doctoral study at the Centre for Biopharmaceutical Sciences, University of Leiden, The Netherlands, and a period as visiting scientist at the Faculty of Pharmacy, University of Groningen, The Netherlands. He then became a Mass Spectrometry and Chromatography Specialist at Varian Associates, Palo Alto, USA, before returning to Italy in 1991 to become principal administrator at the Italian Ministry of Agriculture. Since 1992 he has been a scientific officer at the DG Joint Research Centre, European Commission in Ispra, Italy. He worked as head of the Organic Reference Laboratory in the Environment Institute between 1992 and 1995, and he headed the European Soil Bureau project in the period 1995-2000. Since 2003 he is the head of the new JRC action 'Monitoring the State of European Soils' (MOSES). His main interests are currently soil databases, GIS, soil protection and land degradation. Contact: *luca.montanarella@jrc.it*.

Maria Luisa Paracchini obtained a degree in Civil Engineering with a specialisation in Land Planning at the Politecnico in Milan, Italy. She then moved to the Faculty of Agricultural Sciences in Milan where she received the PhD on the use of remote sensing and GIS for the analysis of the rural landscape at the Institute of Rural Engineering in 1994. In her thesis she carried out digital terrain analysis for the geometric correction of satellite images and run-off modelling at the catchment scale. She continued her post-doctoral studies at the DG Joint Research Centre, European Commission in Ispra, Italy, Space Applications Institute, where she developed a two-year project on soil erosion modelling in Mediterranean Environments, followed by three-years involvement in projects on land degradation and desertification modelling. At the same time she was supporting the development of a methodology for the characterization of watersheds at the European scale according to run-off conditions, land cover dynamics, fluvial processes and socio-economic conditions. In the years 2000-2001 she was employed at the Finnish Forest Research Institute in Helsinki, as senior scientist carrying out research on biodiversity in boreal forests and she was also teaching the course 'Advanced applications of GIS analysis within natural resource management at Forest Management Department, Helsinki University. Since 2001 she has been employed at Institute for Environment and Sustainability, JRC. She is a member of the executive board of Landscape Europe, an international network of landscape research. She works in the Rural, Water and Ecosystem Resources Unit, being in charge of High Nature Value farmland mapping and the analysis of the impact of Rural Development policies on biodiversity, and is involved in research on agricultural land use modelling. Contact: *maria-luisa.paracchini@jrc.it.*

Robert Peckham obtained a First Class Honours degree in Physics from Imperial College, London and then a PhD in Radio Interferometry from University of Manchester, U.K. After a period of Postdoctoral Research in Radio Interferometry he worked on Energy Analysis for the European Community at the University of Strathclyde, Glasgow, before transferring to DG Joint Research Centre, European Commission in Ispra, Italy, to work on Energy Policies for Europe. Projects on solar energy and three dimensional modelling led to the requirement for GIS and for the last 15 years he worked on a wide range of GIS projects and applications to support European policies. His work involved development and use of digital terrain models in applications such as three dimensional site modelling, harmonisation of European codes for building snow loads, and the mapping of countries using orthorectification of satellite imagery. He was co-ordinator of the JRC's Special Interest Group for GIS and was responsible for initiating and promoting a successful series of EC-GIS workshops which have now become the EC-GI&GIS Workshop - a major European GI event now in its 12th edition. Contact: *rjpeckham@hotmail.com.*

Pierre Soille received an Engineering Degree from the Université Catholique de Louvain (Belgium) in 1988, and his PhD from the same university in collaboration with the Ecole des Mines de Paris in 1992. He received the Habilitation Degree from the Université Montpellier~II, France in 1997. During the period 1992-1999, he pursued research on image analysis and mathematical morphology with the CSIRO Mathematical and Information Sciences in Sydney, the Centre de Morphologie Mathématique of the Ecole des Mines de Paris, Fontainebleau, France, the Pattern Recognition Department of the Fraunhofer-Institut IPK in Berlin, the Ecole des Mines d'Alès and EERIE in Nîmes, France, and the Biotechnology and

Biological Sciences Research Council in Silsoe, U.K. Since 1999 he has been with the DG Joint Research Centre, European Commission in Ispra, Italy, focusing on the research of advanced methodologies and algorithms for the generation of enhanced pan-European geospatial data sets. Pierre is author of the reference monograph `Morphological Image Analysis: Principles and Applications' published by Springer. Contact: *pierre.soille@jrc.it*.

Marcel Šúri holds a MSc and PhD in Geography, Cartography and Geoinformatics. Since 2001 he has been working at the DG Joint Research Centre, European Commission in Ispra, Italy, the Institute for Environment and Sustainability. His main activities are geographical analysis of renewable energy resources, energy meteorology, and technical, environmental and economic assessment of solar energy and bioenergy technologies. He is one of the authors of the web-based Photovoltaic Geographical Information System (PVGIS) for estimation of solar electricity potential. Contact: *marcel.suri@jrc.it*.

János Adolf Szabó is a senior researcher of applied mathematics in the field of mathematical modelling of hydrological processes. He completed his MSc studies in Mathematics with specialised in operations research at the Eötvös Loránd University, Budapest, Hungary in 1985. He carried out research in the field of multi-extremal (global) optimisation methods, parameter calibration for mathematical models, real-time modelling and forecasting, error-analysis and modelling, numerical methods and their application for simulation. He worked for the Research Group for Water Management of the Hungarian Academy of Sciences, and as a lecturer at University of Technology and Economics in Budapest, Hungary until 2003. He joined DG Joint Research Centre, European Commission in Ispra, Italy in 2003. He developed the physically-based large-scale and high-resolution hydrological model DIWA (Distributed Watershed) for modelling rainfall-runoff processes, flood events analysis, real-time hydrological forecasting and control. During his professional career he also obtained experience in using calibrated precipitation-radar, and ECMWF forecasted meteorological data distributions for real-time hydrological forecasting on the Tisza river basin with the DIWA model. Currently, he works for the Weather Driven Natural Hazards Action on Floods at the Institute of Environment and Sustainability (IES) of the JRC. The main objective of his present research is the supporting, evaluation, developing and real-time testing of a pre-operational European flood alert system with 1 to 10 day lead-time, focussing on the major trans-national river basins. Contact: *janos.szabo@jrc.it*.

Jürgen Vogt obtained a Diploma in Rural Survey from the International Institute for Aerospace Survey and Earth Sciences (ITC) in Enschede, the Netherlands in 1984 and an MSc degree in Geography and Remote Sensing from Trier University, Germany in 1986. From 1986 to 1988 he was employed at the Geography Department of Trier University, where he conducted research projects and taught physical geography as well as photo interpretation and remote sensing. He then moved to DG Joint Research Centre, European Commission in Ispra, Italy to perform research on the spatio-temporal variability of climatologically relevant surface parameters. He received his PhD with highest honors from Trier University in 1992. Since 1993 he has conducted research projects at the JRC in the fields of agrometeorology, yield forecasting, drought monitoring, terrain analysis and surface energy balance modeling. Jürgen Vogt is further affiliated with Laval University in Québec, Canada, where he has been an Associate Researcher at the Centre de Recherche en Géomatique since 1996 and an Associate Professor at the Faculté de Foresterie et de Géomatique since 1997. Jürgen Vogt initiated and led the development of the CCM River and Catchment Database for Europe, which should become the European reference network for the Water Framework Directive (WFD). From 2001 to 2002 he chaired the GIS Working Group for the Implementation of the WFD and he has been a member of the Steering Committee of the EU Landscape Tomorrow network since 2003. Since 2003 he has been working in the Rural, Water and Ecosystem Resources unit of Institute for Environment and Sustainability at JRC, in charge of CCM and the development of a European Land Information System for Agriculture and Environment (ELISA). His current research interests focus on terrain analysis and modeling the impact of agricultural policies on land use and landscapes in Europe as well as on the analysis of remote sensing and digital elevation data for modeling land surface processes. Contact: *jurgen.vogt@jrc.it*.

Borut Vrščaj completed his BSc studies at University of Ljubljana, Slovenia and obtained an Agricultural Engineering Degree and an MSc Agriculture. Currently, he is carrying out PhD research at the Department of Exploitation and Protection of the Agricultural and Forestry Resources, University of Torino, Italy on urban soil quality evaluation and modelling. Borut Vrščaj has been employed as a Senior Researcher at the Centre for Soil and Environmental Science, University of Ljubljana since 1984. Currently he works as a Senior Researcher for the Agricultural Institute of Slovenia, Ljubljana on national and international projects. His main research focus is the establishment of Soil Information System of Slovenia, the elaboration, improvement and use of the thematic GIS soil datasets, soil-capability-landscape modelling, soil-environmental protection and land use planning. He was USDA National Resource Conservation Service Cochran fellowship holder in Lincoln, Nebraska and Fort Worth, Texas where he worked on the structures and content of the US digital soil and land information system in 2000. He has actively participated in several international projects, including establishment of the Corine Land Cover Slovenia 1995, elaboration of the Geographical Soil Database of Europe at 1:1M for Slovenian, modelling soil and DEM parameters in FP5 Project SPIN, soil pollution evaluation in FP5 Project URBSOIL, developing cross-border soil databases within Interreg III ECALP Project and soil quality evaluation systems in the EU Interreg III TUSEC-IP Project. He worked for seven months at DG Joint Research Centre, European Commission in Ispra, Italy as a visiting scientist on testing DEMs for vector soil datasets and on environmental modelling with DEMs in 2003 and 2004. He is a member of the Digital Soil Mapping Working Group, European Soil Bureau. Contact: *Borut.Vrscaj@kis.si*.

Index

MIX
Papier aus verantwortungsvollen Quellen
Paper from responsible sources
FSC® C105338

If you have any concerns about our products,
you can contact us on
ProductSafety@springernature.com

In case Publisher is established outside the EU,
the EU authorized representative is:
Springer Nature Customer Service Center GmbH
Europaplatz 3, 69115 Heidelberg, Germany

Printed by Libri Plureos GmbH
in Hamburg, Germany